Finger Prints Examination

Finger Prints Examination

Er. Devesh Kumar Dixit

RANDOM PUBLICATIONS
NEW DELHI (INDIA)

Finger Prints Examination

ISBN 978-93-5111-254-9

Published in 2014 in India by

Reprint 2019

RANDOM PUBLICATIONS

4376-A/4B, Gali Murari Lal, Ansari Road
New Delhi-110 002
Phone : +91-11-43580356, +91-11-23289044
e-mail: randomexports@gmail.com, sales@randompublications.com, info@randompublications.com

Type Setting by : Keystoneprintads, Delhi-110051
Printed at : Mehra Printers, Delhi-110 092

Preface

Fingerprint Examination has clearly stood the test of time and experience. It has been established over 100 years ago that friction ridge patterns remain unchanged naturally in their ridge detail during the lifetime of an individual and that fact makes it possible for individualization of fingerprints leading to absolute personal identification.

Fingerprint identification, known as dactyloscopy, or hand print identification, is the process of comparing two instances of friction ridge skin impressions, from human fingers or toes, or even the palm of the hand or sole of the foot, to determine whether these impressions could have come from the same individual. The flexibility of friction ridge skin means that no two finger or palm prints are ever exactly alike in every detail; even two impressions recorded immediately after each other from the same hand may be slightly different. Fingerprint identification, also referred to as individualization, involves an expert, or an expert computer system operating under threshold scoring rules, determining whether two friction ridge impressions are likely to have originated from the same finger or palm (or toe or sole).

Fingerprinting has served all governments worldwide during the past 100 years or so to provide accurate identification of criminals. No two fingerprints have ever been found identical in many billions of human and automated computer comparisons. Fingerprints are the fundamental tool for the identification of people with a criminal history in every police agency. It remains the most commonly gathered forensic evidence worldwide and in most jurisdictions fingerprint examination outnumbers all other forensic examination casework combined. Moreover, it continues to expand as the premier method for identifying persons, with tens of thousands of people added to fingerprint repositories daily in America alone — far more than other forensic databases.

Fingerprint image quality is an important factor in the performance of minutiae extraction and matching algorithms. A good quality fingerprint image has high contrast between ridges and valleys. A poor quality fingerprint image is low in contrast, noisy, broken, or smudgy, causing spurious and missing minutiae. Poor quality can be due to cuts, creases, or bruises on the surface of a finger tip, excessively wet or dry skin condition, uncooperative attitude of subjects, damaged and unclean scanner devices, low quality fingers (elderly people, manual workers), and other factors.

The goal of an enhancement algorithm is to improve the clarity (contrast) of the ridge structures in a fingerprint. General-purpose image enhancement techniques are not very useful due to the non-stationary nature of a fingerprint image. However, techniques such as gray-level smoothing, contrast stretching, histogram equalization, and Wiener filtering can be used as preprocessing steps before a sophisticated fingerprint enhancement algorithm is applied.

Techniques that use single filter convolutions on the entire image are not suitable. Usually, a fingerprint image is divided into sub regions and then a filter whose parameters are pre-tuned according to the region's characteristics is applied. Each local region of a fingerprint can be seen as a surface wave of a particular wave (ridge) orientation (perpendicular to the flow direction) and frequency.

The fingerprints have been traditionally classified into categories based on information in the global pattern of ridges. A recognition procedure consists in retrieving one or more fingerprints in a large database corresponding to a given fingerprint, whereas a classification procedure consists in assigning a fingerprint to a pre-defined class. Fingerprint recognition is the basic task of the identification systems of the most famous policy agencies. If all the fingerprints within the database are a priori classified, the recognition procedure can be performed more efficiently, since the given fingerprint has to be compared only with the database items belonging to the same class.

The book will be useful to police officers, lawyers, researchers and students working or studying to acquire knowledge of fingerprints.

I thank all members of my team who have helped in the preparation of the book. My special thanks go to "Random Publications" who have published the book.

–Er. Devesh Kumar Dixit

Contents

1

Introduction to Fingerprint

A fingerprint in its narrow sense is an impression left by the friction ridges of a human finger. In a wider use of the term, fingerprints are the traces of an impression from the friction ridges of any part of a human or other primate hand. A print from the foot can also leave an impression of friction ridges. A friction ridge is a raised portion of the epidermis on the digits (fingers and toes), the palm of the hand or thesole of the foot, consisting of one or more connected ridge units of friction ridge skin. These are sometimes known as "epidermal ridges" which are caused by the underlying interface between the dermal papillae of the dermis and the interpapillary (rete) pegs of the epidermis. These epidermal ridges serve to amplify vibrations triggered, for example, when fingertips brush across an uneven surface, better transmitting the signals to sensory nerves involved in fine texture perception. These ridges may also assist in gripping rough surfaces and may improve surface contact in wet conditions.

Impressions of fingerprints may be left behind on a surface by the natural secretions of sweat from the eccrine glands that are present in friction ridge skin, or they may be made by ink or other substances transferred from the peaks of friction ridges on the skin to a relatively smooth surface such as a fingerprint card. Fingerprint records normally contain impressions from the pad on the last joint of fingers and thumbs, although fingerprint cards also typically record portions of lower joint areas of the fingers.

Since the early 20th century, fingerprint detection and analysis has been one of the most common and important forms of crime sceneforensic investigation. More crimes have been solved with fingerprint evidence than for any other reason.

FOR IDENTIFICATION

Fingerprint identification, known as dactyloscopy, or hand print identification, is the process of comparing two instances of friction ridge skin impressions, from human fingers or toes, or even the palm of the hand or sole of the foot, to determine whether these impressions could have come from the same individual. The flexibility of friction ridge skin means that no two finger or palm prints are ever exactly alike in every detail; even two impressions recorded immediately after each other from the same hand may be slightly different. Fingerprint identification, also referred to as individualization, involves an expert, or an expert computer system operating under threshold scoring rules, determining whether two friction ridge impressions are likely to have originated from the same finger or palm (or toe or sole).

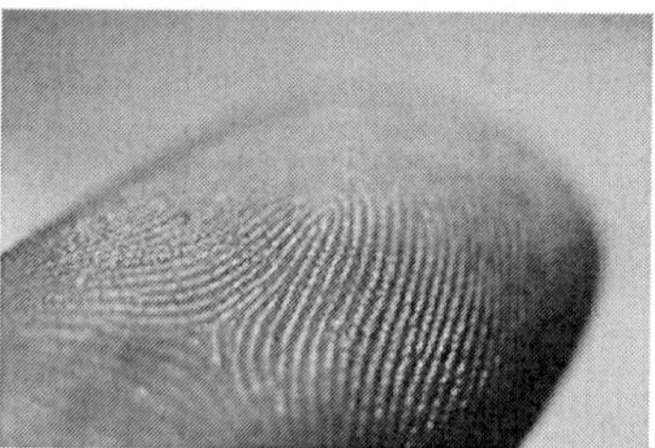

Fig. The friction ridges on a finger

An intentional recording of friction ridges is usually made with black printer's ink rolled across a contrasting white background, typically a white card. Friction ridges can also be recorded digitally, usually on a glass plate, using a technique called Live Scan. A "latent print" is the chance recording of friction ridges deposited on the surface of an object or a wall. Latent prints are invisible to the naked eye, whereas "patent prints" or "plastic prints" are viewable with the un-aided eye. Latent prints are often fragmentary and require the use of chemical methods, powder, or alternative light sources in order to be made clear. Sometimes an ordinary bright flashlight will make a latent print visible.

Fig. An image of a fingerprint created by the friction ridge structure

When friction ridges come into contact with a surface that will take a print, material that is on the friction ridges such as perspiration, oil, grease, ink or blood, will be transferred to the surface. Factors which affect the quality of friction ridge impressions are numerous. Pliability of the skin, deposition pressure, slippage, the material from which the surface is made, the roughness of the surface and the substance deposited are just some of the various factors which can cause a latent print to appear differently from any known recording of the same friction ridges. Indeed, the conditions surrounding every instance of friction ridge deposition are unique and never duplicated. For these reasons, fingerprint examiners are required to undergo extensive training. The scientific study of fingerprints is called dermatoglyphics.

TYPES

Exemplar

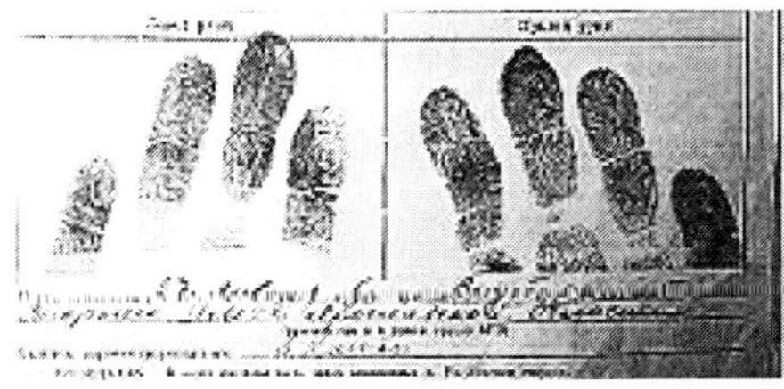

Fig. Exemplar prints on paper using ink

Exemplar prints, or known prints, is the name given to fingerprints deliberately collected from a subject, whether for purposes of enrollment in a system or when under arrest for a suspected criminal offense. During criminal arrests, a set of exemplar prints will normally include one print taken from each finger that has been rolled from one edge of the nail to the other, plain (or slap) impressions of each of the four fingers of each hand, and plain impressions of each thumb. Exemplar prints can be collected usingLive Scan or by using ink on paper cards.

Latent

Fig. Barely visible latent prints on a knife

Although the word latent means hidden or invisible, in modern usage for forensic science the term latent prints means any chance or accidental impression left by friction ridge skin on a surface, regardless of whether it is visible or invisible at the time of deposition. Electronic, chemical and physical processing techniques permit visualization of invisible latent print residues whether they are from natural sweat on the skin or from a contaminant such as motor oil, blood, ink, paint or some other form of dirt. The different types of fingerprint patterns, such as arch, loop and whorl, will be described below.

Latent prints may exhibit only a small portion of the surface of a finger and this may be smudged, distorted, overlapped by other prints from the same or from different individuals, or any or all of these in combination. For this reason, latent prints usually present an "inevitable source of error in making comparisons," as they generally "contain less clarity, less content, and less undistorted information than a fingerprint taken under controlled conditions, and much, much less detail compared to the actual patterns of ridges and grooves of a finger."

Patent

Patent prints are chance friction ridge impressions which are obvious to the human eye and which have been caused by the transfer of foreign material from a finger onto a surface. Some obvious examples would be impressions from flour and wet clay. Because they are already visible and have no need of enhancement they are generally photographed rather than being lifted in the way that latent prints are. An attempt to preserve the actual print is always made for later presentation in court, and there are many techniques used to do this. Patent prints can be left on a surface by materials such as ink, dirt, or blood.

Plastic

A plastic print is a friction ridge impression left in a material that retains the shape of the ridge detail. Although very few criminals would be careless enough to leave their prints in a lump of wet clay, this would make a perfect plastic print. Commonly encountered examples are melted candle wax, putty removed from the perimeter of window panes and thick grease deposits on car parts. Such prints are already visible and need no enhancement, but investigators must not overlook the potential that invisible latent prints deposited by accomplices may also be on such surfaces. After photographically recording such prints, attempts should be made to develop other non-plastic impressions deposited from sweat or other contaminants.

Electronic recording

There has been a newspaper report of a man selling stolen watches sending images of them on a mobile phone, and those images included parts of his hands in enough detail for police to be able to identify fingerprint patterns.

CLASSIFYING

Before computerisation replaced manual filing systems in large fingerprint operations, manual fingerprint classification systems were used to categorize fingerprints based on general ridge formations (such as the presence or absence of circular patterns on various fingers), thus permitting filing and retrieval of paper records in large collections based on friction ridge patterns alone. The most popular ten-print classification systems include the Roscher system, the Juan Vucetich system, and the Henry Classification System. Of these systems, the Roscher system was developed in Germany and implemented in both Germany and Japan, the Vucetich system (developed by a Croatian-born Buenos Aires Police Officer) was developed in Argentina and implemented throughout South America, and the Henry system was developed in India and implemented in most English-speaking countries.

In the Henry system of classification, there are three basic fingerprint patterns: loop, whorl and arch, which constitute 60–65%, 30–35% and 5% of all fingerprints respectively. There are also more complex classification systems that break down patterns even further, into plain arches or tented arches, and into loops that may be radial or ulnar, depending on the side of the hand toward which the tail points. Ulnar loops start on the pinky-side of the finger, the side closer to the ulna, the lower arm bone. Radial loops start on the thumb-side of the finger, the side closer to the radius. Whorls may also have sub-group classifications including plain whorls, accidental whorls, double loop whorls, peacock's eye, composite, and central pocket loop whorls.

Other common fingerprint patterns include the tented arch, the plain arch, and the central pocket loop. The system used by most experts, although complex, is similar to the Henry System of Classification. It consists of five fractions, in which R stands for right, L for left, i for index finger, m for middle finger, t for thumb, r for ring finger and p(pinky) for little finger. The fractions are as follows: Ri/Rt + Rr/Rm + Lt/Rp + Lm/Li + Lp/Lr. The numbers assigned to each print are based on whether or not they are whorls. A whorl in the first fraction is given a 16, the second an 8, the third a 4, the fourth a 2, and 0 to the last fraction. Arches and loops are assigned values of 0. Lastly, the numbers in the numerator and denominator are added up, using the scheme:

(Ri + Rr + Lt + Lm + Lp)/(Rt + Rm + Rp + Li + Lr)

and a 1 is added to both top and bottom, to exclude any possibility of division by zero. For example, if the right ring finger and the left index finger have whorls, the fractions would look like this:

0/0 + 8/0 + 0/0 + 0/2 + 0/0 + 1/1, and the calculation: (0 + 8 + 0 + 0 + 0 + 1)/(0 + 0 + 0 + 2 + 0 + 1) = 9/3 = 3.

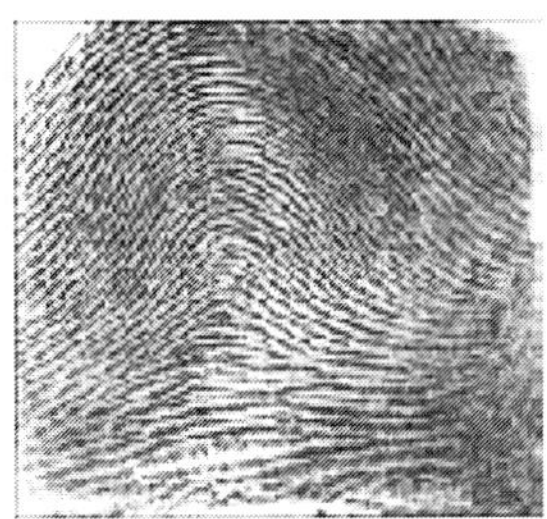

Using this system reduces the number of prints that the print in question needs to be compared to. For example, the above set of prints would only need to be compared to other sets of fingerprints with a value of 3.

Arch

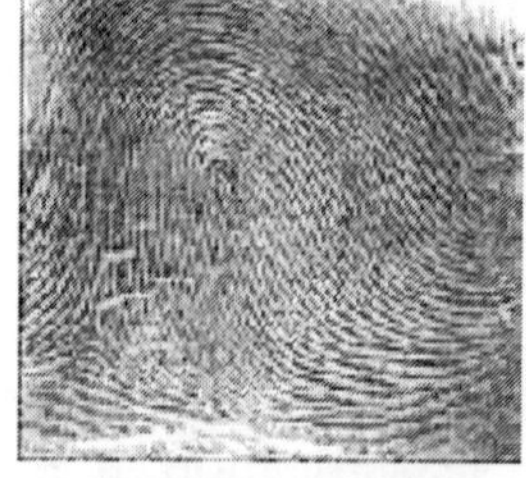

Loop (Right Loop)

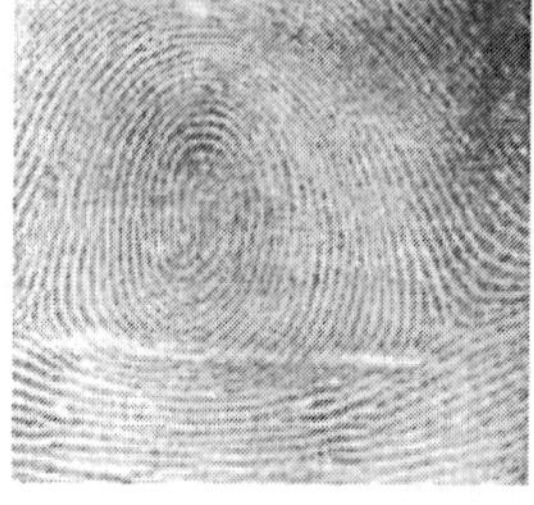

Whorl

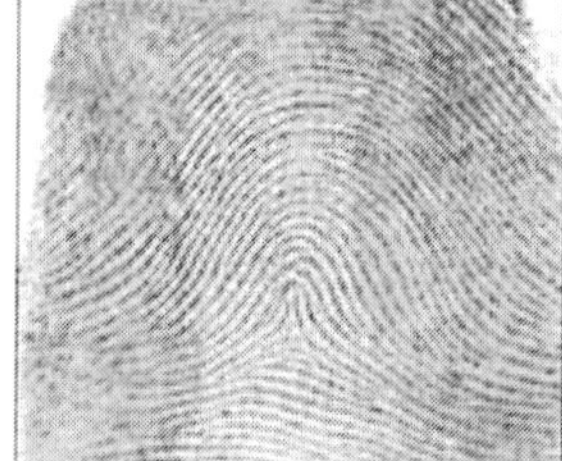

Arch (Tented Arch)

FOOTPRINTS

Friction ridge skin present on the soles of the feet and toes (plantar surfaces) is as unique in its ridge detail as are the fingers and palms (palmar surfaces). When recovered at crime scenes or on items of evidence, sole and toe impressions can be used in the same manner as finger and palm prints to effect identifications. Footprint (toe and sole friction ridge skin) evidence has been admitted in courts in the United States since 1934.

The footprints of infants, along with the thumb or index finger prints of mothers, are still commonly recorded in hospitals to assist in verifying the identity of infants. Often, the only identifiable ridge detail that can be seen on a baby's foot is from the large toe or adjacent to the large toe.

It is not uncommon for military records of flight personnel to include bare foot inked impressions. Friction ridge skin protected inside flight boots tends to survive the trauma of a plane crash (and accompanying fire) better than fingers. Even though the US Armed Forces DNA Identification Laboratory (AFDIL), as of 2010, stored refrigerated DNA samples from all active duty and reserve personnel, almost all casualty identifications are effected using fingerprints from military ID card records (live scan fingerprints are recorded at the time such cards are issued). When friction ridge skin is not available from military personnel's remains, DNA and dental records are used to confirm identity.

CAPTURE AND DETECTION

Livescan devices

Fig. Fingerprint being scanned

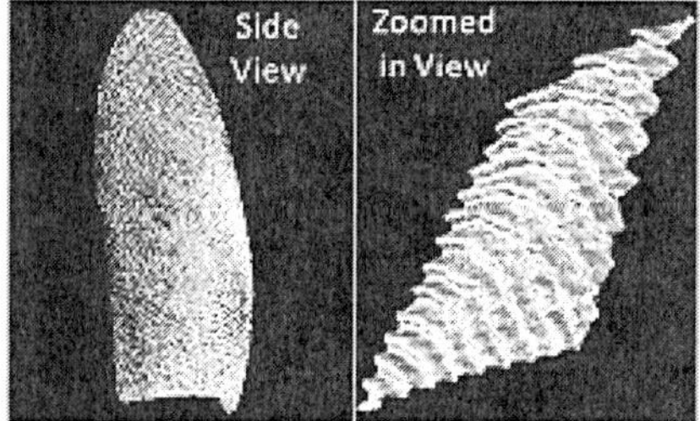

Fig.3D fingerprint

Fingerprint image acquisition is considered to be the most critical step in an automated fingerprint authentication system, as it determines the final fingerprint image quality, which has a drastic effect on the overall system performance. There are different types of fingerprint readers on the market, but the basic idea behind each is to measure the physical difference between ridges and valleys.

All the proposed methods can be grouped into two major families: solid-state fingerprint readers and optical fingerprint readers. The procedure for capturing a fingerprint using a sensor consists of rolling or touching with the finger onto a sensing area, which according to the physical principle in use (optical, ultrasonic, capacitive or thermal) captures the difference between valleys and ridges. When a finger touches or rolls onto a surface, the elastic skin deforms. The quantity and direction of the pressure applied by the user, the skin conditions and the projection of an irregular 3D object (the finger) onto a 2D flat plane introduce distortions, noise and inconsistencies in the captured fingerprint image. These problems result in inconsistent, irreproducible and non-uniform irregularities in the image. During each acquisition, therefore, the results of the imaging are different and uncontrollable. The representation of the same fingerprint changes every time the finger is placed on the sensor plate, increasing the complexity of any attempt to match fingerprints, impairing the system performance and consequently, limiting the widespread use of thisbiometric technology.

In order to overcome these problems, as of 2010, non-contact or touchless 3D fingerprint scanners have been developed.Acquiring detailed 3D information, 3D fingerprint scanners take a digital approach to the analog process of pressing or rolling the finger. By modelling the distance between neighboring points, the fingerprint can be imaged at a resolution high enough to record all the necessary detail.

Scanning dead or unconscious people

Placing the hand of a dead or unconscious person on a scanner to gain unauthorized access has become a common plot device. However, a Mythbusters episode revealed that this doesn't work (at least with the scanners available to the program). But Adam Savage and Jamie Hyneman found a way to convert fingerprints lifted from the hand to a photographic form that the sensor would accept. For obvious reasons, they refuse to reveal the technique.

Latent detection

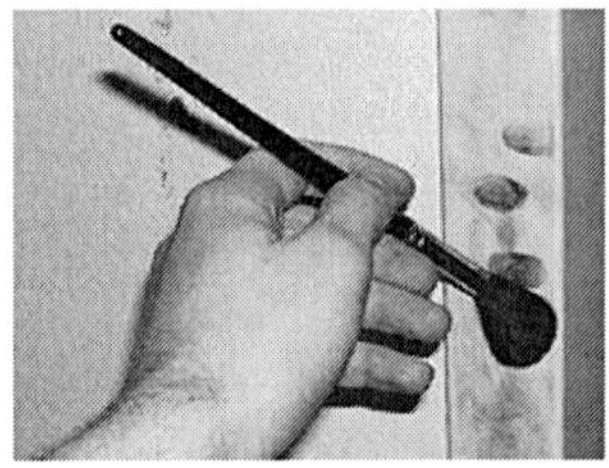

Fig. Use of fine powder and brush to reveal latent fingerprints

Since the late nineteenth century, fingerprint identification methods have been used by police agencies around the world to identify suspected criminals as well as the victims of crime. The basis of the traditional fingerprinting technique is simple. The skin on the palmar surface of the hands and feet forms ridges, so-called papillary ridges, in patterns that are unique to each individual and which do not change over time. Even identical twins (who share their DNA) do not have identical fingerprints. The best way to render latent fingerprints visible, so that they can be photographed, can be complex and may depend, for example, on the type of surfaces on which they have been left. It is generally necessary to use a 'developer', usually a powder or chemical reagent, to produce a high degree of visual contrast between the ridge patterns and the surface on which a fingerprint has been deposited.

Developing agents depend on the presence of organic materials or inorganic salts for their effectiveness, although the water deposited may also take a key role. Fingerprints are typically formed from the aqueous-based secretions of the eccrine glands of the fingers and palms with additional material from sebaceous glands primarily from the forehead. This latter contamination results from the common human behaviors of touching the face and hair. The resulting latent fingerprints consist usually of a substantial proportion of water with small traces of amino acids and chlorides mixed with a fatty, sebaceous component which contains a number of fatty acids and triglycerides. Detection of a small proportion of reactive organic substances such as urea and amino acids is far from easy.

Fingerprints at a crime scene may be detected by simple powders, or by chemicals applied in situ. More complex techniques, usually involving chemicals, can be applied in specialist laboratories to appropriate articles removed from a crime scene. With advances in these more sophisticated techniques, some of the more advanced crime scene investigation services from around the world were, as of 2010, reporting that 50% or more of the fingerprints recovered from a crime scene had been identified as a result of laboratory-based techniques.

Fig.A city fingerprint identification room.

Laboratory techniques

Although there are hundreds of reported techniques for fingerprint detection, many of these are only of academic interest and there are only around 20 really effective methods which are currently in use in the more advanced fingerprint laboratories around the world. Some of these techniques, such as ninhydrin, diazafluorenone and vacuum metal deposition, show great sensitivity and are used operationally. Some fingerprint reagents are specific, for example ninhydrin or diazafluorenone reacting with amino acids. Others such as ethyl cyanoacrylate polymerisation, work apparently by water-based catalysis and polymer growth. Vacuum metal deposition using gold and zinc has been shown to be non-specific, but can detect fat layers as thin as one molecule. More mundane methods, such as the application of fine powders, work by adhesion to sebaceous deposits and possibly aqueous deposits in the case of fresh fingerprints. The aqueous component of a fingerprint, whilst initially sometimes making up over 90% of the weight of the fingerprint, can evaporate quite quickly and may have mostly gone after 24 hours. Following work on the use of argon ion lasers for fingerprint detection, a wide range of fluorescence techniques have been introduced, primarily for the enhancement of chemically-developed fingerprints; the inherent fluorescence of some latent fingerprints may also be detected. The most comprehensive manual of the operational methods of fingerprint enhancement is published by the UK Home Office Scientific Development Branch and is used widely around the world.

Research

The International Fingerprint Research Group (IFRG) which meets biennially, consists of members of the leading fingerprint research groups from Europe, the US, Canada, Australia and Israel and leads the way in the development, assessment and implementation of new techniques for operational fingerprint detection.

One problem for the early twenty-first century is the fact that the organic component of any deposited material is readily destroyed by heat, such as occurs when a gun is fired or a bomb is detonated, when the temperature may reach as high as 500°C. Encouragingly, however, the non-volatile inorganic component of eccrine secretion has been shown to remain intact even when exposed to temperatures as high as 600°C.

A technique has been developed that enables fingerprints to be visualised on metallic and electrically conductive surfaces without the need to develop the prints first. This technique involves the use of an instrument called a scanning Kelvin probe (SKP), which measures the voltage, or electrical potential, at pre-set intervals over the surface of an object on which a fingerprint may have been deposited.

These measurements can then be mapped to produce an image of the fingerprint. A higher resolution image can be obtained by increasing the number of points sampled, but at the expense of the time taken for the process. A sampling frequency of 20 points per mm is high enough to visualise a fingerprint in sufficient detail for identification purposes and produces a voltage map in 2–3 hours. As of 2010, this technique had been shown to work effectively on a wide range of forensically important metal surfaces including iron, steel and aluminium. While initial experiments were performed on flat surfaces, the technique has been further developed to cope with irregular or curved surfaces, such as the warped cylindrical surface of fired cartridge cases. Research during 2010 at Swansea University has found that physically removing a fingerprint from a metal surface, for example by rubbing with a tissue, does not necessarily result in the loss of all fingerprint information from that surface. The reason for this is that the differences in potential that are the basis of the visualisation are caused by the interaction of inorganic salts in the fingerprint deposit and the metal surface and begin to occur as soon as the finger comes into contact with the metal, resulting in the formation of metal-ion complexes that cannot easily be removed.

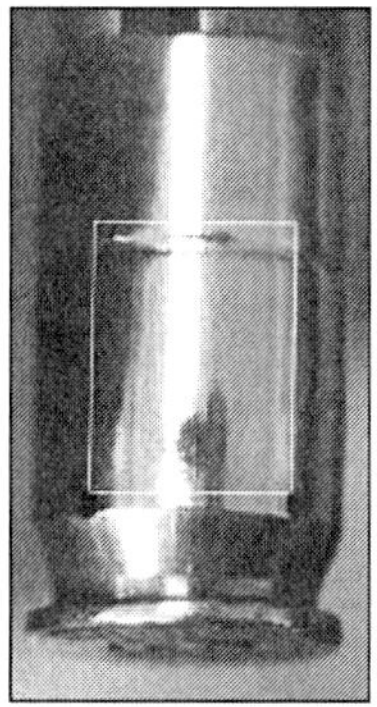

Fig. Cartridge case with an applied fingerprint

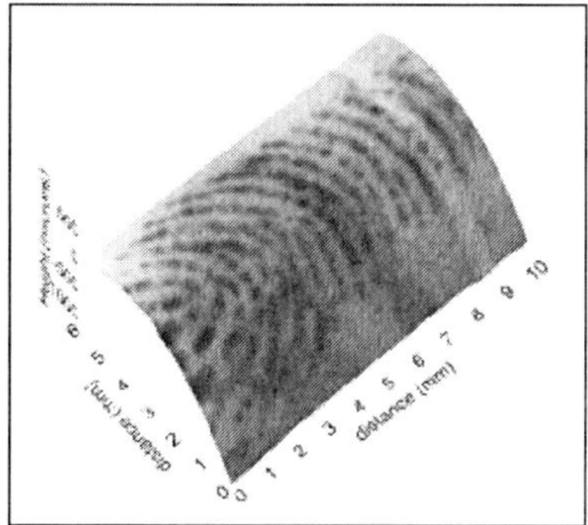

Fig. Scanning Kelvin probe scan of the same cartridge case with the fingerprint detected. The Kelvin probe can easily cope with the 3D curvature of the cartridge case, increasing the versatility of the technique.

Another problem for the early twenty-first century is that during crime scene investigations, a decision has to be made at an early stage whether to attempt to retrieve fingerprints through the use of developers or whether to swab surfaces in an attempt to salvage material for DNA profiling. The two processes are mutually incompatible, as fingerprint developers destroy material that could potentially be used for DNA analysis, and swabbing is likely to make fingerprint identification impossible.

The application of the new scanning Kelvin probe (SKP) fingerprinting technique, which makes no physical contact with the fingerprint and does not require the use of developers, has the potential to allow fingerprints to be recorded whilst still leaving intact material that could subsequently be subjected to DNA analysis. A forensically usable prototype was under development at Swansea University during 2010, in research that was generating significant interest from the British Home Office and a number of different police forces across the UK, as well as internationally. The hope is that this instrument could eventually be manufactured in sufficiently large numbers to be widely used by forensic teams worldwide.

DISAPPEARANCE OF CHILDREN'S LATENT PRINTS

In 1995, researchers at the Oak Ridge National Laboratory, at the instigation of Detective Art Bohanan of the Knoxville Police Department, discovered that children's fingerprints are considerably more short-lived than adult fingerprints. The rapid disappearance of children's fingerprints was attributed to a lack of the more waxy oils that become present at the onset of puberty. The lighter fatty acids of children's fingerprints evaporate within a few hours. As of 2010, researchers at Oak Ridge National Laboratory are investigating techniques to capture these lost fingerprints.

Detection of drug use

The secretions, skin oils and dead cells in a human fingerprint contain residues of various chemicals and their metabolites present in the body. These can be detected and used for forensic purposes. For example, the fingerprints of tobacco smokers contain traces of cotinine, a nicotine metabolite; they also contain traces of nicotine itself. Caution should be used, as its presence may be caused by mere contact of the finger with a tobacco product. By treating the fingerprint with gold nanoparticles with attached cotinine antibodies, and then subsequently with a fluorescent agent attached to cotinine antibodies, the fingerprint of a smoker becomes fluorescent; non-smokers' fingerprints stay dark. The same approach, as of 2010, is being tested for use in identifying heavy coffee drinkers, cannabis smokers, and users of various other drugs.In 2008, British researchers developed methods of identifying users of marijuana, cocaine and methadone from their fingerprint residues.

United States databases and compression

In the United States, the FBI manages a fingerprint identification system and database called the Integrated Automated Fingerprint Identification System, or IAFIS, which currently holds the fingerprints and criminal records of over 51 million criminal record subjects and over 1.5 million civil (non-criminal) fingerprint records. US Visit currently holds a repository of the fingerprints of over 50 million people, primarily in the form of two-finger records. In 2008, US Visit hoped to have changed over to a system recording FBI-standard ten-print records.

Most American law enforcement agencies use Wavelet Scalar Quantization (WSQ), a wavelet-based system for efficient storage of compressed fingerprint images at 500pixels per inch (ppi). WSQ was developed by the FBI, the Los Alamos National Lab, and the National Institute for Standards and Technology (NIST). For fingerprints recorded at 1000 ppi spatial resolution, law enforcement (including the FBI) uses JPEG 2000 instead of WSQ.

Fig.A city fingerprint identification office

Validity

The validity of forensic fingerprint evidence has been challenged by academics, judges and the media. While fingerprint identification was an improvement on earlier anthropometric systems, the subjective nature of matching, despite a very low error rate, has made this forensic practice controversial. Certain specific criticisms are now being accepted by some leaders of the forensic fingerprint community, providing an incentive to improve training and procedures.

Criticism

The words "reliability" and "validity" have specific meanings to the scientific community. Reliability means that successive tests bring the same results. Validity means that these results are judged to accurately reflect the external criteria being measured. Although experts are often more comfortable relying on their instincts, this reliance does not always translate into superior predictive ability. For example, in the popular Analysis, Comparison, Evaluation, and Verification (ACE-V) paradigm for fingerprint identification, the verification stage, in which a second examiner confirms the assessment of the original examiner, may increase the consistency of the assessments. But while the verification stage has implications for the reliability of latent print comparisons, it does not assure their validity.

The few tests that have been made of the validity of forensic fingerprinting have not been supportive of the method. "Despite the absence of objective standards, scientific validation, and adequate statistical studies, a natural question to ask is how well fingerprint examiners actually perform. Proficiency tests do not validate a procedure per se, but they can provide some insight into error rates. In 1995, the Collaborative Testing Service (CTS) administered a proficiency test that, for the first time, was “designed, assembled, and reviewed” by the International Association for Identification(IAI).The results were disappointing. Four suspect cards with prints of all ten fingers were provided together with seven latents. Of 156 people taking the test, only 68

(44%) correctly classified all seven latents. Overall, the tests contained a total of 48 incorrect identifications. David Grieve, the editor of the Journal of Forensic Identification, describes the reaction of the forensic community to the results of the CTS test as ranging from "shock to disbelief," and added:

'Errors of this magnitude within a discipline singularly admired and respected for its touted absolute certainty as an identification process have produced chilling and mind- numbing realities. Thirty-four participants, an incredible 22% of those involved, substituted presumed but false certainty for truth. By any measure, this represents a profile of practice that is unacceptable and thus demands positive action by the entire community.' What is striking about these comments is that they do not come from a critic of the fingerprint community, but from the editor of one of its premier publications."

Investigations have been conducted into whether experts can objectively focus on feature information in fingerprints without being misled by extraneous information, such as context. Fingerprints that have previously been examined and assessed by latent print experts to make a positive identification of suspects have then been re-presented to those same experts in a new context which makes it likely that there will be no match. Within this new context, most of the fingerprint experts made different judgments, thus contradicting their own previous identification decisions.

Complaints have been made that there have been no published, peer-reviewed studies directly examining the extent to which people can correctly match fingerprints to one another. Experiments have been carried out using naïve undergraduates to match images of fingerprints. The results of these experiments demonstrate that people can identify fingerprints quite well, and that matching accuracy can vary as a function of both source finger type and image similarity.

Defense

Fingerprints collected at a crime scene, or on items of evidence from a crime, have been used in forensic science to identify suspects, victims and other persons who touched a surface. Fingerprint identification emerged as an important system within police agencies in the late 19th century, when it replaced anthropometric measurements as a more reliable method for identifying persons having a prior record, often under a false name, in a criminal record repository. The science of fingerprint identification has been able to assert its standing amongst forensic sciences for many reasons.

Track record

Fingerprinting has served all governments worldwide during the past 100 years or so to provide accurate identification of criminals. No two fingerprints have ever been found identical in many billions of human and automated computer comparisons. Fingerprints are the fundamental tool for the identification of people with a criminal history in every police agency. It remains the most commonly gathered forensic evidence worldwide and in most jurisdictions fingerprint examination outnumbers all other forensic examination casework combined. Moreover, it continues to expand as the premier method for identifying persons, with tens of thousands of people added to fingerprint repositories daily in America alone — far more than other forensic databases.

Professional standing and certification

Fingerprinting was the basis upon which the first forensic professional organization was formed, the International Association for Identification (IAI), in 1915. The first professional certification program for forensic scientists was established in 1977, the IAI's Certified Latent Print Examiner program, which issued certificates to those meeting stringent criteria and had the power to revoke certification where an individual's performance warranted it. Other forensic disciplines have followed suit and established their own certification programs.

INSTANCES OF ERROR

Brandon Mayfield and the Madrid bombing

Brandon Mayfield is an Oregon lawyer who was identified as a participant in the 2004 Madrid train bombings based on a fingerprint match by the FBI. The FBI Latent Print Unit processed a fingerprint collected in Madrid and reported a "100 percent positive" match against one of the 20 fingerprint candidates returned in a search response from their IAFIS — Integrated Automated Fingerprint Identification System. The FBI initially called it an "absolutely incontrovertible match". Subsequently, however, Spanish National Police examiners suggested that the print did not match Mayfield and after two weeks, identified another man whom they claimed the fingerprint did belong to. The FBIacknowledged their error, and a judge released Mayfield, who had spent two weeks in police custody, in May 2004. In January 2006, a U.S. Justice Department report was released which criticized the FBI for sloppy work but exonerated them of some more serious allegations. The report found that the misidentification had been due to a misapplication of methodology by the examiners involved: Mayfield is an American-born convert to Islam and his wife is an Egyptian immigrant, but these are not factors that should have affected fingerprint search technology.

On 29 November 2006, the FBI agreed to pay Brandon Mayfield US$2 million in compensation. The judicial settlement allowed Mayfield to continue a suit regarding certain other government practices surrounding his arrest and detention. The formal apology stated that the FBI, which erroneously linked him to the 2004 Madrid bombing through a fingerprinting mistake, had taken steps to "ensure that what happened to Mr. Mayfield and the Mayfield family does not happen again."

René Ramón Sánchez

René Ramón Sánchez, a legal Dominican Republic immigrant to the US was arrested on July 15, 1995, on a charge of driving while intoxicated (Driving Under the Influence, or DUI). His fingerprints, however, were placed on a card containing the name, Social Security number and other data for one Leo Rosario, who was being processed at the same time. Leo Rosario had been arrested for selling cocaine to an undercover police officer. On October 11, 2000, while returning from a visit to relatives in the Dominican Republic, René was misidentified as Leo Rosario at John F. Kennedy International Airport in New York and arrested. Even though he did not match the physical description of Rosario, the erroneously-cataloged fingerprints were considered to be more reliable.

Shirley McKie

Shirley McKie was a police detective in 1997 when she was accused of leaving her thumb print inside a house in Kilmarnock, Scotland where Marion Ross had been murdered. Although McKie denied having been inside the house, she was arrested in a dawn raid the following year and charged with perjury. The only evidence the prosecution had was this thumb print allegedly found at the murder scene. Two American experts testified on her behalf at her trial in May 1999 and she was found not guilty. The Scottish Criminal Record Office (SCRO) would not admit any error, although Scottish first minister Jack McConnell later said it had been an "honest mistake".

On February 7, 2006, McKie was awarded £750,000 in compensation from the Scottish Executive and the Scottish Criminal Record Office. Controversy continued to surround the McKie case and the Fingerprint Inquiry into the affair finished taking evidence in November 2009 and is awaiting publication of the final report.

Stephan Cowans

Stephan Cowans was convicted of attempted murder in 1997 after he was accused of shooting a police officer whilst fleeing a robbery in Roxbury, Massachusetts. He was implicated in the crime by the testimony of two witnesses, one of whom was the victim. There was also a fingerprint on a glass mug from which the assailant had drunk some water and experts testified that the fingerprint belonged to Cowans. He was found guilty and sent to prison for 35 years. Whilst in prison, Cowans earned money cleaning up biohazards until he could afford to have the evidence against him tested for DNA. The DNA did not match his and he was released. He had already served six years in prison when he was released on January 23, 2004. Cowans died on October 25, 2007.

Craig D. Harvey

In April 1993, in the New York State Police Troop C scandal, Craig D. Harvey, a New York State Police trooper was charged with fabricating evidence. Harvey admitted he and another trooper lifted fingerprints from items the suspect, John Spencer, touched while in Troop C headquarters during booking. He attached the fingerprints to evidence cards and later claimed that he had pulled the fingerprints from the scene of the murder. The forged evidence was presented during John Spencer's trial and his subsequent conviction resulted in a term of 50 years to life in prison at his sentencing. Three state troopers were found guilty of fabricating fingerprint evidence and served prison sentences.

HISTORY

Antiquity and the medieval period

Fingerprints have been found on ancient Babylonian clay tablets, seals, and pottery. They have also been found on the walls of Egyptian tombs and on Minoan, Greek, and Chinese pottery, as well as on bricks and tiles from ancient Babylon and Rome. Some of these fingerprints were deposited unintentionally by the potters and masons as a natural consequence of their work, and others were made in the process of adding decoration. However, on some pottery, fingerprints have been impressed so deeply into the clay that they were possibly intended to serve as an identifying mark by the maker.

Fingerprints were used as signatures in ancient Babylon in the second millennium BCE. In order to protect against forgery, parties to a legal contract would impress their fingerprints into a clay tablet on which the contract had been written. By 246 BCE, Chinese officials were impressing their fingerprints into the clay seals used to seal documents. With the advent of silk and paper in China, parties to a legal contract impressed their handprints on the document. Sometime before 851 CE, an Arab merchant in China, Abu Zayd Hasan, witnessed Chinese merchants using fingerprints to authenticate loans. By 702, Japan allowed illiterate petitioners seeking a divorce to "sign" their petitions with a fingerprint.

Although ancient peoples probably did not realize that fingerprints could uniquely identify individuals, references from the age of the Babylonian king Hammurabi (1792-1750 BCE) indicate that law officials would take the fingerprints of people who had been arrested. During China's Qin Dynasty, records have shown that officials took hand prints, foot prints as well as finger prints as evidence from a crime scene. In China, around 300 CE, handprints were used as evidence in a trial for theft. By 650, the Chinese historian Kia Kung-Yen remarked that fingerprints could be used as a means of authentication. In his Jami al-Tawarikh (Universal History), the Persian physicianRashid-al-Din Hamadani (also known as "Rashideddin", 1247–1318) refers to the Chinese practice of identifying people via their fingerprints, commenting: "Experience shows that no two individuals have fingers exactly alike." In Persia at this time, government documents may have been authenticated with thumbprints.

Europe in the 17th and 18th centuries

In 1684, the English physician, botanist, and microscopist Nehemiah Grew (1641–1712) published the first scientific paper to describe the ridge structure of the skin covering the fingers and palms. In 1685, the Dutch physician Govard Bidloo (1649–1713) and the Italian physician Marcello Malpighi (1628–1694) published books on anatomy which also illustrated the ridge structure of the fingers. A century later, in 1788, the German anatomist Johann Christoph Andreas Mayer (1747–1801) recognized that fingerprints are unique to each individual.

Modern era

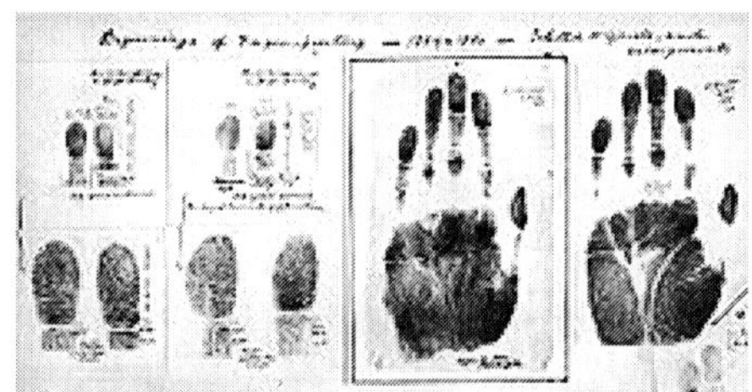

Fig. Fingerprints taken by William Herschel 1859/60

Jan Evangelista Purkyne or Purkinje (1787–1869), a Czech physiologist and professor of anatomy at the University of Breslau, published a thesis in 1823 discussing 9 fingerprint patterns, but he did not mention any possibility of using fingerprints to identify people. Some years later, the German anatomist Georg von Meissner (1829–1905) studied friction ridges, and five years after this, in 1858, Sir William James Herschel initiated fingerprinting in India. In 1877 at Hooghly (near Calcutta) he instituted the use of fingerprints on contracts and deeds to prevent the then-rampant repudiation of signatures and he registered government pensioners' fingerprints to prevent the collection of money by relatives after a pensioner's death. Herschel also fingerprinted prisoners upon sentencing to prevent various frauds that were attempted in order to avoid serving a prison sentence.

In 1863, Paul-Jean Coulier (1824–1890), professor for chemistry and hygiene at the medical and pharmaceutical school of the Val de Grâce military hospital in Paris, discovered that iodine fumes can reveal fingerprints on paper. In 1880, Dr. Henry Faulds, a Scottish surgeon in a Tokyo hospital, published his first paper on the subject in the scientific journal Nature, discussing the usefulness of fingerprints for identification and proposing a method to record them with printing ink. He also established their first classification and was also the first to identify fingerprints left on a vial. Returning to the UK in 1886, he offered the concept to the Metropolitan Police in London but it was dismissed at that time. Faulds wrote to Charles Darwinwith a description of his method but, too old and ill to work on it, Darwin gave the information to his cousin, Francis Galton, who was interested in anthropology. Having been thus inspired to study fingerprints for ten years, Galton published a detailed statistical model of fingerprint analysis and identification and encouraged its use in forensic science in his book Finger Prints. He had calculated that the chance of a "false positive" (two different individuals having the same fingerprints) was about 1 in 64 billion.

Juan Vucetich, an Argentine chief police officer, created the first method of recording the fingerprints of individuals on file, associating these fingerprints to the anthropometric system of Alphonse Bertillon, who had created, in 1879, a system to identify individuals by anthropometric photographs and associated quantitative descriptions. In 1892, after studying Galton's pattern types, Vucetich set up the world's first fingerprint bureau. In that same year, Francisca Rojas of Necochea, was found in a house with neck injuries, whilst her two sons were found dead with their throats cut. Rojas accused a neighbour, but despite brutal interrogation, this neighbour would not confess to the crimes. Inspector Alvarez, a colleague of Vucetich, went to the scene and found a bloody thumb mark on a door. When it was compared with Rojas' prints, it was found to be identical with her right thumb. She then confessed to the murder of her sons.

Fig. Women clerical employees of LA Police Department getting fingerprinted and photographed in 1928.

A Fingerprint Bureau was established in Calcutta (Kolkata), India, in 1897, after the Council of the Governor General approved a committee report that fingerprints should be used for the classification of criminal records. Working in the Calcutta Anthropometric Bureau, before it became the Fingerprint Bureau, were Azizul Haque and Hem Chandra Bose. Haque and Bose were Indian fingerprint experts who have been credited with the primary development of a fingerprint classification system eventually named after their supervisor, Sir Edward Richard Henry. The Henry Classification System, co-devised by Haque and Bose, was accepted in England and Wales when the first United Kingdom Fingerprint Bureau was founded in Scotland Yard, the Metropolitan Police headquarters, London, in 1901. Sir Edward Richard Henry subsequently achieved improvements in dactyloscopy.

In the United States, Dr. Henry P. DeForrest used fingerprinting in the New York Civil Service in 1902, and by 1906, New York City Police Department Deputy Commissioner Joseph A. Faurot, an expert in the Bertillon system and a finger print advocate at Police Headquarters, introduced the fingerprinting of criminals to the United States.

The Scheffer case of 1902 is the first case of the identification, arrest and conviction of a murderer based upon fingerprint evidence. Alphonse Bertillon identified the thief and murderer Scheffer, who had previously been arrested and his fingerprints filed some months before, from the fingerprints found on a fractured glass showcase, after a theft in a dentist's apartment where the dentist's employee was found dead. It was able to be proved in court that the fingerprints had been made after the showcase was broken. A year later, Alphonse Bertilloncreated a method of getting fingerprints off smooth surfaces and took a further step in the advance of dactyloscopy.

Since the advent of fingerprint detection, many criminals have resorted to the wearing of gloves in order to avoid leaving fingerprints, which thus makes the crime investigation more difficult. However, the gloves themselves can leave prints that are just as unique as human fingerprints. After collecting glove prints, law enforcement can then match them to gloves that they have collected as evidence. In many jurisdictions the act of wearing gloves itself while committing a crime can be prosecuted as an inchoate offense.

As many offenses are crimes of opportunity, many assailants are not in the possession of gloves when they commit their illegal activities. Thus, assailants have been viewed using pulled-down sleeves and other pieces of clothing and fabric to handle objects and touch surfaces during the commission of their crimes.

PRIVACY ISSUES

Fingerprinting of children

Various schools have implemented fingerprint locks or made a record of children's fingerprints. In the United Kingdom there have been fingerprint locks in Holland Park Schoolin London, and children's fingerprints are stored on databases. There have also been instances in Belgium, at the école Marie-José in Liège, in France and in Italy. The non-governmental organization (NGO) Privacy International in 2002 made the cautionary announcement that tens of thousands of UK school children were being fingerprinted by schools, often without the knowledge or consent of their parents. That same year, the supplier Micro Librarian Systems, which uses a technology similar to that used in US prisons and the German military, estimated that 350 schools throughout Britain were using such systems to replace library cards. By 2007, it was estimated that 3,500 schools were using such systems. Under the United Kingdom Data Protection Act, schools in the UK do not have to ask parental consent to allow such practices to take place. Parents opposed to fingerprinting may only bring individual complaints against schools. In response to a complaint which they are continuing to pursue, in 2010 the European Commission expressed 'significant concerns' over the proportionality and necessity of the practice and the lack of judicial redress, indicating that the practice may break the European Union data protection directive.

In Belgium, the practice of taking fingerprints from children gave rise to a question in Parliament on February 6, 2007 by Michel de La Motte (Humanist Democratic Centre) to the Education Minister Marie Arena, who replied that it was legal provided that the school did not use them for external purposes, or to survey the private life of children. AtAngers in France, Carqueiranne College in the Var won the Big Brother Award for 2005 and the Commission nationale de l'informatique et des libertés (CNIL), the official organisation in charge of the protection of privacy in France, declared the measures it had introduced "disproportionate."

In March 2007, the British government was considering fingerprinting all children aged 11 to 15 and adding the prints to a government database as part of a new passportand ID card scheme and disallowing opposition for privacy concerns. All fingerprints taken would be cross-checked against prints from 900,000 unsolved crimes. Shadow Home secretary David Davis called the plan "sinister". An Early Day Motion which called on the UK Government to conduct a full and open consultation with stakeholders about the use of biometrics in schools, secured the support of 85 Members of Parliament (Early Day Motion 686). Following the establishment in the United Kingdom of a Conservative and Liberal Democratic coalition government in May 2010, the ID card scheme was scrapped.

Serious concerns about the security implications of using conventional biometric templates in schools have been raised by a number of leading IT security experts, one of whom has voiced the opinion that "it is absolutely premature to begin using 'conventional biometrics' in schools". The vendors of biometric systems claim that their products bring benefits to schools such as improved reading skills, decreased wait times in lunch lines and increased revenues.

They do not cite independent research to support this view. One education specialist wrote in 2007: "I have not been able to find a single piece of published research which suggests that the use of biometrics in schools promotes healthy eating or improves reading skills amongst children... There is absolutely no evidence for such claims". The Ottawa Police in Canada have advised parents who fear their children may be kidnapped, to fingerprint their children.

OTHER USES

Welfare claimants

It has been alleged that taking the fingerprints of welfare recipients as identification serves as a social stigma that evokes cultural images associated with the processing of criminals.

Log-in authentication and other locks

Since 2000, electronic fingerprint readers have been introduced for security applications such as log-in authentication for the identification of computer users. However, some less sophisticated devices have been discovered to be vulnerable to quite simple methods of deception, such as fake fingerprints cast in gels. In 2006, fingerprint sensors gained popularity in the notebook PC market. Built-in sensors in ThinkPads, VAIO, HP Pavilion laptops, and others also double as motion detectors for document scrolling, like the scroll wheel.

Following the release of the iPhone 5s model, a group of German hackers announced on September 21, 2013 that they had bypassed Apple's new Touch ID fingerprint sensor by photographing a fingerprint from a glass surface and using that captured image as verification. The spokesman for the group stated: "We hope that this finally puts to rest the illusions people have about fingerprint biometrics. It is plain stupid to use something that you can't change and that you leave everywhere every day as a security token."

Electronic registration and library access

Fingerprints and, to a lesser extent, iris scans can be used to validate electronic registration, cashless catering, and library access. By 2007, this practice was particularly widespread in UK schools, and it was also starting to be adopted in some states in the US.

ABSENCE OR MUTILATION OF FINGERPRINTS

A very rare medical condition, adermatoglyphia, is characterized by the absence of fingerprints. Affected persons have completely smooth fingertips, palms, toes and soles, but no other medical signs or symptoms. A 2011 study indicated that adermatoglyphia is caused by the improper expression of the protein SMARCAD1. The condition has been called immigration delay disease by the researchers describing it, because the congenital lack of fingerprints causes delays when affected persons attempt to prove their identity while traveling. Only five families with this condition have been described as of 2011.

People with Naegeli–Franceschetti–Jadassohn syndrome and dermatopathia pigmentosa reticularis, which are both forms of ectodermal dysplasia, also have no fingerprints. Both of these rare genetic syndromes produce other signs and symptoms as well, such as thin, brittle hair. The anti-cancer medication capecitabine may cause the loss of fingerprints. Swelling of the fingers, such as that caused by bee stings, will in some cases cause the temporary disappearance of fingerprints, though they will return when the swelling recedes. Since the elasticity of skin decreases with age, many senior citizens have fingerprints that are difficult to capture. The ridges get thicker; the height between the top of the ridge and the bottom of the furrow gets narrow, so there is less prominence.

Fingerprints can be erased permanently and this can potentially be used by criminals to reduce their chance of conviction. Erasure can be achieved in a variety of ways including simply burning the fingertips, using acids and advanced techniques such as plastic surgery. John Dillinger burned his fingers with acid, but prints taken during a previous arrest and upon death still exhibited almost complete relation to one another.

In other species

Some other animals have evolved their own unique prints, especially those whose lifestyle involves climbing or grasping wet objects; these include many primates, such as gorillas and chimpanzees, Australian koalas and aquatic mammal species such as the North American fisher. According to one study, even with an electron microscope, it can be quite difficult to distinguish between the fingerprints of a koala and a human. Koala's independent development of fingerprints is an example of convergent evolution.

IN FICTION

Mark Twain

Mark Twain's memoir Life on the Mississippi (1883), notable mainly for its account of the author's time on the river, also recounts parts of his later life, and includes tall talesand stories allegedly told to him. Among them is an involved, melodramatic account of a murder in which the killer is identified by a thumbprint. Twain's novel Pudd'nhead Wilson, published in 1893, includes a courtroom drama that turns on fingerprint identification.

Crime fiction

The use of fingerprints in crime fiction has, of course, kept pace with its use in real-life detection. Sir Arthur Conan Doyle wrote a short story about his celebrated sleuthSherlock Holmes which features a fingerprint: "The Norwood Builder" is a 1903 short story set in 1894 and involves the discovery of a bloody fingerprint which helps Holmes to expose the real criminal and free his client.

The British detective writer R. Austin Freeman's first Thorndyke novel The Red Thumb-Mark was published in 1907 and features a bloody fingerprint left on a piece of paper together with a parcel of diamonds inside a safe-box. These become the center of a medico-legal investigation led by Dr. Thorndyke, who defends the accused whose fingerprint matches that on the paper, after the diamonds are stolen.

Movies

The movie Men in Black, a popular 1997 science fiction thriller, required Agent J, played by Will Smith, to remove his ten fingerprints by putting his hands on a metal ball, an action deemed necessary by the MIB agency to remove the identity of its agents. And in a 2009 science fiction movie starring Paul Giamatti, Cold Souls, a mule who is paid to smuggle souls across borders, wears latex fingerprints to frustrate airport security terminals. She can change her identity by changing her wig, and switching latex fingerprints from the privacy of a restroom, always storing extra fingerprints in a ziploc bag, so she can always assume an alias that is suitable to her undertaking.

OTHER RELIABLE IDENTIFIERS

Other forms of biometric identification utilizing a physical attribute that is unique to every human include iris recognition, the use of dental records in forensic dentistry, thetongue and DNA profiling, also known as genetic fingerprinting.

INTRODUCTION TO FINGERPRINTS AND FINGERPRINTING

Most human skin is quite smooth and covered with hair follicles and oil glands. The finger, palm and sole areas, however have no hair or oil glands but instead have sweat pores and friction ridges that take various forms and shapes. The function of the friction ridges is to increase grip and the sense of touch. The study of friction ridge patterns is known as dermatoglyphics.

The pattern of fingerprint ridges and pores is different in each person; no two people have the exact same pattern of ridges. It seems that the general pattern of friction ridges may be genetic, however the specific pattern or fine detail is unique. Even for identical twins this is true: they may have similar general patterns but the fine details or 'minutiae' (my-new-shay) are different.

Fingerprints are often used to identify or eliminate suspects from a crime. These days, with security being so important, the field of biometrics (the process by which distinguishing human anatomy is used for identification and verification) is becoming increasingly important. Fingerprints are an important element of biometrics.

HOW ARE FINGERPRINTS FORMED?

During early embryonic development (four to five weeks) there is swelling of (mesenchymal) tissue on areas of the sole, palm and digits. These areas are known as volar pads. The pads stop growing at about 10 weeks of development but the hand continues to grow. The volar pads are then absorbed back into the hand and as the pads shrink the skin folds to produce the ridges of fingerprints. The first ridges begin to appear at around 10 weeks.

Based on the pattern of pad absorption and timing, various combinations of fingerprint can occur:

- If ridges appear when the volar pads are quite pronounced, the ridge pattern is a whorl.
- If ridges appear when the volar pads are less pronounced, the ridge pattern is a loop.
- If ridges appear when the volar pads are nearly absorbed, the ridge pattern is an arch.

Timing and absorption during development is genetically influenced; this is why identical twins have similar ridge patterns. However, within the general friction ridge pattern there are many small variations known as minutiae. The development of minutiae is a result of the environment and external stresses and pressures such as growth. So while identical twins may have very close similarities in their fingerprints, they do not have identical fingerprints as they are subject to different stresses and pressures while they are in the womb.

As well as ridges, there are other elements that make up a fingerprint. These are sweat pores which can be located along the ridges. Sweat pores are spaced almost evenly along the ridges and they are responsible for secretions which can sometimes leave finger-prints at the scene of a crime.

What is a fingerprint?

Fingerprints are the marks left behind when someone touches an object with their finger. There are three types of fingerprints that can be left behind.

1. An impression left in something soft (such as butter, putty, soap or wet paint)
2. A print left by a finger that is covered in something that is left behind such as dirt, blood, paint or ink.
3. An invisible deposit left by secretions from the skin. Everyone's fingers have small amounts of oil and perspiration which come out of microscopic pores on the tiny ridges of the fingerprints. These secretions also come out of different parts of the body too.

Why are fingerprints used in forensics?

There are three basic principles about why forensics scientists use fingerprints as evidence:

1. After almost a century of fingerprint existence, no two fingers have ever been found to possess identical ridge characteristics.
2. A fingerprint remains unchanged during a persons' lifetime. (Unless they are involved in a accident which affects their hands.)

3. Fingerprints have ridge characteristics that allow for efficient classification and examination.

History

The FBI Identification Division started in 1924 with 8 million fingerprint files. Many of the fingerprints were taken from criminals in Leavenworth Prison. By 2001 the FBI in the United States had well over 250 million records from both criminals and civilians. Civilian prints were taken from government employees and applicants for federal jobs.

First major case in the US

United States 1911. People vs Jennings: The first case in the US that involved fingerprints as evidence was that of Thomas Jennings. He was convicted of murder of Mr Hiller, the owner of a home that Jennings illegally entered. Clarence Hiller always left a gas-light burning at the head of the stairs near the door leading to his daughter's room. Shortly after 2am on a Monday morning Mrs Hiller woke and noticed that the light was out and roused Mr Hiller. Hiller got up and encountered an intruder at the head of the stairs. They struggled and both fell which eventuated in Hiller being shot twice. Hiller died from the gunshot wounds. Hiller's house had recently been painted. The back veranda had been painted two days earlier, on Saturday. Jennings entered the house through a rear window in the kitchen, which was close to the back veranda railing. During the investigation of the crime scene, police found the imprint of four fingers of a left hand (which were recorded in the fresh paint on the railing). The police removed the railing so that it could be used as evidence. Enlarged photographs of the prints were made.

Jennings was arrested after several eyewitnesses identified him. A coincidence during this case was that he had earlier had been fingerprinted during a stint in prison in 1910. Jennings was again fingerprinted after his arrest and when compared, the prints produced a match.

The defendant argued that the evidence of the comparison of his fingerprints to those found on the railing was improperly admitted. The accuracy of the photographs, the analysis method, the taking of fingerprints or the correctness of the photographic enlargement was not questioned in court. Various witnesses for the prosecution testified about their expertise in regard to fingerprints. This case established that there was a scientific basis for the system of fingerprint identification and that expert witnesses were qualified to testify. Thomas Jennings was sentenced to death and executed on February 16, 1912 for the murder of Clarence B Hiller.

FINGERPRINT CLASSIFICATION

There are three main structures that make up fingerprints. These are loops, whorls and arches.

Loops

Loops are comprised of one or more ridges entering from one side, curving, and then going out the same side it entered. The ridges in loops double back on themselves. All loops have elements called a delta and a core. The delta is a triangular area usually shaped like a T-junction, while a core is the centre of the pattern. About 65% of fingerprints have loops. Loops can be divided into two groups:

- Radial loops – these flow downward and toward the radius (or the thumb side)
- Ulnar loops – which flow toward the ulnar (or the little finger side). The ulnar loop is more common.

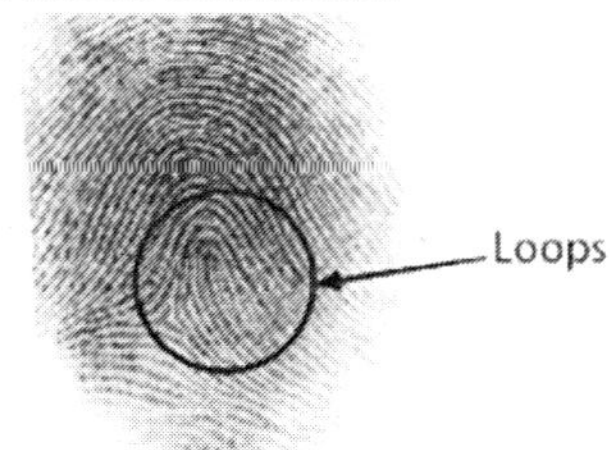

NOTE: you need to know which hand (right or left) made the print before your can tell if it is an ulnar or a radial loop.

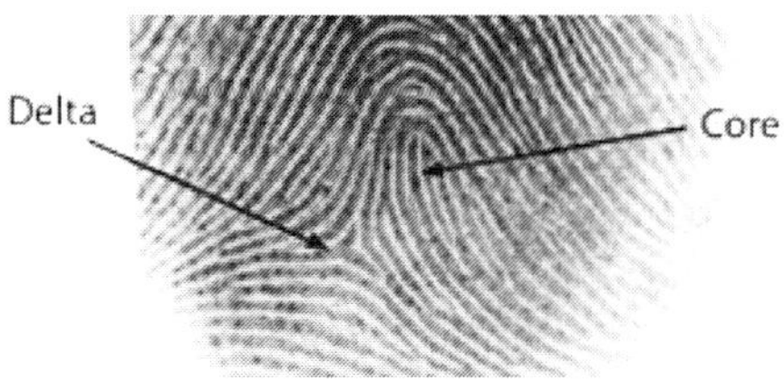

Whorls

Whorls have a circular pattern and have at least two deltas and a core. Whorls look a little like target shapes or whirlpools – circles within circles. Whorls make up 35% of patterns seen in human fingerprints and can be sub-grouped into four categories:

- Plain whorls – which are either concentric circles like a bull's eye or spirals like a wound spring.
- Central pocket loop whorls – these resemble a loop with a whorl at its end.

- Double loop whorls – these occur when two loops collide to produce an 'S'-shaped pattern.
- Accidental loop whorls – these are slightly different from other whorls and are irregular.

Arches

Arches are the least common pattern making up only 5% of all pattern types. Arches are ridgelines that rise in the centre and create a wave like pattern. The ridges enter from one side and exit the other side with a rise in the middle. They do not have a delta or a core and can be broken into two sub-groups:

- Plain arch – which has a gentle rise.
- Tented arches - have a steeper rise than plain arches.

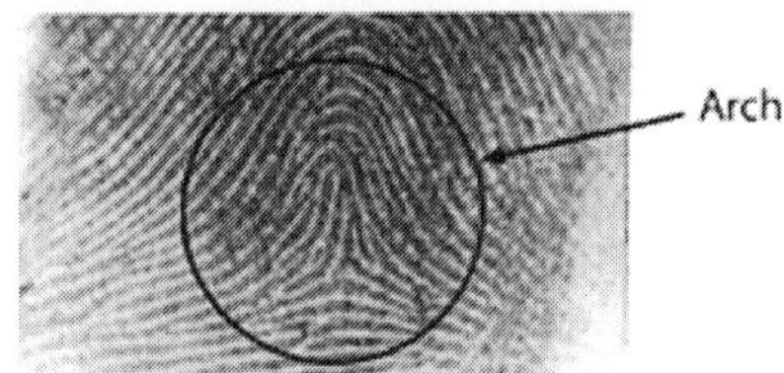

The finer structures that make up fingerprints are called minutiae. There are several types of minutiae – these are displayed in the table below. The dots found in the structures below represent the pores on the ridges where sweat is secreted.

Fine structures – Minutiae ('min-oo-shay')

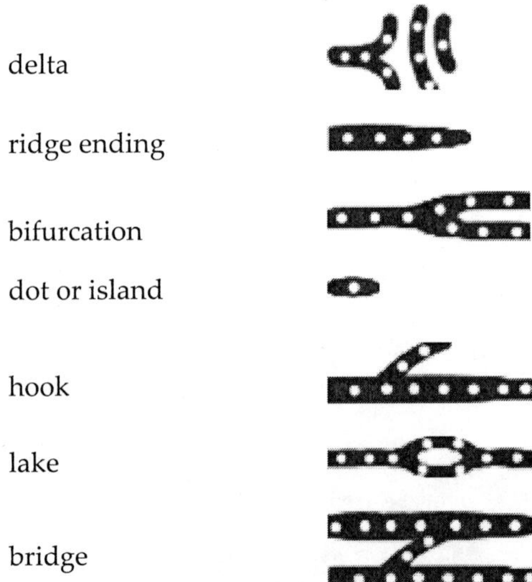

An example of identifying minutiae in a fingerprint is in the labelled picture.

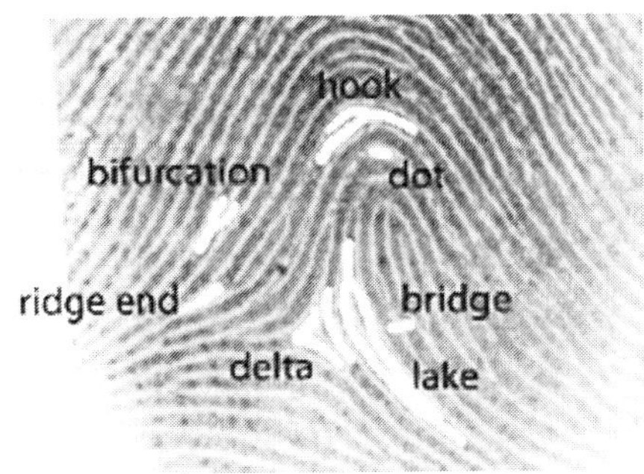

Latent, patent or plastic prints

There are three general types of fingerprints that can be left behind at a crime scene. These types depend on how and where the fingerprints were left. Some fingerprints are easier to find than others. For example, a print left in wet paint or on a greasy wall is easier to find than one left with quite clean hands on a garbage bag.

These are the three types of prints forensic scientists are likely to find:

- Latent prints are invisible and cannot be seen without special lighting or processing. In latent fingerprints, the ridge structure is reproduced on an object in sweat, or other substances naturally present (or added there by accident) on the fingers. Although there are no oil secretions from the fingers themselves we often have oils on our hands from touching our hair and face and this oil is enough to leave a print.
- Patent prints are visible and occur when substances such as blood, ink, paint, grease or dirt on the fingers of the perpetrator leave behind an easily seen print. Visible or patent prints are visible without any particular treatment
- Plastic prints have a three-dimensional quality and occur when the finger is pressed into something that leaves an indentation or impression print. Soft substances that patent prints may occur in include wax, putty, soap, cold butter and dust.

SIMPLE DETECTION TECHNIQUES USED FOR DEVELOPING FINGERPRINTING LATENT PRINTS

There are two main ways that latent prints can be detected; powders and chemical fuming.

Powders and dusting

Powders are used for detecting latent marks on non-porous surfaces such as glass, plastic, metal surfaces, glazed tiles and glossy paint. The fine powder attaches to any greasy/oily substances in the fingerprint deposit. Powdering is only effective on quite fresh marks as the mark dries out over time and looses its stickiness. Black power (made from carbon black or charcoal) and grey powder (made from aluminium or titanium powder) are used most often. Police who dust at a scene will use a powder that contrasts with the object colour they are dusting on so that prints are easier to see. Once the fingerprint has been dusted, the latent impressions are preserved as evidence either by photography or by lifting powdered prints using special sticky tape.

If you wish to try dusting with your students you will find that cocoa powder or talcum powder work well. Fine powders work best.

Chemical fuming:

Latent prints can also be developed using chemicals to expose prints. This is called chemical fuming. Chemical fuming is when the surface with suspected prints is exposed to chemical fumes. When fingerprints are difficult to see or if forensic scientists are not sure if fingerprints are present, they use this technique to 'process' fingerprints and make them visible.

1. Cyanoacrylate vapour. More commonly known by its trade name – Super Glue (which is 98% cyanoacrylate). This is an extremely useful forensic tool. When heated and mixed with sodium hydroxide (another common household chemical), cyanoacrylate releases vapours that bind to amino acids. There are amino acids present in print residue, so voila, when they bind a hard, white latent print is left behind. The print can then be photographed as is, or treated with a florescent dye that will make the print glow under UV light.

The evidence that is to be checked for fingerprints is often exposed to the vapour in something called a fuming chamber. This can be done at the crime scene in a fuming box, but these days police frequently use hand held wands. These gadgets heat a small cartridge of cyanoacrylate mixed with florescent dye which can be directed at latent prints at a crime scene. When prints react with the cyanoacrylate, they are also fixed at the same time with florescent dye making the process much faster and easier.

Super Glue vapour works well on surfaces such as glass, plastics and metal objects

2. Iodine fuming. When heated the solid crystals of iodine release iodine vapours. This is done in a fuming chamber and when the iodine vapour combines with oils in the latent print a brownish print is produced. This sort of print does fade quickly though, so photographs must be taken straight away or it must be fixed by spraying the print with a solution of starch and water. The starch and water will preserve the print for several weeks or months.
3. Ninhydrin. Ninhydrin (triketohydridene hydrate) has been used for years to reveal latent prints and it is an important and regularly used technique today. The evidence that the latent print is meant to be on is dipped in or sprayed with ninhydrin solution. This

solution reacts with the oils and proteins in the print. This process is extremely slow and it may take several hours for the print to appear as a purple-blue colour. Heating the object to around 26 to 38° C will speed up the process.

Ninhydrin is used on porous surfaces such as paper, cardboard, fabrics and untreated wood. As protein molecules are quite stable, old latent marks and prints can be developed with ninhydrin.

4. Silver nitrate. Silver nitrate is a component that is used in black and white photographic film. This process works to produce a black or reddish brown print when viewed under ultra violet (UV) light. Sweat is often a component found in latent prints. Sweat contains salts, one of which is sodium chloride (like the salt you put on your dinner). When investigators expose the latent print to silver nitrate, the chloride in the salt reacts with the silver nitrate to form a new compound called silver chloride. This colourless compound will develop and become visible when exposed to UV light.

ANALYSIS OF FINGERPRINTS

The assessor must determine if the recovered fingerprint is clear enough to enable a full analysis. If so, there are three steps that the scientists must follow to be able to accurately assess the print. The print is always photographed.

Step 1 – does not require magnification: Assess the overall pattern of the friction ridges. Whorl, loop or arch.

Step 2 – requires x5 to x10 magnification: Major ridge pattern deviations or minutiae or points of identification are established: ridge endings, bifurcations and dots (and variations of these basic patterns). Features such as scars, creases etc are also included.

Step 3 – higher magnification: Details such as the alignment or shape of ridges, pore shape and position. Fingerprints are then compared between the unknown mark (the fingerprint found at the scene) and a known print (a print found in police records or collected from a suspect).

CLEANING UP THE PRINT USING DIGITAL TECHNIQUES

More often than not a print or a partial print found at the scene of a crime is unclear. It is uncommon to get a perfect print! The minute details of a print may be fuzzy, missing or difficult to see. With the development of digital technology, the problems with fingerprints can be solved more easily than before. Prints can be scanned into a computer and then transferred into one of many computer programs. These programs can enhance, improve, and clean up the computer generated image of the print. By changing the light, clarity, contrast and background patterns electronically, certain details of the print can become more obvious giving the investigators more clues than previously available, providing more accurate evidence and also speeding up the matching process.

INTERESTING INFORMATION

It is possible to collect DNA from non-blood latent fingerprints. Skin cells are continuously being shed and can be deposited in a fingerprint. There a many more techniques used to detect latent fingerprints. Among them are: vacuum metal deposition, small particle reagent, gentian violet and use of luminescence and fluorescence. Latent fingerprints can be detected on human skin. It is however a very difficult surface to get prints off mainly because the same secretory compounds that are in the fingerprints are also naturally on the skin. If a print is less than a few hours old it can be possible. All primates also have friction ridges on their hands and feet. Some new world monkeys also have friction ridges on their tails.

FINGERPRINT IMAGE REPRESENTATION

In this chapter various fingerprint representations are introduced and general review of image enhancement, feature extraction, and matching techniques that are used in fingerprint recognition systems are provided. A fingerprint is the impression resulting from the friction ridges on the outer surface of the skin on a finger or thumb. While an in depth analysis of the way that fingerprints are formed is not within the scope of this thesis, it is commonly assumed within fingerprint biometric circles that no two people have the same fingerprints. A corollary to this assumption is that given a fingerprint, the information contained within is sufficient to uniquely identify a single individual. The validity of these assumptions is also outside the scope of this thesis. The ridges and interleaving valleys that constitute a fingerprint create two levels of detail that can be observed. The high level detail is the overall shape that is formed by the ridges.

FINGERPRINT REPRESENTATION

There are mainly three different kinds of fingerprint representations that are used in fingerprint recognition systems and each has its own advantages

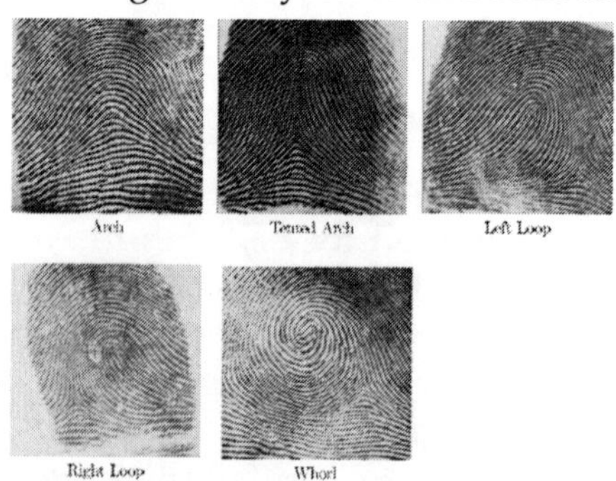

Fig. Sample fingerprints with their associated shapes

and drawbacks. When observing the patterns that the ridges of a fingerprint form together, Sir Edward Henry created a classification of fingerprints into five classes. These classes are, arch, tented arch, left loop, right loop and whorl. Samples of these fingerprint shapes can be seen in Fig.

There are two main features that define the shape of a fingerprint. These are cores and deltas (also collectively known as macro-singularities). A core is often described as a point where a single ridge line turns through 180 degrees. Similarly, a delta is described as a point where three ridge lines form a triangle. It can be seen in Fig 0 2 where the cores and deltas are marked. These core and delta points characterize the overall shape. Arches can be easily identified through the lack of any delta or core points. Also, whorls can be easily identified through the presence of two core and two delta points. Differentiating the right loop, left loop and tented arch is slightly more difficult, as all three have one core and one delta point.

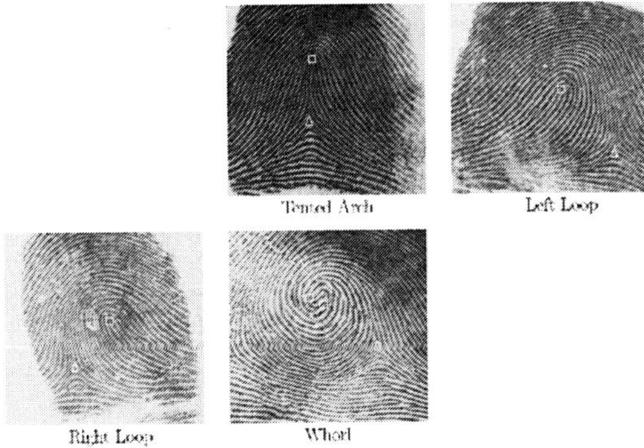

Fig. 0 2 Sample fingerprints, with core points marked with a square, and delta points marked with a triangle

Image-based representation

In image based representation, the fingerprint image itself is used as a template. There is no need for a specific feature extracting algorithm, and the raw intensity pixel values are directly used. This representation retains the most information about a fingerprint since fewer assumptions are made about the application. However, a fingerprint recognition system that uses the image-based representation requires tremendous storage space. For example, a 0.8mm×1.0mm (400 ×500 pixels) fingerprint is obtained by a scanner at 500 dots per inch (DPI) with 8 bits gray-scale resolution. The resulting fingerprint image is 400 × 500 × 8 = 0.2 Mbytes. A system with large amount of fingerprint data may have difficulty storing all the templates. For example, FBI has collected more than 200 million fingerprints since 1924 which require more than 250 terabytes storage space . Traditional compression techniques, such as JPEG tend to lose the highest frequency details, which contain discriminating information and the blocking artifacts also affect the performance of automatic fingerprint recognition systems. FBI recommends a compression method based on WSQ (Wavelet Scalar Quantization) , which can preserve the discriminating information without blocking artifacts while achieving a high compression ratio (around 20:1). However, it still requires about 20 Kbytes to store a compressed fingerprint image.

Global Ridge Pattern

This representation relies on the ridge structure, global landmarks and ridge pattern characteristic, such as the singular points, ridge orientation map, and the ridge frequency map. This representation is sensitive to the quality of the fingerprint images. However, the discriminative abilities of this representation are limited due to absence of singular points.

Local Ridge Detail

This is the most widely used and studied fingerprint representation. Local ridge details are the discontinuities of local ridge structure referred to as minutiae. Sir Francis Galton (1822-1922) was the first person who observed the structures and permanence of minutiae. Therefore, minutiae are also called "Galton details". They are used by forensic exports to match two fingerprints.

There are about 150 different types of minutiae categorized based on their configuration. Among these minutia types, "ridge ending" and "ridge bifurcation" are the most used, since all other types of minutiae can be seen as the combinations of "ridge endings" and "ridge bifurcations".

After the fingerprint ridge thinning, marking minutia points is relatively easy. In general, for each 3x3 window, if the central pixel is 1 and has exactly 3 one-value neighbors, then the central pixel is a ridge branch as shown in figure. If the central pixel is 1 and has only 1 one-value neighbor, then the central pixel is a ridge ending as shown in Fig 0 3

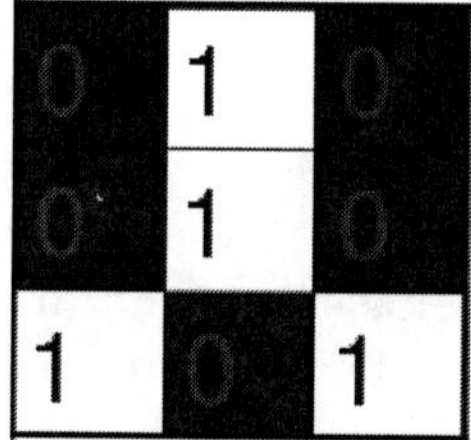

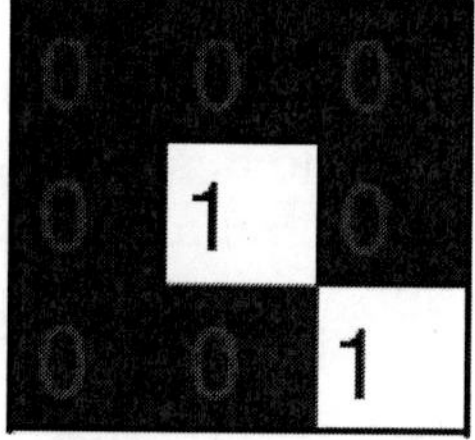

Fig. 0 3 (a) Bifurcation (b) Termination

The American National Standards Institute-National Institute of Standards and Technology (ANSI-NIST) proposed a minutiae-based fingerprint representation. It includes minutiae location and orientation . The minutiae orientation is defined as the direction of the underlying ridge at the minutiae location. Minutiae-based fingerprint representation also has an advantage in helping privacy issues since one cannot reconstruct the original image from using only minutiae information. Minutiae is relatively stable and robust to contrast, image resolutions, and global distortion when compared to other representations. However, to extract the minutiae from a poor quality image is not an easy task. At present, most of the automatic fingerprint recognition systems are designed to use minutiae as their fingerprint representations.

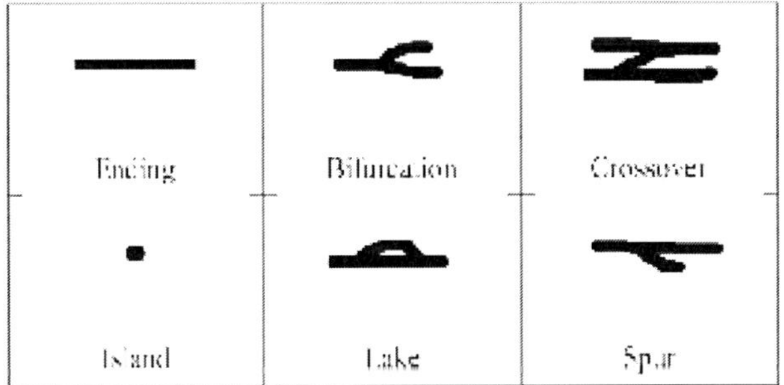

Fig. Some of the common minutiae types

Intra-ridge Detail

On every ridge of the finger epidermis, there are many tiny sweat pores. Pores are considered to be highly distinctive in terms of their numbers, positions, and shapes. However, extracting pores is feasible only in high-resolution fingerprint images (for example 1000 DPI) and with good image quality. Therefore, this kind of representation is not practical for most applications.

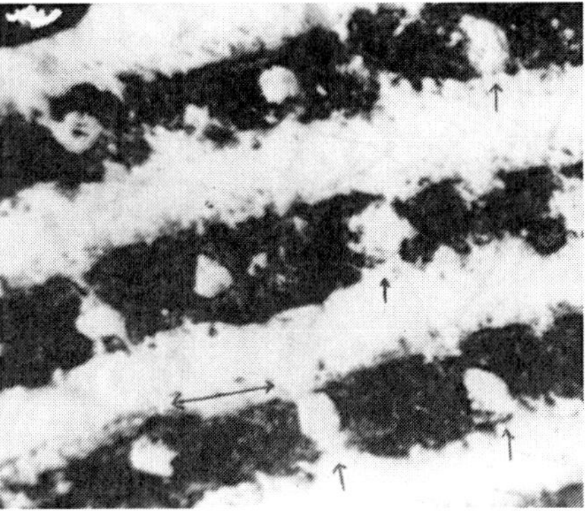

Fig. 0 5 Pore and ridge edge contour

MINUTIAE-BASED FINGERPRINT RECOGNITION

Minutiae-based fingerprint representation to design the systems due to the advantages of wide accessibility and stability. Minutiae-based fingerprint representation and matching are widely used by both machine and human experts. Minutiae representation has several advantages compared to other fingerprint representations. Minutiae have been (historically) used as key features in fingerprint recognition tasks. Its configuration is highly distinctive and several theoretical models [13, 14, and 15] use it to provide an approximation of the individuality of fingerprints. Minutiae-based systems are more accurate than correlation based systems and the template size of minutiae-based fingerprint representation is small. Forensic experts use this representation which has now become part of several standards for exchange of information between different systems across the world.

FINGERPRINT IMAGE ENHANCEMENT

Fingerprint image quality is an important factor in the performance of minutiae extraction and matching algorithms. A good quality fingerprint image has high contrast between ridges and valleys. A poor quality fingerprint image is low in contrast, noisy, broken, or smudgy, causing spurious and missing minutiae. Poor quality can be due to cuts, creases, or bruises on the surface of finger tip, excessively wet or dry skin condition, uncooperative attitude of subjects, damaged and unclean scanner devices, low quality fingers (elderly people, manual workers), and other factors.

The goal of an enhancement algorithm is to improve the clarity (contrast) of the ridge structures in a fingerprint. General-purpose image enhancement techniques are not very useful due to the non-stationary nature of a fingerprint image. However, techniques such as gray-level smoothing, contrast stretching, histogram equalization, and Wiener filtering can be used as preprocessing steps before a sophisticated fingerprint enhancement algorithm is applied.

Techniques that use single filter convolutions on the entire image are not suitable. Usually, a fingerprint image is divided into sub regions and then a filter whose parameters are pre-tuned according to the region's characteristics is applied. Each local region of a fingerprint can be seen as a surface wave of a particular wave (ridge) orientation (perpendicular to the flow direction) and frequency. Several types of contextual filters in both spatial and frequency domains have been proposed in the literature.

Fig. Good quality fingerprint image

The purpose of the filters is to fill small gaps (low-pass effect) in the direction of a ridge and to increase the discrimination (band-pass effect) between ridges and valleys in the direction orthogonal to the ridge . O'Gorman and Nikerson were the first to propose the use of contextual filtering.

Recently, Greenberg et al proposed the use of an anisotropic filter that adapts its parameters to the structure of the underlying sub region. Wu, Shi, and Govindaraju proposed to convolve a fingerprint image with an anisotropic filter to remove Gaussian noise and then apply directional median filters (DMF) to remove impulse noise.

On visual inspection, enhancement results of Wu et al appear to be superior to those obtained by Greenberg et al. Anisotropic filters remove Gaussian noise and smoothen the fingerprint image along the local ridge direction. The standard rectangular-shaped median filters produce artifacts in fingerprint images. Wu et al use directional median filters whose shapes vary by their direction.

Note that the shape of the filter changes along with its direction. Sherlock, Monro, and Millard proposed a fingerprint enhancement method in the Fourier domain. In this approach, a fingerprint image is convolved with pre-computed filters which results in a set of filtered images. The enhanced fingerprint image is constructed by selecting each pixel from the filtered image whose orientation is the closest to that of the original pixel. However, their assumption of constant ridge frequency limits the performance of the approach. Willis and Myers presented an FFT based fingerprint enhancement method. Instead of explicitly computing the local ridge direction and frequency, enhancement is achieved by multiplying the Fourier transforms of the block by magnitude of power, k. Chikkerur proposed an algorithm based on short time Fourier transform (STFT), and a probabilistic approximation of dominant ridge orientation and frequency was used instead of the maximum response of the Fourier spectrum to remove impulse noise (small gaps on bridge or dots in valleys). The ridge orientation image, ridge frequency image, and foreground region image are generated simultaneously while performing the STFT analysis.

A wavelet-based method is proposed by Hsieh et al . It uses both local ridge orientation and global texture information. Fingerprint image is first wavelet-decomposed into "approximation" and "detail" sub-images. A series of texture filters and a directional compensation process based on a voting technique are applied on those sub-images. The enhanced fingerprint image is then obtained by the reconstructing process of wavelet transform.

MINUTIAE EXTRACTION

The reliability of minutiae features plays a key role in automatic fingerprint recognition. Generally, the minutiae representation of a fingerprint consists of simply a list of minutia points associated with their spatial coordinates and orientation. Some methods also include the types and quality of minutiae in the representation. Minutiae extraction algorithms are of two types: (i) binarization-based extraction and (ii) gray-scale based extraction.

Binarization-based Minutiae Extraction

Most of the proposed minutiae extraction methods are binarization-based approaches. They require conversion of the gray-scale fingerprint image (8 bits per pixel, 256 gray levels) into a binary form (1 bit per pixel, black or white). Various binarization techniques have been presented in the image processing literature . One intuitive approach is to use a global threshold (th) and assign each pixel a value as follows:

$$IB(x, y) = \begin{cases} 1 \text{ if } I(x, y) > th \\ 0 \text{ if } I(x, y) <= th \end{cases}$$

where I (x, y) is the intensity value of the pixel at (x, y) in a gray-scale image. Otsu's method describes a technique to obtain the global threshold (th) from a statistical viewpoint. Dong and Yu use a data clustering approach which is equivalent to Otsu's method but is more efficient. The contrast variation in a fingerprint image makes it impossible to find an optimal global threshold. Adaptive techniques are preferred in general but they fail on poor quality images.

Several methods have been proposed to utilize the flow texture of a fingerprint image in binarization tasks. Stock and Swonger observed that the average local intensity of a ridge line along its flow direction is highest and used it in binarization. Ratha, Chen and Jain use a 16 × 16 window centered and oriented along the local ridge direction on each pixel. Ridge lines are recognized as peaks of the gray-level profile of pixel intensities projected on the central segment of the window. Coetzee and Botha use a local binarization technique. The area between two edges of a local block is blob-colored and then logical ORed with the result of local binarization of the same local block to produce the final binarized image. Garris et al and Watson et al propose a directional binarization technique. In this approach, each pixel is examined successively and assigned to black (0) or white (1). By consulting the intrinsic orientation map, a pixel is assigned to white if there is no detectable ridge flow for the local block. If the flow is well defined in the pixel's local block, then an orientated window (7 × 9) is used to analyze the neighboring pixel intensity of the pixel. The rows of the window are aligned with the local ridge and the central row sum is compared against the average row sum of the entire window. A pixel is white if the central row sum is less than the window's average row sum; otherwise, it is black.

Usually, the binarization-based minutiae extraction methods apply a thinning algorithm after the binarization step to obtain the skeletons of fingerprint ridges. Once a binary skeleton of a fingerprint is obtained, minutiae extraction becomes a trivial task. It is assumed that the foreground and background pixel values of a fingerprint skeleton are 1 and 0, respectively.

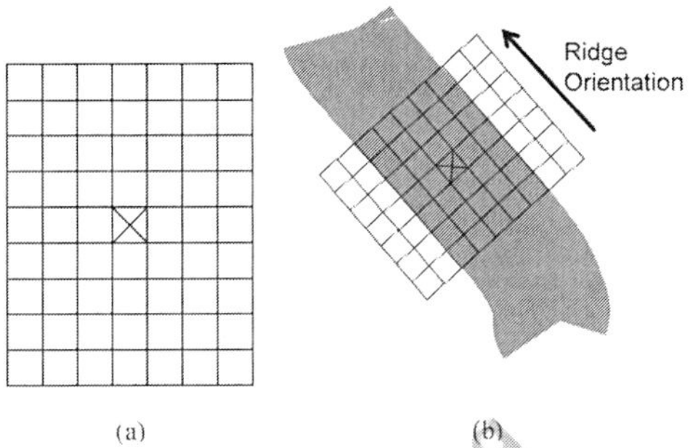

Fig. (a) The window used for analyzing the surrounding pixel intensity (b) the window oriented along the local ridge direction

Minutia can be detected by examining the 8-neighborhood of a ridge skeleton pixel at (x, y).

Many thinning approaches have been proposed. However, thinning tends to introduce hair-like artifacts along the one-pixel wide skeleton, which leads to detection of spurious minutiae. Various techniques are introduced between the stages of binarization and thinning to improve the quality of binarized fingerprint images by filling holes, smoothing ridges, and removing small gaps and other artifacts.

Several approaches have been also proposed to extract minutiae directly from the binarized fingerprint image to avoid the computationally intensive thinning process. Weber proposed a method that extracts minutiae from the thick binary ridges using a rule based ridge tracking algorithm. Garris et al (see also Watson) use a series of pixel patterns to detect minutiae on binraized fingerprint images. A method based on chaincode is proposed by Govindaraju et al . It is a lossless representation of an object contour and is widely used in document analysis and recognition research . It is generated by tracing the exterior contours of a binary object counterclockwise (clockwise for interior contours) and stored in contour lists. In the contour list, each contour element contains the x, y coordinates of the pixel, the direction of the contour into the pixel, and curvature information. The chaincode representation of fingerprint ridge contours provides several advantages in minutia detection:

- It is a lossless representation, thus, most of fingerprint information is retained.
- It is easy to remove small objects and holes from the ridge contours. Therefore, the number of spurious minutiae is few.
- It works directly on binarized image and eliminates the need for thinning.
- Minutiae are the significant turns in the ridge contour.

2

Working Principle of Existing Minutiae Based Fingerprints Matching

A feature extractor finds the ridge endings and ridge bifurcations from the input fingerprint images, If ridges can be perfectly located in an input fingerprint image, then minutiae extraction is just a trivial task of extracting singular points in a thinned ridge map. However, in practice, it is not always possible to obtain a perfect ridge map . The performance of currently available minutiae extraction algorithms depends heavily on the quality and orientation of the input fingerprint images, fingerprint images may not always have well-defined ridge structures. A reliable minutiae extraction algorithm is critical to the performance of an automatic identity authentication system using fingerprints. The overall working of existing minutiae based fingerprint representation is depicted below

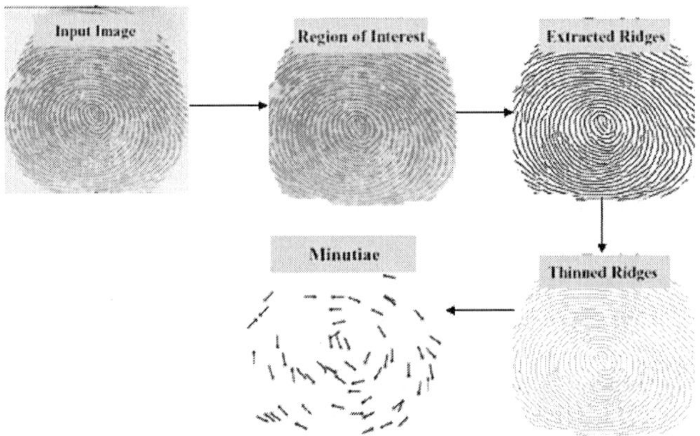

Once the minutiae are extracted from existing fingerprint, the matching process starts. It mainly consists of four components user interface, enrollment module, authentication module and system database. The user interface provides mechanisms for a user to input his fingerprints into the system. The system database consists of metadata, information about all minutiae points extracted. In this mechanism a lot of memory and effort required to store the information of all minutiae in the database, further it has its limitations during matching operation.

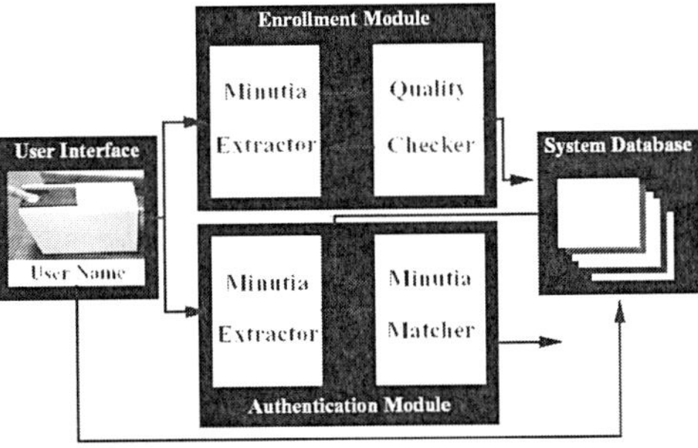

Fig. Architecture of an automatic identity authentication system.

PROPOSED SCHEME

The most popular technology to obtain a live-scan fingerprint image is based on optical frustrated total internal reflection (FTIR) concept . When a finger is placed on one side of a glass platen (prism), ridges of the finger are in contact with the platen, while the valleys of the finger are not in contact with the platen. The rest of the imaging system essentially consists of an assembly of an LED light source and a CCD (charged coupled device) placed on the other side of the glass platen.

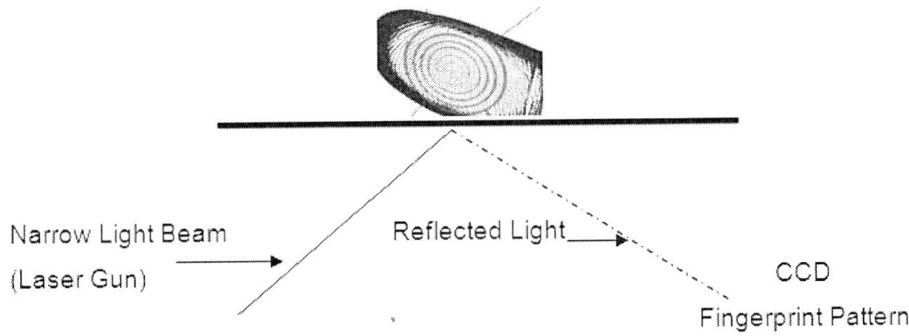

Fig. (Optical frustrated total internal reflection)

The laser light source illuminates the glass at a certain angle and the camera is placed such that it can capture the laser light reflected from the glass. The light incidenting on the platen at the glass surface touched by the ridges is randomly scattered while the light incidenting at the glass surface corresponding to valleys suffers total internal reflection. Consequently, a portion of the image formed on the imaging plane of the CCD corresponding to ridges is dark and those corresponding to valleys are bright. In live scanning fingerprint, Optical Frustrated Total Internal Reflection and Digital Optical Frustrated Total Internal Reflection has its own aspects, here off-line fingerprint representation is presented instead of live finger.

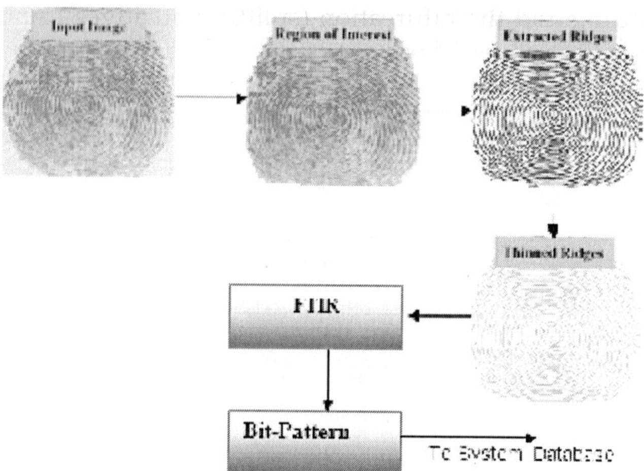

Fig. Proposed Scheme for off-line FTIR

During off-line representation the final step is minutiae detection as shown in figure 2. But in proposed scheme we bypass this step of detecting minutiae points and replacing it with FTIR, here the thinned image will input to FTIR and corresponds to that image as bit pattern will be generated . This bit pattern will now work as system database rather metadata as shown in figure 3. It will naturally improve the system speed and accuracy.

THE PROCESS OF MATCHING

There are multiple approaches found in literature to match fingerprint images. The one that is well known and used often is called minutiae-based matching. In this chapter is first described the process of fingerprint matching, in case of minutiae-based matching is used.

Fingerprints are unique, there are no two individuals that have exactly the same pattern. In principle, every finger is suitable to give prints for authentication purposes. However, there are differences between the ten fingers. There is no clear evidence as to which specific finger should be used for identification. The thumb provides a bigger surface area but there is not much association of the thumb with criminality. Forefingers have been typically used in civilian applications. In most cases one can assume that the index finger obtains the best performance. Since the majority of the people is right handed, the best choice would be to take the right hand index finger.

After capturing the fingerprint, for example in crime, it is compared to other fingerprints in the database to find a matching pair. Nowadays this whole process goes automatically via identification marks. A fingerprint has various identification marks. On the global level there are the ridges that make a particular pattern. Moreover there are singular points to detect, like the delta and core. At the local level you find minutiae details. The two most occurring are a ridge ending and the bifurcation (splitting of ridges). At the very fine level there are sweat pores. These can only be used at images from very high quality and are not discussed in this paper.

The comparing of two prints can only take place if the fingerprints are of reasonable quality and have values that are measured the same way and mean the same thing. A few preparations must precede before the actual matching takes place. These steps are called preprocessing. Preprocessing improves the quality of the fingerprint (mostly a digital image) and removes scars and other noise in the image. It also makes the image better readable for the matching step by making the image black and white for instance.

The matching of fingerprints has been studied a lot, resulting in multiple approaches. One of these approaches is minutiae matching. It is the most well known and often used method that makes use of small details in the fingerprint, called minutiae, as ridge endings and split points. Each minutia has its own information, like the angle and position. Extracting this information is one of the steps of the minutiae matching approach, matching these extracted minutiae is another problem. This matching of minutiae can be seen as a point matching problem. A suitable algorithm to solve this problem is the Hough transform-based algorithm. It calculates the optimal transformation for matching minutiae. If there exists a matching fingerprint in the database, the template with the most matching minutiae is probably the same as the input.

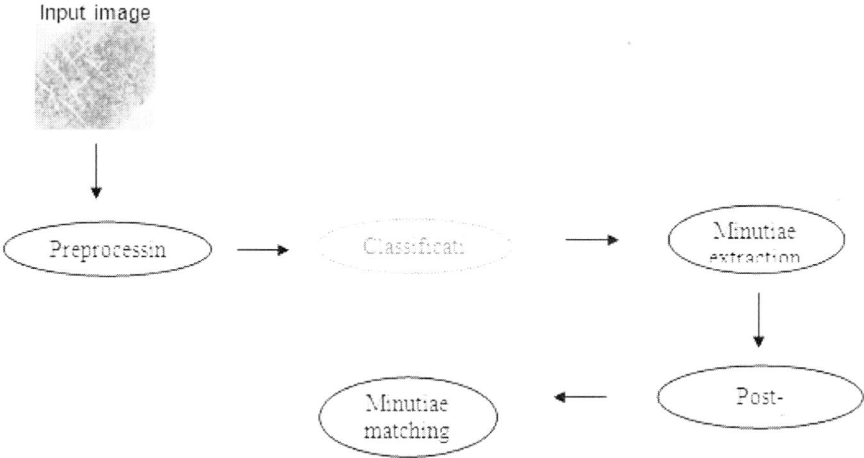

Fig. The process of matching an input fingerprint with a template from the database

Between the extraction of minutiae points and the matching, there is often a post-processing step to filter out false minutiae, caused by scars, sweat, dirt or even by the preprocessing step. In the end, using the minutiae can lead to a unique match of fingerprints. The procedure described above can be extended with one important time saving step, namely classification.

Classification is used in cases where the database containing fingerprints is so large that matching a fingerprint with all the images is too time-consuming. Fingerprints can be categorized by their patterns. The classification is based on the following patterns: loops, whorls and arches. In this way an input fingerprint does not have to be matched with all the fingerprints in the database, but only a part of it. The steps to take for the whole process of matching an input fingerprint with one of the templates from the database. The assumption is made that the templates in the database have already gone through this process and so only the input fingerprint has to be prepared for matching.

PREPROCESSING

When a fingerprint image is captured, nowadays through a scan, it contains a lot of redundant information. Problems with scars, too dry or too moist fingers, or incorrect pressure must also be overcome to get an acceptable image. Therefore, preprocessing, consisting of enhancement and segmentation is applied to the image. It is widely acknowledged that at least two to five percent of target population has fingerprints of poor quality. These fingerprints that cannot be reliably processed using automatic image processing methods. This fraction is even higher when the target population consists of older people, people doing manual work, people living in dry weather conditions or having skin problems, and people who have poor fingerprints due to their genetic and racial attributes.

A fingerprint can contain regions of different quality:

- A well-defined region, where ridges are clearly differentiated from each another;
- A recoverable region, where ridges are corrupted by a small amount of gaps, creases and smudges, but they are still visible and the neighboring regions provide sufficient information about their true structure;
- An unrecoverable region, where ridges are corrupted by such a severe amount of noise and distortion that no ridges are visible and the neighboring regions do not allow them to be reconstructed.

STEPS

A critical step in automatic fingerprint matching is to automatically and reliably extract minutiae from the input fingerprint images. However, the performance of a minutiae extraction algorithm relies heavily on the quality of the input fingerprint images. To ensure that the performance of an automatic fingerprint identification system will be robust with respect to the quality of the fingerprint images, it is essential to implement a fingerprint enhancement algorithm in the minutiae extraction module. In the literature there are several methods to improve the quality of an image and make it ready for matching details. The steps that are present in almost every process are:

1. Normalization,
2. Filtering,
3. Binarization,
4. Skeletonization.

In the first step the input image from the sensor is normalized. This is important since image parameters may differ significantly with varying sensors, fingers and finger conditions. By normalizing an image, the colors of the image are spread evenly throughout the gray scale. The filtering step is the one that changes the most. There are a lot of filters to smooth the ridges, take away scars, noise and irrelevant segments. Low pass filters are used, just as Gaussian masks, Gabor filters or orientation filters. The third step is making the image binary; transform the gray scale image into a binary image (black and white). The ridges are then made thinner from five to eight pixels in width down to one pixel, for precise location of endings and bifurcations.

Normalization

Normalization is a good first step for improving image quality. To normalize an image is to spread the gray scale in a way that it is spread evenly and fill all available values instead of just a part of the available gray scale, see figure 3-1. The normal way to plot the distribution of pixels with a certain amount of gray (the intensity) is via a histogram. To be able to normalize an image, the area which is to normalize within, has to be known. Thus it is necessary to find the highest and the lowest pixel value of the current image. Every pixel is then evenly spread out along this scale. Equation (1) represents the normalization process.

$$I_{norm}(x,y) = \frac{I(x,y) - I_{min}}{I_{max} - I_{min}} \times M,$$

where I is the intensity (gray level) of the image. Imin is the lowest pixel value found in the image, Imax is the highest one found. M represents the new maximum value of the scale, mostly M = 255, resulting in 256 different gray

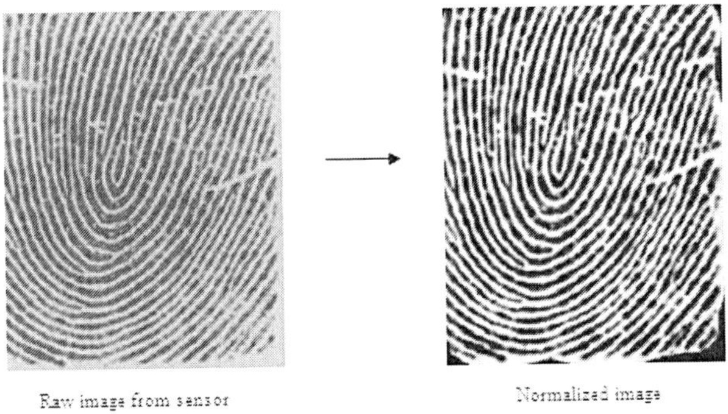

Fig. The normalization step

levels, including black (0) and white (255). Inorm(x, y) is the normalized value of the pixel with coordinates x and y in the original image I(x,y). When images have been normalized it is much easier to compare and determine quality since the spread now has the same scale. Without the normalization it would not be possible to use a global method for comparing quality.

Filtering

It is important to filter out image noise coming from finger consistency and sensor noise. For that purpose the orientation of the ridges can be determined so that it is able to filter the image exactly in the direction of the ridges.

Fig. An orientation field overlayed on a fingerprint

By this filter method the ridge noise is greatly reduced without affecting the ridge structure itself. One approach to ridge orientation estimation relies on the local image gradient. A gray scale gradient is a vector whose orientation indicates the direction of the steepest change in the gray values and whose magnitude depends upon the amount of change of the gray values in the direction of the gradient. The local orientation in a block can be determined from the pixel gradient orientations of the block.

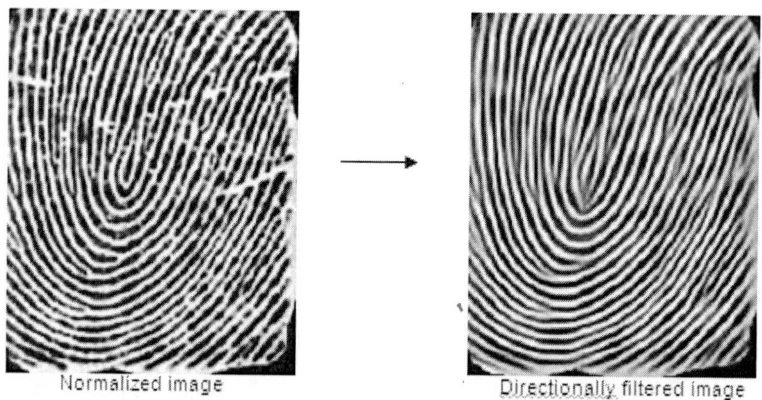

Fig. The filtering step (in this case an orientation field filter)

Binarization

Binarization can be seen as the separation of the object and background. It turns a gray scale picture into a binary picture. A binary picture has only two different values. The values 0 and 1 are represented by the colors black and white, respectively. To perform binarization on an image, a threshold value in the gray scale image is picked. Everything darker (lower in value) than this threshold value is converted to black and everything lighter (higher in value) is converted to white. This process is performed to facilitate finding identification marks in the fingerprints such as singularity points or minutiae.

The difficulty with binarization lies in finding the right threshold value to be able to remove unimportant information and enhance the important one. It is impossible to find a working global threshold value that can be used on every image. The variations can be too large in these types of fingerprint images that the background in one image can be darker than the print in another image. Therefore, algorithms to find the optimal value must be applied separate on each image to get a functional binarization. There are a number of algorithms to perform this, the most simple one uses the mean value or the median of the pixel values in the image. This algorithm is based on global thresholds.

What often are used nowadays are local thresholds. The image is separated into smaller parts and threshold values are then calculated for each of these parts. This enables adjustments that are not possible with global calculations. Local thresholds demand a lot more calculations but mostly compensate it with a better result.

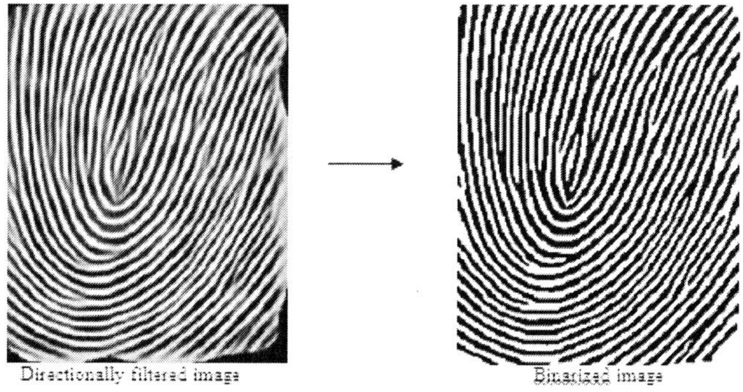

Fig. The binarization step

Skeleton modeling

One way to make a skeleton is with thinning algorithms. The technique takes a binary image of a fingerprint and makes the ridges that appear in the print just one pixel wide without changing the overall pattern and leaving gaps in the ridges creating a sort of "skeleton" of the image. An example of skeletonization. The form ▪✚▪ is used as structural element, consisting of five blocks that each present a pixel. The pixel in the center of that element is called the origin. When the structural element overlays the object pixels in its entirety, only the pixels of the origin remain. The others are deleted.

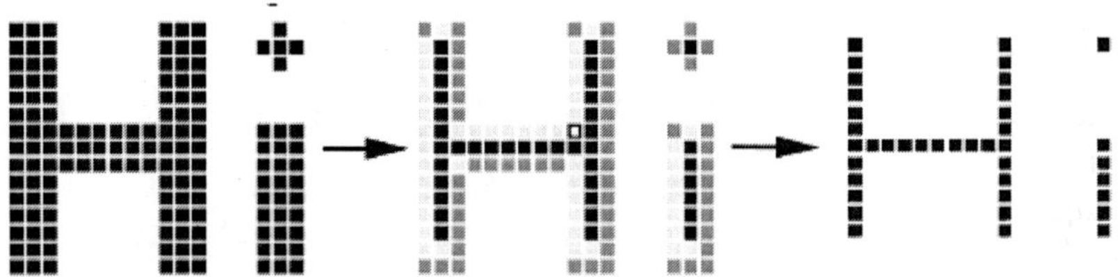

Fig. An example of skeletonization

Skeleton modeling makes it easier to find minutiae and removes a lot of redundant data, which would have resulted in longer process time and sometimes different results. There are a lot of different algorithms for skeleton modeling that differ slightly.

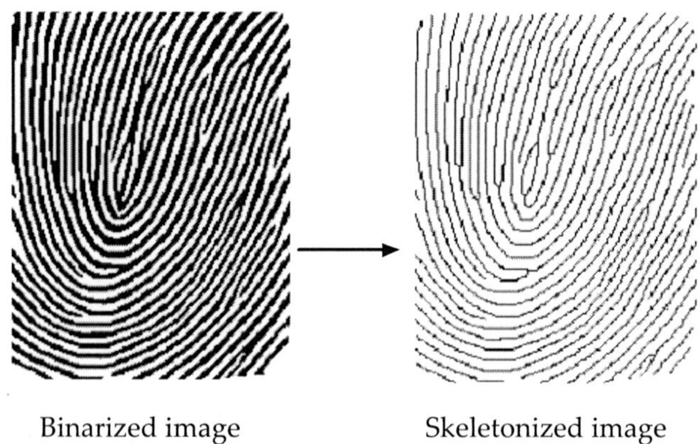

Binarized image Skeletonized image

Fig. The skeletonizing step

DISCUSSION

Binarization reduces the information in the image from gray scale to black and white. While this simplifies for further algorithms to decode the fingerprint image into information useful for identification, it also reduces the complexity of the image and removes information that might have been necessary for the identification of the fingerprint. Some authors have proposed minutiae extraction approaches that work directly on the gray-scale images without binarization and skeleton modeling. This choice is motivated by these considerations :

- A significant amount of information may be lost during the binarization process;

- Binarization and thinning are time consuming;
- Thinning may introduce a large number of spurious minutiae;
- In the absence of an a priori enhancement step, most of the binarization techniques do not provide satisfactory results when applied to low-quality images.

A reduction of information is not necessarily a negative thing though.

Still the binarization is a necessary step for many of the algorithms used for minutiae analysis. Advanced algorithms like skeletonization will only work if this process is performed. The thinning performed in skeleton modeling enables point identification via simple counting of the nearby pixels. When this process has been performed, a mapping of available minutiae in the image is made, used for minutiae-based matching. It is now comparable to earlier stored templates. However, if pattern recognition is used instead, it is not necessary to perform binarization (or even normalization). Instead every pixel direction is calculated and a direction field of the total image is created.

CLASSIFICATION

The fingerprints have been traditionally classified into categories based on information in the global pattern of ridges. A recognition procedure consists in retrieving one or more fingerprints in a large database corresponding to a given fingerprint, whereas a classification procedure consists in assigning a fingerprint to a pre-defined class.

WHY CLASSIFICATION

Fingerprint recognition is the basic task of the identification systems of the most famous policy agencies. If all the fingerprints within the database are a priori classified, the recognition procedure can be performed more efficiently, since the given fingerprint has to be compared only with the database items belonging to the same class.

The first (semi-)automatic systems for the fingerprint recognition were developed in the '70s by the US Federal Bureau of Investigation (FBI) in collaboration with the National Bureau of Standards, Cornell Aeronautics Laboratory and Rockwell International Corporation. Since then the volumes of the fingerprint databases and the amount of requests of identification increased constantly, so that it was necessary very soon to classify the fingerprints to improve the recognition efficiency. The FBI now has about forty to fifty million fingerprints stored in their database.

Although the first approaches to automatic fingerprint classification were proposed many years ago , policy agencies still go on performing classification manually with a great expense of time. On the other hand, the problem of fingerprint classification is very hard: a good classification system should be very reliable (it must not misclassify fingerprints), it should be selective (the fingerprint database has to be partitioned in a number of non-overlapping classes with about the same properties) and it should be efficient (each fingerprint must be processed in a short time).

PATTERNS

A fingerprint can be looked at from different levels; the global level, the local level, and the very fine level. At the global level, you find the singularity points, called core and delta points. These singularity points are very important for fingerprint classification, but they are not sufficient for accurate matching.

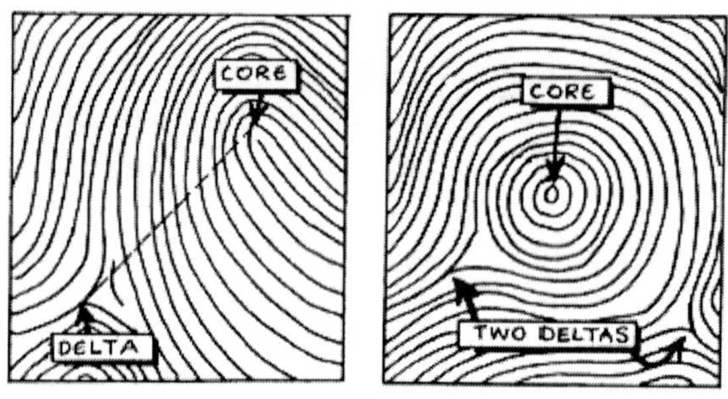

Fig. Core and delta points marked on sketches of the two fingerprint patterns loop and whorl

The core is the inner point, normally in the middle of the print, around which whorls, loops, or arches center. It is frequently characterized by a ridge ending and several curved ridges. Deltas are the points, normally at the lower left and right hand of the fingerprint, around which a triangular series of ridges center.

Fingerprint classification nowadays is usually based on the Galton-Henry classification scheme. Galton divided the fingerprints into three major classes (arches, loops and whorls) and further divided each category into subcategories . Edward Henry refined Galton's classification by increasing the number of classes . The five most common classes of the Galton-Henry classification scheme are: plain arches, tented arches, left loop, right loop, and whorl. Of all fingerprint patterns 65% consist of loops, whorls make up about 30%, and arches the remaining 5%.

Loops

The loop is the most common fingerprint pattern. They are usually separated into right loops and left loops. The difference between these two is the direction that the ridges turn to. If the ridges turn to the left it is a left loop and vice versa. Figure 4-2 shows a right loop. In a loop pattern, one or more of the ridges enter on either side of the fingerprint, recurve, touch or cross the line running from the delta to the core and terminate or tend to terminate on or toward the same side of the fingerprint from which such ridge or ridges entered. There is one delta.

Fig. A right loop

Whorls

Whorls are the second most common pattern. Here the ridges form circular patterns around the core. Most often they form to spirals, but they can also appear as concentric circles. In a whorl some of the ridges make a turn through at least one circuit. There are two loops (or a whorl) and two deltas.

Fig. A whorl

Arches

Arches are more uncommon than loops and whorls. In the arch type the ridges run from one side to the other, making no backward turn. Usually arches are classified into plain (simple) and tented (narrow) arches. A plain arch does not have loops or deltas. The tented arch often has a loop and a delta point below.

Fig. A plain arch

Fig. A tented arch

CLASSIFICATION TECHNIQUES

It is important to note that the distribution of fingers into the five classes is highly skewed. A fingerprint classification system should take that into account and should be invariant to rotation, translation and elastic distortion of the skin.

Although a wide variety of classification algorithms has been developed for this problem, a relatively small number of features extracted from fingerprint images have been used by most of the authors. In particular, almost all the methods are based on one or more of the following features: ridge line flow, orientation image, singular points and Gabor filter responses. Ridge line flow is usually represented as a set of curves running parallel to the ridge lines. An orientation image is used in most classification approaches because it contains all the information required for the classification. Usually the orientation image is registered with respect to the core point, one of the singular points. A Gabor filter is a common directional filter which has both frequency-selective and orientation-selective properties. Several approaches have been developed for automatic fingerprint classification. The best approaches can be broadly categorized into the following categories :

Rule-based

These approaches mainly depend on the number and position of the singular points of the fingerprint. This is the approach commonly used by human experts for manual classification. A plain arch has no singular points. A tented arch, left loop and right loop have one loop and one delta. A whorl has two loops (or a whorl) and two deltas. The result of this technique is a scheme to follow, which tells in which class the input image belongs to.

Structural

Structural approaches are based on the relational organization in the structure of the fingerprints. They are often based on the orientation field. Cappelli et al proposed partitioning the orientation image into homogeneous regions and subsequently classify the image based on the pattern of the lines between the regions. See also Maio and Maltoni .

Statistical

In statistical approaches, a fixed-size numerical feature vector is derived from each fingerprint and a general-purpose statistical classifier is used for the classification. One of the most widely adopted statistical classifiers is the k-nearest neighbor. Many approaches directly use the orientation image as a feature vector.

Neural network-based

Most of the proposed neural network approaches are based on multilayer perceptrons and use the elements of the orientation image as input features. The success of neural networks was mainly in the 1990s.

Multi-classifier

Different classifiers offer complementary information about the patterns to be classified, which may improve performance. This approach is especially studied in recent years. For example Jain, Prabhakar and Hong adopt a two-stage classification strategy: a k-nearest neighbor classifier is used to find the two most likely classes. Then a specific neural network, trained to distinguish between the two classes, is exploited to obtain the final decision. The techniques have an error rate between 6.5 and 11.9% when fingerprints are assigned to five different classes. Many errors are due to the misclassification of tented arch fingerprints as plain arch. That is the reason why some authors only use four classes; tented arch and plain arch as one. The classification errors drop to between 5.1 and 9.4%.

FINGERPRINT MATCHING

An input fingerprint is compared with one of the template fingerprints, stored in a database. In the first paragraph some methods for fingerprint matching. Subsequently, the most well known method of these, minutiae-based matching, is described in detail.

METHODS

Many algorithms have been developed to match two different fingerprints and they can be divided into the following groups:

Correlation-based matching: Two fingerprint images are laid on top of each other and the correlation between corresponding pixels is computed for different alignments (various displacements and rotations). This technique has some disadvantages. Correlation-based techniques require the precise location of a registration point and are affected by non-linear distortion. Minutiae-based matching: A fingerprint pattern is full of minutiae points, which characterize the fingerprint. In minutiae-based matching, these points are extracted from the image, stored as sets of points in the two-dimensional plane, and then compared with the same points extracted from the template. With this technique, it is difficult to extract the minutiae points accurately when the fingerprint is of low quality.

Ridge feature-based matching: This matching can be conceived as a superfamily of minutiae-based matching and correlation-based matching, since the pixel intensity and the minutiae positions are themselves features of the finger ridge pattern. The matching method uses features like local orientation and frequency, ridge shape, and structure information. Even though minutiae-based matching is considered more reliable, there are cases where ridge feature-based matching is better to use. In very low-quality fingerprint images, it can be difficult to extract the minutiae points, and using the ridge pattern for matching is then preferred.

Minutiae-based matching is certainly the most well known and widely used method for fingerprint matching, thanks to its strict analogy with the way forensic experts compare fingerprints and its acceptance as a proof of identity in the courts of law in almost all countries. For this reason specific attention is paid to this method.

MINUTIAE EXTRACTION

At the local level, a total of 150 different local ridge characteristics, called minutiae details, have been identified. Most of them depend heavily on the impression conditions and quality of fingerprints and are rarely observed in fingerprints. The seven most prominent ridge characteristics.

RIDGE TERMINATION
BIFURCATION
INDEPENDENT RIDGE
DOT OR ISLAND
LAKE
SPUR
CROSSOVER

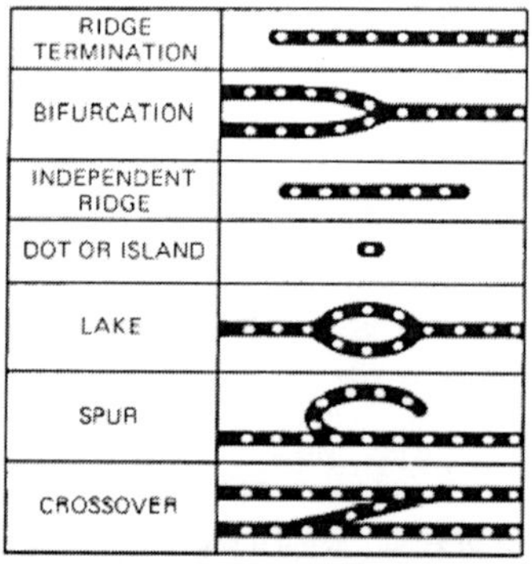

Fig. Minutiae details

The measured fingerprint area consists in average of about thirty to sixty minutiae points depending on the finger and on the sensor area. These can be extracted from the image after the image processing step (and possibly the classification step) is performed. The point at which a ridge ends, and the point where a bifurcation begins, are the most rudimentary minutiae, and are used in most applications. Once the thinned ridge map is available, the ridge pixels with three ridge pixel neighbors are identified as ridge bifurcations, and those with one ridge pixel neighbor identified as ridge endings. However, all the minutiae detected are no facts yet because of image processing and the noise in the fingerprint image. For each extracted minutia a couple of features are stored: the absolute position (x,y), the direction (θ), and if necessary the scale (*s*).

The location of the minutiae are commonly indicated by the distance from the core, with the core serving as the (0,0) on an x,y-axis. Some authors use the far left and bottom boundaries of the image as the axes, correcting for misplacement by locating and adjusting from the core. In addition to the placement of the minutia, the angle of the minutia is normally used. When a ridge ends, its direction at the point of termination establishes the angle. This angle is taken from a horizontal line extending rightward from the core.

At the very fine level, intra-ridge details can be detected. These are essentially the finger sweat pores whose position and shape are considered highly distinctive. However, extracting pores is usable only in high-resolution fingerprint images of good quality and therefore this kind of representation is not practical for most applications.

POST-PROCESSING

Minutiae localization begins with a preprocessed image. At this point, even a very precise image will have distortions and false minutiae that need to be filtered out. For example, an algorithm may search the image and eliminate one of two adjacent minutiae, since minutiae are very rarely adjacent. Irregularities caused by scars, sweat or dirt appear as false minutiae, and algorithms locate any points or patterns that do not make sense, such as a spur on an island (probably false) or a ridge crossing at right angles to two or three others (probably a scar or dirt). A large percentage of false minutiae are discarded in this post-processing stage.

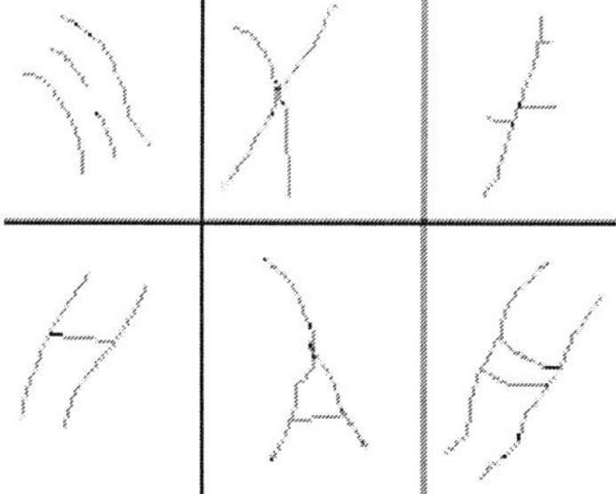

Fig. False minutiae: interrupted ridges, forks, spurs, structure ladders, triangles and bridges, in clockwise order

The several examples of false minutiae can be observed. In clockwise order: interrupted ridges, forks, spurs, structure ladders, triangles and bridges are depicted in the figure. The interrupted ridges are two very close lines with the same direction. Two lines connected by a noisy line compose a fork. The spurs are short lines whose direction is orthogonal to ridges direction. The structure ladders are pseudo-rectangle between two ridges. The triangles are formed by a real bifurcation with a noisy line between two ridges. Finally, the bridge is a noisy line between two ridges. All these characteristics generate several false minutiae. The algorithm is divided into several steps, executed in a pre-arranged order: elimination of the spurs, union of the endpoints, elimination of the bridges, elimination of the triangles, elimination of the structure ladders.

Pre-alignment

An intuitive more logical choice would be to pre-align all the templates in the database and the input image before the matching procedure. In this way the alignment takes place only once for every image. A great advantage is that it significantly reduces the computational time. Pre-alignment cannot compare images to one another so only depends on the properties of itself. The most common pre-alignment technique convert the fingerprint according to the position of the core point. Unfortunately, reliable detection of the core is very difficult in noisy images and in arch type patterns. For adjusting the rotation the shape of the silhouette, the orientation of the core delta segment, the average orientation in some regions around the core, and the orientations of the singularities can be used. But still this is a complex problem and causes often errors in the matching procedure. That is the main reason why embedded alignment in the minutiae matching stage is often used.

Conclusion

The main steps that a automatic fingerprint identification system should take when performing minutiae-based matching. However, some experts repeat steps or speed the process up by carrying out multiple steps at a time. The preprocessing and classification step can partly be executed at the same time for instance. The reason for this is that some preprocessing steps are necessary for the classification and some are necessary for minutiae extraction. As said before, weather to carry out the classification step depends on de size of the database and the time in which the identification has to be fulfilled.

Almost each author has his own preprocessing steps. Especially the filtering step varies by author. Moreover, some steps of the preprocessing can be skipped to shorten the processing time and even improve quality in some cases.

Improvement

Still there is a lot to improve in fingerprint identification technology. The improvement in performance is mainly to achieve in the preprocessing step. Fingerprints of poor quality, are still difficult to classify and match. Most of the misclassified or mismatched fingerprints are the result of bad quality. The preprocessing can improve the quality, but also discard useful information. Improvements are still achieved in this area.

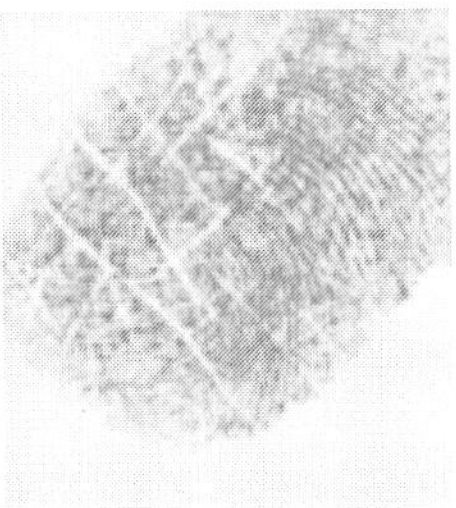

Fig. Fingerprint image of poor quality

Furthermore there can be improvements in time. A user does not want to wait for his results. Identification has to be real-time when going through security at an airport. Recently a new way of classification is developed, that speeds up the process considerably. In stead of using the classes of the Galton-Henry scheme, the images are classified based on the angle of the ridges. Each image is divided into four main directions: 0, 45, 90 and 135 degrees. There arise four scatter plots of the fingerprint, where the gray scale indicates the concentration of present angles. A white spot on the plot of 45 degrees indicates only angles of 45 degrees, black spots indicate no angles of 45 degrees. The result is a very fast classification. This goes about 100 times more rapidly than so far possible, and the error percentage is smaller than 0.5%.

Future of fingerprint identification

A dirty skin, scars, sweat and bruises can easily distort fingerprint recognition. A false matching can be the result of this. This factor does not play a role in iris recognition, a very reliable form of biometrics. Experts say that iris scans are therefore superior to fingerprint recognition systems. Also because an iris scan will produce results more quickly than a scan of a fingerprint. A check against 100.000 iris codes in a database takes two seconds. A fingerprint search will take fifteen seconds to perform the same task . On the other hand, the fingerprint is the most practical biometric recognition method. It only takes a simple sensor and a smart link to a database. These necessities can easily be built in in small equipment like PDA's and mobile phones. Iris detection on the contrary needs a high resolution camera. Besides, fingerprints are used in criminology where iris scans can not be used. Fingerprint identification will therefore be indispensable in the future.

3

Ballistic Fingerprinting Identification and Criminal Justice System

Ballistic fingerprinting refers to a set of forensic techniques that rely on marks that firearms leave on bullets to match a bullet to the gun it was fired with. It is a subset of forensic ballistics (the application of ballistics to legal questions) and internal ballistics (the study of events between the firing of a gun and the bullet leaving the barrel).

HISTORY

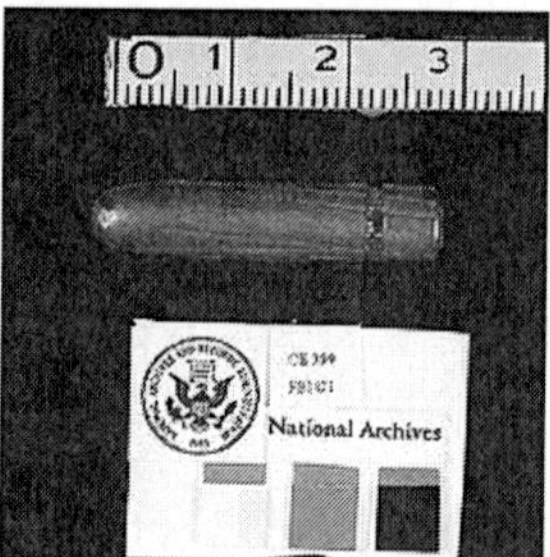

Fig. Controversial bullet from the John F. Kennedy assassination.

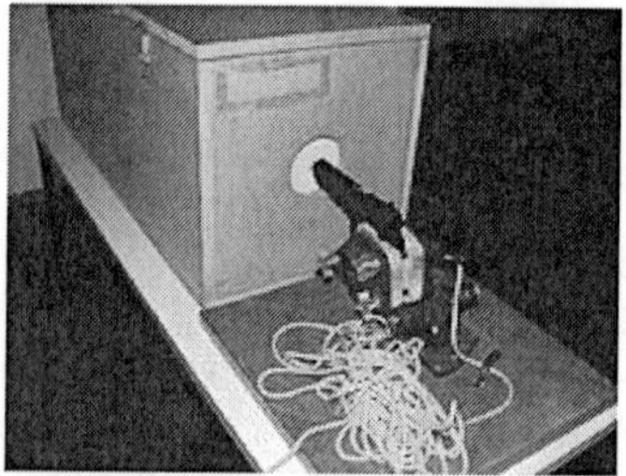

Fig.A forensic ballistics experiment

Rifling, which first made an appearance in the 15th century, is the process of making grooves in gun barrels that imparts a spin to the projectile for increased accuracy and range. Bullets fired from rifled weapons acquire a distinct signature of grooves, scratches, and indentations which are of value for matching a fired projectile to a firearm.

The first firearms evidence identification can be traced back to England in 1835 when the unique markings on a bullet taken from a victim were matched with a bullet mould belonging to the suspect. When confronted with the damning evidence, the suspect confessed to the crime. Alexandre Lacassagne was the first scientist to try to match an individual bullet to a gun barrel. The first court case involving firearms evidence took place in 1902 when a specific gun was proven to be the murder weapon. The expert in the case, Oliver Wendell Holmes, had read about firearm identification, and had a gunsmith test-fire the alleged murder weapon into a wad of cotton wool. A magnifying glass was used to match the bullet from the victim with the test bullet. Calvin Goddard, physician and ex-army officer, acquired data from all known gun manufacturers in order to develop a comprehensive database. With his partner, Charles Waite, he catalogued the results of test-firings from every type of handgun made by 12 manufacturers. Waite also invented the comparison microscope. With this instrument, two bullets could be laid adjacent to one another for comparative examination.

In 1925 Goddard wrote an article for the Army Ordnance titled "Forensic Ballistics" in which he described the use of the comparison microscope regarding firearms investigations. He is generally credited with the conception of the term "forensic ballistics," though he later admitted it to be an inadequate name for the science. In 1929 the Saint Valentine's Day Massacre led to the opening of the first independent scientific crime detection laboratory in the United States.

TECHNIQUES

Ballistic fingerprinting techniques are based on the principle that all firearms have inevitable variations due to marks left by the machining process, leaving shallow impressions in the metal which are rarely completely polished out. Also, normal wear and tear from use can cause each firearm to acquire distinct characteristics over time.

Fig.A firearm identification room. This room includes microscopes, a water tank for firing bullets and, an Automated firearms identification system. This room is used to identify and test firearms picked up asevidence.

Gross differences

The simplest considerations are the gross differences. A 10 mm bullet, for example, could not have been fired from a 9 mm barrel.

Striations

When a bullet is fired through a rifled barrel, the raised and lowered spirals of the rifling etch fine grooves called "striations" into the bullet. These can be matched with the barrel through which the bullet was fired. Examiners distinguish between striations common to all guns of a particular type ("class characteristics") and those unique to a particular gun ("individual characteristics").

The class characteristics depend upon the type of rifling in the barrel, which varies among manufacturers and models in number and shape of the grooves, twist rate, and direction. Colt, for example, traditionally uses a left-hand twist, while Smith and Wesson uses a right hand twist; a current production M16 rifle uses a 1 in 7 inch twist, while most civilian AR-15s and the current Mini-14 use a 1 in 9 inch twist. Marlin Firearms use a distinctive 16-groove Micro-Groove rifling in many of their firearms, while the M1903 Springfield rifle had two, four, or six grooves depending on the manufacturer. Polygonal rifling may leave striations that are difficult to match to a particular barrel.

Individual characteristics are caused by imperfections in the rifling process and tools, but also by the wear and tear caused by regular use, and can therefore change over time. Criminals or those concerned with government intrusion in privacy sometimes attempt to alter a gun's individual characteristics by changing or shortening the barrel, or by rubbing its interior with a steel brush.

Breech markings

Marks on the cartridge case can be matched to marks in the chamber and breech. For a number of reasons, Cartridge cases are often easier to identify than bullets. First, the parts of a firearm that produce marks on cartridge cases are less subject to long-term wear, and second, bullets are often severely deformed on impact, destroying much of the markings they acquire.

Shotguns

Ballistic fingerprinting of bullets does not work with firearms such as shotguns that fire shot-containing cartridges. In many cases the shot rides inside a plastic sleeve that prevents it from ever touching the barrel, and even in cases where the shot does touch the barrel, the random movement of the shot down the barrel will not leave any consistent marks. But shotgun cases can still be examined for firing pin marks and the like.

BALLISTIC FINGERPRINTING AIDS

Databases

Some localities, particularly Maryland, have attempted to build up a large database of "fingerprints"; in the case of the Maryland law, all new firearms sales must provide a fired case from the firearm in question to the Maryland State Police, who photograph it and log the information in a database. The Maryland State Police wrote a report critical of the program and asking the Maryland General Assembly to disband it, since it was expensive and had not contributed to solving a single crime. Subsequently however, the database did provide evidence used to obtain one murder conviction at an estimated cost of 2.6 million dollars per conviction.

A California Department of Justice survey, using 742 guns used by the California Highway Patrol as a test bed, showed very poor results; even with such a limited database, less than 70% of cases of the same make as the "fingerprint" case yielded the correct gun in the top 15 matches; when a different make of ammunition was used, the success rate dropped to less than 40%.

Bullet marking

There have been several proposals for the mandated marking of bullets to aid in ballistic fingerprinting, and some jurisdictions have passed legislation to that effect. California, for instance, passed a bill AB 1471 which requires all new models of handguns to be equipped with microstamping technology by 2010. However, due to the way the California law was written, microstamping cannot be implemented currently due to patent encumberment.

Several techniques have been proposed:

- Firearm microstamping is a process that engraves the make, model, and serial number on the cartridge and on the face of the firing pin, which stamps the primer as the firing pin impacts it.
- A British researcher proposed in a 2008 report that ammunition manufacturers coat their bullets with pollen, or with a pollen deposit coated with a metal oxide. Pollen grains are sticky enough and have a sufficiently hard outer case to survive being fired. They also attach themselves to the clothing and hands of people who handle the ammunition and the gun, providing an additional forensics clue (the pollen is extremely difficult to wash off completely, according to the researchers). If manufacturers used unique pollen varieties or unique mixtures of pollen and oxide coatings, the manufacturing database could be used to quickly identify a bullet found at a crime scene, assuming the investigating bodies equip themselves with the necessary pollen-identification equipment.

FINGERPRINT IDENTIFICATION AND THE CRIMINAL JUSTICE SYSTEM

There is a great deal of debate these days about the impact of the newest identification technology, DNA typing, on the criminal justice system. The introduction and rapid diffusion of this powerful technique over the past two decades or so has raised a host of important question including: How accurate, discriminating, and reliable is DNA, and how do we measure these attributes? How do we police the application of DNA typing to minimize errors? How inclusive should DNA databases be? What kind of threat do they pose to individual privacy and to civil liberties? What is the relationship between the criminal justice application of DNA typing and other applications in areas like health care, immigration control, and scientific research? Do genetic databases raise the specter of a resurgence of eugenics?

While there has been extensive debate over many of these issues for at least a decade now, much of this debate has been – and still is – conducted as if these issues have been raised anew by DNA typing. In fact, biometric systems of criminal identification have been with us for more than a century and half. Although other identification technologies are important, I am chiefly referring here to fingerprint identification, which has reigned as the world's dominant method of criminal identification since around the 1920s. Many of the most urgent issues now being debated with reference to DNA have been debated before with reference to fingerprinting. Indeed, our current discourse over DNA typing in many ways uncannily echoes the discourse in the early twentieth century when fingerprinting was a powerful new criminal identification technology sweeping the world. To the extent that our historical experience with fingerprint identification has been cited in the DNA debate, it has largely been based on a superficial, and largely mythical, understanding of the history of fingerprinting. The current debate will be better informed if we dig a bit more deeply into our past experiences with other biometric technologies of criminal identification.

A BRIEF HISTORY OF CRIMINAL IDENTIFICATION

In order to assess the impact of criminal identification databases on criminal justice systems, it is first necessary to have some understanding of why biometric criminal identification databases were built in the first place. Contrary to the popular image of fingerprinting as tool for forensic investigation, fingerprint identification was developed for purposes of criminal record keeping, rather than forensics. Specifically, fingerprinting was developed in order to facilitate the storage and retrieval of criminal histories by the state.

The impetus behind the development of biometric criminal identification technologies in the late nineteenth century was complex, including such factors as rapid urbanization; the increasing anonymity of urban life; and the dissolving of local networks of familiarity in which individuals were "known" by their neighbors; growing migration of individuals from city to city, country to country, and continent to continent; and the necessity of governing imperial possessions populated by large numbers of people whom it was necessary to monitor, control, and identify. Perhaps most important, however, was the shift in the philosophy of punishment, documented most famously by the French historian Michel Foucault, from classical to reformist jurisprudence. Under classical jurisprudence, punishments were meted out in strict proportion to the severity of the crime for which the offender had been convicted. Criminal histories were largely irrelevant under classical jurisprudence: a robbery was punished the same whether it was the offender's first or his fifth. Reformist jurisprudence, in contrast, sought to tailor punishment to the character of the offender. Under reformist jurisprudence, therefore, a criminal history was desirable – crucial, even -- because it enabled the state to draw sharp distinctions between first-time offenders and what were variously known as "habitual criminals," "incorrigibles," "repeat offenders," or "recidivists," a term which was coined in both English and French in the late nineteenth century.

Attempts to retrieve criminal histories filed according to names could be evaded by the simple expedient of adopting an alias. Beginning in the 1880s, two new technologies emerged which promised to solve the problem of aliases by linking criminal records, not to names, but to some representation of the criminal's body. Such a system required that criminal records be filed according some bodily property rather than by name. One of these systems, fingerprinting, is familiar to us today. The other, anthropometry – the measurement of the human body -- has largely been forgotten. Nonetheless, the two systems battled for dominance until well into the 1920s.

The chief architect of anthropometric identification was Alphonse Bertillon, an official at the Paris police prefecture. Bertillon's father was a demographer, and Alphonse was familiar with anthropometry as a social scientific tool. In Bertillon's system, developed during the 1880s, eleven anthropometric measurements were recorded using specially calibrated calipers and rulers, and the results were recorded on a printed card. Based on empirical data, Bertillon devised three equally populated categories for each measurement: small, medium, and large. Cards were filed according to which category they fell into for all eleven measurements. Faced with a suspect, Bertillon's operators could record all eleven measurements and search the existing criminal records for a card containing very similar measurements. If operators suspected a match, they confirmed it by reference to "peculiar marks," such as birthmarks, scars, and tattoos. The dimensions, locations, orientations, and descriptions of these peculiar marks were also recorded on the Bertillon card.

Though it sounds cumbersome to our ears, by contemporary accounts, the system was extraordinarily effective. Bertillon awed visitors to the Paris police department with his ability to retrieving matching cards in minutes from a vast archive containing tens of thousands of criminal records. Meanwhile, government bureaucrats in colonial India, Japan, and the United States were experimenting with using inked impressions of the papillary ridges on the tip of the finger for the less challenging task of authenticating identity – that is, verifying that an individual was indeed who he claimed to be. The use of fingerprints for indexing criminal records, however, was stymied by the lack of a system for filing identification cards according to fingerprint patterns. By the 1890s, anthropometric identification was rapidly being adopted by prisons and police departments worldwide.

Today, of course, anthropometric identification, with its meticulous skull measurements and attention to body size, evokes the pseudo-sciences of phrenology, craniometry, and somatotyping and their contributions to racist science. What is less known, however, is that both identification systems – fingerprinting as well as anthropometry – were closely tied to biologically determinist efforts to find bodily markers of character traits like intelligence and criminality. Cesare Lombroso, the founder of criminal anthropology, a discipline which purported to read signs of criminality in skulls sizes and shapes, facial features, and body types, called the Bertillon system "an ark of salvation" for the nascent field, although Bertillon himself maintained a agnostic stance toward the use of anthropometric data to diagnose criminal propensity. And, though it is now largely forgotten, a thriving research program, which began in the 1890s and extended – albeit in greatly diminished and increasingly marginalized form – through the present day, sought to correlate fingerprint pattern types with race, ethnicity, and character traits, such as insanity and criminality.

To the late nineteenth century scientific mind, it seemed patently obvious that fingerprint patterns were probably inherited and therefore should correlate with race, ethnicity, disease propensity, abilities, and various behavioral characteristics. Indeed, one of the pioneers of the modern system of fingerprint identification was Sir Francis Galton, who is better known as the statistician who founded the "eugenics" movement (and coined the term). The convergence of fingerprinting and eugenics in the person of Galton is often treated as a mere coincidence, but in fact Galton's ideas about eugenics were closely bound up in the development of criminal identification. Indeed, Galton's chief contribution to the development of the fingerprint system – his tripartite classification scheme for sorting all fingerprint patterns into three groups: arches, loops, and whorls – was devised chiefly for the purpose of using fingerprint patterns as bodily markers of heredity and character.

In his landmark book Finger Prints (1892), Galton published a study of the frequency with which these three pattern types appeared among various races. He was disappointed, however, to find almost no significant variations, other than slightly fewer arches among Jews. This was what the anthropologist Paul Rabinow has called "Galton's regret" in his provocative essay of that title. Despite Galton's "regret," however, a scientific research program arose beginning in the 1890s that did attach biological significance to Galton's three pattern types. In 1892, the same year Galton published his results, the French medico-legalist René Forgeot found an overrepresentation of arch patterns among prisoners at the Bologne penal colony. The following year, the French psychiatrist Charles Féré found more arches among both epileptics (epilepsy was thought to be a manifestation of evolutionary "degeneration," which also caused madness, disease, and criminality) and monkeys. Similarly, David Hepburn, an Irish anthropologist, found more loops and whorls among the "higher" primate species like chimpanzees and orangutans than in the "lower" monkeys. "Degeneration theory," a gloss on human evolution fashionable at the time, explained these results intuitively: the arch was the simplest pattern and, therefore, the least evolved. However, there was also a rival school of thought. In the United States, the Smith College anatomist Harris Wilder and his assistant (and later wife) Inez Whipple argued that the arch was the most evolved pattern because it was least functional as a tread for swinging from tree branches. Thus, the arch denoted the greatest distance from our primate ancestors.

This sort of research continued well into the 1920s. In Galton's laboratory, researchers studied the inheritance of fingerprint patterns. Other researchers measured the frequency with which pattern types appeared in different ethnic groups. The most ambitious study of this kind, published by the Norwegian biologist Kristine Bonnevie in 1924, found that Asians had a higher proportion of whorls, and fewer arches, than Europeans. In 1922, the New York Times reported, a German professor, Heinrich Poll predicted that life insurance companies would soon be able to "tell from finger prints what will be the insured's career." This research never died out completely. The most recent publication claiming to be able to diagnose criminality from fingerprint pattern types dates from 1991.

In short, the discourse surrounding fingerprinting in the early part of this century was strikingly similar to the discourse surrounding DNA today. It was widely assumed that pattern types appeared with different frequencies among different racial and ethnic populations, that heredity could be traced using fingerprint patterns (there was even the occasional paternity case in which fingerprint patterns were introduced as evidence), and that fingerprint patterns contained information that would soon be able to predict individuals' propensity for certain diseases and even their behavioral characteristics, including criminal propensity.

Paradoxically, Galton's fingerprint research had a much greater impact in an area – criminal record keeping – which he would have viewed as far less important than hereditary research. Galton's tripartite division of fingerprint patterns into arches, loops, and whorls provided the foundation for the creation of system for using fingerprint patterns to index criminal records in a manner similar to Bertillon's system based on measurements. The earliest workable fingerprint classification systems were developed almost simultaneously during the mid-1890s by Juan Vucetich, a police official in La Plata, Argentina, and Edward Henry, Azizul Haque, and Chandra Bose of the British colonial police in the Bengal province of India. The crucial inventive steps consisted of extending Galton's tripartite classification system by subdividing loops by means of "ridge counting" (measuring the number of intervening ridges between the "delta," the point at which the transverse ridges separated to flow around the central pattern and the center of the print) and whorls by means of "ridge tracing" (following a ridge from the delta and determining whether it passed inside or outside the center of the print). In this way, loops and whorls could be assigned to subcategories. By classifying each individual according to the pattern types and subtypes on all ten fingers, fingerprint classifiers could, like Bertillon operators, sort even very large collections of identification cards into relatively small groups. Examiners could classify the fingerprints of an unknown suspect and refer to a small number of potentially matching cards. They could then determine whether all the detail between the two sets of fingerprints was consistent.

The two rival fingerprint classification systems, the Argentine "Vucetich system" and the British Indian "Henry system," diffused during the 1900s and began to compete with anthropometric identification systems. Although it seems counterintuitive to us today, the prevailing wisdom at the time held that anthropometry was the scientifically superior system because anthropometry had a basis in mainstream academic sciences, anthropology and ethnology. The chief advantages of fingerprinting were practical: it was cheaper and faster than anthropometry, and, crucially, the recording of data – the taking of inked finger impressions – required only minimal training. The quality and consistency of anthropometric data, in contrast, depended greatly upon the diligence with which operators adhered to specified procedures for taking measurements. Obtaining consistent measurements required thorough training, supervision, and disciplining of operators.

Contrary to what we might think today, the potential forensic application of fingerprinting exerted only minimal influence over identification bureau chiefs choosing between fingerprinting and anthropometry. Bloody or "latent" fingerprints – invisible finger impressions made visible by "dusting" with powder – had been used to investigate crimes as early as 1892 in Argentina, 1897 in India, and 1903 in Britain (as well as, forgotten to history, in the late 1850s in Albany, New York). While these cases did attract some favorable publicity for fingerprinting, they did not necessarily convince identification bureau chiefs to discard their anthropometric identification files. Identification bureaus could stick with anthropometric identification systems and still use fingerprinting for forensic investigation, as Bertillon himself did in a 1902 murder case. Fingerprints could be used as a check on anthropometric identification, and space for fingerprints was added to the bottom of anthropometric identification cards, even in France. The Bertillon system, meanwhile, had its own bonus application that could assist in the investigation of crime: the portrait parlé, or "spoken likeness," a system of codes, suitable for transmission via telegraph, for describing the physical appearance of a suspect so that officials in another city could apprehend a suspect even without a photograph. While it sounds fanciful to us today, at the time the portrait parlé was viewed as a crime-fighting tool equal in utility to latent fingerprint identification. In any case, identification bureau chiefs were more concerned with maintaining systems of criminal records that would expose recidivists and help remove them from society than with solving specific crimes.

Nor were there many cases in which fingerprint evidence also exonerated the innocent, analogous to the recent spate of post-conviction exonerations generated by forensic DNA evidence. The closest parallel was probably one of the earliest forensic fingerprint cases, the Rojas case in Argentina in 1892. The initial suspect in the murder of two young children in the village of Necochea was the mother's suitor, one Velasquez. Standard tactics, including torture and forcing Velasquez to sleep with the corpses, failed to elicit a confession. Only then, did detectives, familiar with Vucetich's work with fingerprints, examine a bloody fingerprint found at the crime scene. This print matched, not Velasquez, but the victims' mother Francesca Rojas. Confronted with this evidence, Rojas confessed to the crime. While Velasquez had not yet been convicted when the fingerprint evidence was discovered, one might reasonably infer that, as the prime suspect in the grisly murder of two young children, he would have been. At the same time, the bloody fingerprint does not definitively prove the Rojas committed the crime since she may have touched the corpses post-mortem.

In early criminal trials, fingerprint classifiers were qualified to testify as experts in the analysis and interpretation of fingerprint patterns. They testified to the identity of latent fingerprint impressions by matching "minutiae" or "Galton details" or "points of similarity" – generally ridge endings or bifurcations – between the latent print and the inked print of the suspect. Based on these similarities, they proffered testimony that the latent print and the suspect's inked print were "identical" and that it was a "fact" that they came from the same source finger, to the exclusion of all other fingerprints in the world. This extraordinarily strong conclusion was justified on the basis that "there are no two fingerprints alike." This fundamentally unprovable assertion was in turn justified by reference to treatises in the field that declared it to be so and by the fact that identification bureaus had not yet discovered any identical fingerprint patterns on two different fingers. Occasional reference was also made to an unspecified law of nature that "nature never repeats" or to a statistical calculation by Galton that held that the chance of two different complete fingerprints being exactly alike was 1 in 64 billion.

None of these arguments shed much light on the validity or reliability of what it was these expert witnesses were claiming to be able to do: use similarities of detail between two fingerprint impressions to substantiate the conclusion that the two must come from a common source finger, to the exclusion of all other fingers in the world. This would have required some sort of measurement of how much corresponding detail warranted such a conclusion and on how good fingerprint experts were at making the interpretive decisions they claimed to make with absolute certainty. The arguments that no two fingers contained exactly identical papillary ridge formations begged the question of whether different fingers might contain areas of ridge detail similar enough that they might leave latent impressions which might give a fingerprint examiner a false impression of identicality. Forensic fingerprint identification's lone informed critic, the Scottish physician Henry Faulds, meanwhile, pointed out that identification bureaus' failure to find any identical fingerprints from two different fingers was hardly surprising considering that their records were filed according to the aggregate patterns on all ten fingers.

Although in early cases some trial judges and some juries expressed skepticism about warranting a criminal conviction on the basis of a single fingerprint match (while expressing confidence in the system by which criminal histories were authenticated by reference to the full set of ten prints), case law worldwide quickly ratified the scientific nature of forensic fingerprint evidence, and the legitimacy of fingerprint classifiers to testify as expert witness and to declare matches in terms of virtual certainty. Neither the courts nor the defense bar imposed upon fingerprint examiners the kind of demands that forensic DNA experts faced during the "DNA Wars" of the late 1980s and 1990s. While DNA experts would later be required to verify their protocols through testing, fingerprint examiners were never asked to measure how well they could match latent fingerprint impressions with source fingers. While two National Research Council panels would fiercely debate the niceties of the random match probabilities that DNA experts should be allowed to use before juries, fingerprint examiners were permitted to eschew probabilities altogether and phrase forensic fingerprint matches as virtual certainties or "facts." Nonetheless, courts went on to ratify fingerprint identification as a science and as reliable evidence. Eventually, American courts went so far as to reverse the burden of proof, demanding that the defense produce two identical fingerprints from different fingers, a impossible burden and one that was not necessary to support a defense argument that impressions from two different fingers might yet be similar enough to fool a fingerprint examiner.

While these legal decisions certainly helped boost fingerprinting's scientific credentials, they still did not tip the balance in favor of the fingerprint system. Thus, for the first two decades of the twentieth century, most prison identification bureaus outside Argentina, India, and Britain (which were actively proselytizing fingerprinting) relied on anthropometry for the identification of felons. Fingerprint identification filled other applications, such as authenticating the identity of military personnel, civil service applicants, or immigrants. Eventually, fingerprinting found a niche in police departments and courts charged with processing petty offenders, such as prostitutes, drunk-and-disorderlies, and vagrants. In these applications, the consequences of a potential error were less drastic, and the lower cost and greater speed of fingerprinting held great appeal.

The Will West case at the United States Penitentiary in Leavenworth, Kansas in 1903 has been touted by the FBI and by numerous popular and scholarly accounts as having dramatically demonstrated the superiority of fingerprinting to anthropometry. William and Will West were two African-American convicts discovered at Leavenworth who supposedly coincided in their Bertillon measurements and possessed "a facial resemblance like that of twin brothers." Only when the identification clerks fingerprinted them were they able to tell them apart, thus exposing at once the fallibility of identification by name, photography, and anthropometry. In fact, the Wests' anthropometric measurements did not match, and they were not fingerprinted either, at least not in 1903. They were fingerprinted sometime after that, and the dramatic story was concocted by fingerprint advocates in order to give fingerprinting an appealing creation myth. The transition from anthropometry to fingerprint was a more gradual process, the racial dimensions of which are suggested by the construction of the Will West case around two supposedly indistinguishable African-Americans.

Only after the First World War did the eventual triumph of fingerprinting over anthropometry become assured. Fingerprinting upon arrest became standard procedure in law enforcement agencies large and small, and police departments' collections of fingerprint cards began to outstrip prisons' collection of Bertillon cards. Although early legal decisions split on the question of whether the recording of fingerprints upon arrest was constitutional, the weight of legal opinion eventually supported the practice. Courts also allowed law enforcement agencies to retain fingerprint records even when the individual was not convicted or even acquitted, except in cases of illegal arrests. While the bulk of fingerprint collection took place at the local level, in 1924, the newly-founded Federal Bureau of Investigation (FBI) created an Identification Division positioned as a national repository of fingerprint cards, although the original cards were still retained by local law enforcement agencies. At the same time, the FBI also established a civil fingerprint file, which included the prints of civil servants and immigrants. While there are specific federal statues that forbid the use of fingerprints taken for immigration or motor vehicle purposes for criminal investigation, courts have otherwise upheld the searching of civil fingerprint files for criminal investigations.

Proponents of fingerprinting hoped to extend the civil use of fingerprinting even further, calling for "universal identification" of all citizens. As with DNA today, identification entrepreneurs exploited parents' fears of kidnapping by offering to record their children's fingerprints in a private file, for a modest fee. As with DNA today proponents of fingerprinting noted that a universal fingerprint database would enable the authorities to identify disaster victims. They cited the 1904 Slocum ship fire, the deadliest American maritime disaster of the twentieth century, much in the way that DNA proponents today cite the use of DNA to identify the victims of the crash of TWA Flight 800 and the bombings of the World Trade Center and Pentagon. (A key difference, of course, is that by obtaining samples from family members, DNA can be used to identify disaster victims even if the victim does not have a DNA sample on file in a database. Humanitarian use of fingerprints requires that the victim's fingerprints have previously been recorded and stored in a database.)

In the 1930s, the FBI and a variety of conservative civic organization like the American Legion, the American Coalition, the Daughters of the American Revolution, the Merchants' Association and various chambers of commerce got involved in the drive for universal fingerprinting. Local police departments and chambers of commerce jointly organized fingerprint drives to urge citizens to voluntarily submit their fingerprints. Hundreds of thousands of American citizens volunteered their – and their children's – fingerprints in this manner.

Although a universal civilian identifier could have proven extremely useful in many applications both criminal and civil, the stigma that the public already associated with fingerprinting – that being fingerprinted was tantamount to being treated like a criminal – and privacy and civil liberties concerns conspired to doom the universal fingerprint movement. The ultimate death of the universal fingerprint movement can be dated to the period 1935-1943, during which three different bills -- proposing universal fingerprinting or attaching a fingerprint to the newly created social security card – failed to pass Congress.

Despite the FBI's enthusiasm for universal fingerprinting, it is unlikely that the FBI, or anyone else, really had the technical capability to actually handle a universal fingerprint database in the post-war period, even had there been the political will to create one. In fact, fingerprint databases were far from the omniscient surveillance systems early identification pioneers had hoped they would become. The panoptic power of fingerprint databases was limited by several factors: First, the Henry and Vucetich systems had multiplied into a profusion of different systems. Each nation – and, in some countries, each jurisdiction – had modified the Henry or Vucetich system slightly to accord with local preferences. Thus, fingerprint classifications were not compatible across national and jurisdictional boundaries. An effort to search for a matching fingerprint record outside the local jurisdiction, therefore, required copying the fingerprint card and sending a separate copy to each neighboring agency. Efforts to develop universal telegraph codes for fingerprint patterns or to utilize facsimile technology to send fingerprint images to neighboring agencies foundered mainly on lack of cooperation between stubbornly local law enforcement agencies. Second, ten-print filing systems were of limited utility for searching latent prints since cards were filed according to the aggregate patterns on all ten fingers. A manual search of a fingerprint database for a single latent print would, therefore, require multiple searches based on educated guesses about what finger the latent derived from and the pattern types on the absent nine digits. Numerous single-print filing systems devised to remedy this problem proved cumbersome and complicated. The problem of manually searching latent prints was exacerbated by the rapidly increasing size of fingerprint databases. The larger a fingerprint file grew, the more daunting the process of conducting a manual search. Thus, for most of this century fingerprint databases were useful primarily for determining whether a suspect had a criminal record locally. Inquiry could also be made to the FBI, but the response time was slow: usually a matter of weeks. Latent print analysis was useful primarily when a suspect or a set of suspects had been selected by other means and their prints could be compared to the latent. Cracking of crimes solely through latent fingerprint evidence was relatively rare.

Only with the advent of computerized fingerprint identification has routine searching of unidentified latent prints and instantaneous national searching become realistic. Data-processing technology was used to sort fingerprint cards as early as the 1940s, and research into computer imaging of fingerprints began in the 1960s. During the 1970s the FBI developed an automated search and retrieval system. It was not until the mid-1980s, however, that Automated Fingerprint Identification Systems (AFIS) were mature enough for local law enforcement agencies to begin investing in them. Instead of using ink, AFIS record prints using an optical scanner and store them as digital images. Technicians can enter an unidentified latent print into the system and the AFIS will search its files and produce a list of candidate matches. A trained examiner then compares the latent with the candidates and determines whether any of them warrants a conclusion of identity. AFIS are good at winnowing a large database into a small number of likely candidates, but relatively poor at choosing the matching print: the "true" match is often not ranked first.

Optical scanning, digital storage, and computerized search and retrieval now give fingerprinting the potential to at last live up to its popular image: in which crime-scene technicians can routinely solve crimes lacking suspects by searching latent prints against a large database. These matches, called "cold hits" provided anecdotal justification for the procurement of AFIS. Thus, the investigative application of fingerprint technology is just now beginning to catch up to the archival application. This development has already prompted calls to further extend the scope of fingerprint databases. The logic was exemplified by New York City's campaign against "quality of life" offenses. By arresting, booking, and fingerprinting offenders who would previously have been released with a warning or a court date, the New York Police Department (NYPD) intentionally sought to get as many potential offenders "into the system" as possible. The argument was that quality of life offenders were the same individuals that commit more serious crimes. This argument was borne out anecdotally in the John Royster case in which the NYPD cracked a string of linked rapes and a homicide by entering a latent print found at the homicide scene into their AFIS. The print matched a stored print taken from a man named John Royster on the occasion of an arrest for subway fare evasion. The case seemed to reverse the principle that had long held for fingerprint databases: the larger the database the more difficult it was to search it. Now, with searches dependent on computer processing power rather than more costly human resources, the larger the database the greater the chance of success. It appears that this logic is being borne out by "cold case squads" which, using AFIS, are now able to search unidentified latent prints against much larger databases (such as the FBI's) than they were originally searched against. In Los Angeles this recently led to the arrest of a man for the killing of a police officer based on 45-year-old unidentified latent prints.

Digitization also holds great promise for networking disparate local databases into a single national or international database, a long deferred dream of identification advocates. At present, this effort has been hindered by incompatibilities between different AFIS vendors, but the FBI and the National Institute of Standards and Technology are developing a universal standard for exchanging fingerprint information. The recent sniper case, in which it took weeks for the unidentified latent print from a murder scene -- which eventually helped hone in on a suspect -- to be run against the FBI database, demonstrates that we still do not have a seamless, national fingerprint system. This technology, combined with fingerprint scanning devices located in police cars and wirelessly linked to a networked database, has the potential to turn the fingerprint system into the kind of omniscient, global surveillance network envisioned by identification pioneers nearly a century ago.

Eugenics

One of the most frequent objections raised to DNA databases is the threat of eugenics. The eugenics argument is based on what George Annas calls "genetic exceptionalism," distinguishing genetic identification from supposedly harmless biometric identification technologies like fingerprinting. The argument is that genes, unlike fingerprint patterns, contain information about individuals' racial and ethnic heritage, disease susceptibility, and even behavioral propensities. In this volume, Barry Steinhardt sums up this argument most clearly:

Let me start with a point that I hope we can all agree on. Drawing a DNA sample is simply not the same as taking a fingerprint. Fingerprints are two-dimensional representations of the physical attributes of our fingertips. They are useful only as a form of identification. DNA profiling may be used for identification purposes, but the DNA itself represents far more than a fingerprint. Indeed, it trivializes DNA data banking to call it a genetic fingerprint.

Steinhardt and other opponents of DNA databases argue that DNA samples contain sensitive information about race, ethnicity, paternity, disease susceptibility, and possibly behavioral propensities that might be – indeed, inevitably will be -- abused in a number of ways. Insurance companies, employers or other government agencies might raid the data for health-related data, leading to genetic discrimination against individuals or groups. Behavioral researchers will not be able to resist a database of convicted criminals, and most states' current laws do not bar them from accessing this data. Shoddy researchers may easily turn the skewed racial composition of our prisons into supposedly "scientific" evidence of links between crime and race. Most ominously of all, state-controlled DNA databases give the state the means to quickly identify members of racial or ethnic groups or individuals with certain disease, or possibly behavioral propensities. A genetic database in the hands of the Nazis, it is argued, would have made the holocaust easier to execute.

In contrast, Amitai Etzioni responds that the threat of eugenics is political, not technological and that our best defense against a eugenic state is to strengthen democratic institutions, rather than to ban technologies with potential eugenic applications. While I do not share Etzioni's easy confidence in the ability of American democratic institutions to resist eugenic impulses – and history suggests otherwise -- Steinhardt's well-intentioned critique mistakenly locates the threat of eugenics in technology, rather than in ideology. As I described above, the assumption that fingerprint patterns do not contain information that could be used for eugenic purposes is the product of a superficial understanding of history, based primarily on Rabinow's essay "Galton's Regret," which took on great significance because it represented the historical contribution to one of the earliest volumes that attempted to wrestle with the ethical and policy issues surrounding genetic identification, Paul Billings's DNA on Trial (1992).

Rabinow took Galton's failure to find convincing correlations between fingerprint patterns and race (a failure not shared by his colleagues in France, Germany, Ireland, and the United States) to mean that fingerprint patterns do not in fact correlate with race, disease or behavioral propensity. Fingerprint patterns "tell us nothing about individual character or group affiliation." In fact, the most recent studies have confirmed that racial and ethnic discrepancies in the distribution of fingerprint pattern types do exist, and they are just as Bonnevie described in 1924: more whorls among Asians. Similarly, a 1982 article in a reputable scientific journal concludes that fingerprint patterns are indeed to some extent inherited. As for behavioral propensities, the search continues: as recently as 1994, another reputable scientific journal published a study correlating certain fingerprint pattern types with homosexuality. This research is questionable, of course, but no more so than the genetic research on homosexuality.

Rather than concluding that it is not possible to correlate fingerprint patterns with race, ethnicity, disease, or behavior, it would be more accurate to say that the scientific research program that sought to do so did not thrive. The reasons for the decline of the diagnostic fingerprint research program were can be found in history, not in nature. First, biologists found new biological markers to examine, including, of course, the gene, which was re-discovered around 1900. Second, law enforcement officials, who became an increasingly dominant force within the community of people interested in fingerprints, found it more convenient to treat fingerprint patterns as meaningless information. This kept the identification process focused solely on the individuals and uncluttered with distracting theories about whether race, inheritance, or criminal propensity might also be legible in fingerprint patterns. The law enforcement officials who dominated the fingerprint community after the First World War had an interest in erasing the history of diagnostic fingerprint research and muting discussion of the issue. By transforming fingerprint patterns from potentially significant biological markers into merely individualizing information, "used only for identification," law enforcement officials bestowed a "purity" or "neutrality" on fingerprint identification that augmented its credibility as an identification technique among both the general public and the courts. Thus, diagnostic research into fingerprint patterns was marginalized from both sides. Had even a fraction of the scientific resources devoted to researching links between genes and disease, race, and behavior been devoted to researching links between fingerprint patterns and disease, race, and behavior, the latter might seem as significant to us as the former. The conception of fingerprints as useful only for identification is not a natural fact but a historical achievement.

The concerns about the potential resurgence of eugenics in the genetic age. Garland Allen's paper in the this volume clearly shows that the media and a few misguided or unscrupulous researchers are laying the groundwork for the gene to become the latest biological marker to be enlisted in the eugenic program, despite the protests of most geneticists. I do believe, however, that a careful reading of the history of eugenics makes clear that eugenics is an ideological, not a technological, phenomenon. The pernicious aspect of eugenics is the stubborn belief that complex phenomena like race, health, behavior, and ability can be explained by looking at biological markers. The choice of biological marker – whether skull size or shape, fingerprint pattern, or a gene -- is less important than the irrational faith that is bestowed in it. The correlations between biological marker and race, disease, or behavior do not need to be "real" to be viewed as significant by large numbers of otherwise intelligent people. In fact, it matters little whether they are real or not. This, after all, is the lesson of the craniometry episode, related famously by Stephen Jay Gould. Similarly, Troy Duster, in this volume, demonstrates how easily genetic markers can be conflated with genetic causes of race, disease, or behavioral traits. The lesson to be learned from fingerprint identification is not that some biological markers keep us safe from eugenics and others are dangerous, but rather that society can attribute a bogus significance to any biological marker if it chooses to do so. Thus, DNA is not any more dangerous than fingerprinting because the correlation of a single gene with Huntington's disease is "real." What is dangerous is allowing – and indeed encouraging -- people to make the leap from Huntington's to more complex phenomena like schizophrenia, homosexuality, or criminality. Media speculation about genetic causes of such slippery phenomena as shyness, aggressiveness, thrill seeking, altruism, alcoholism, intelligence, and sexual orientation, has drowned out the cautions of professional geneticists that single genes are unlikely to cause any of these behavioral characteristics, nor even of very many diseases. More voices need to be added to the few that are trying to defuse the media hype surrounding genetics. While civil libertarians worry that equating DNA to fingerprints "trivializes" DNA, I worry that genetic exceptionalism dangerously exaggerates the causal power of DNA. The libertarian opposition to DNA databases inadvertently fuels the hype surrounding DNA, lending credence to simplistic assumptions (which these same critics abhor) that genes determine race, disease propensity, and even behavioral characteristics.

We would argue that the principal contribution of genetics to a resurgence of eugenics remains metaphorical, rather than technological. The threat is less that government officials will actually be able to use genetic databases to weed potential lawbreakers out of the population, or target racial minorities in the effort to do so, than that the gene's enormous cultural resonance will serves as a powerful metaphor for convincing a gullible public that complex social behaviors like criminality have biological causes. The amply documented history of what Nicole Rafter calls "eugenic criminology" in this country and in Europe gives some indication of what comes next: the stigmatization of individuals with criminal records as "born criminals," the lifetime warehousing of them as irredeemable, involuntary sterilization, the fueling of racial and ethnic prejudice. Our best defense against this kind of future remains defusing the cultural power of the gene to fuel the resurgence of simplistic biological determinism, rather than principled opposition to the technology of genetic identification.

Ensuring the Reliability of Forensic Evidence

The overhasty anointing of forensic fingerprint identification as reliable evidence deterred all relevant actors from scrutinizing fingerprint evidence. Aside from a handful, "hard" scientists, like anatomists, statisticians, and forensic scientists with biological and mathematical training, did not view fingerprinting as a sexy research problem. Judges relied on a chain of citations founded on the faulty logic that demonstrating that there are no two fingerprint patterns exactly alike establishes the reliability of forensic fingerprint identification that expressed no doubt about the absolute reliability of fingerprint evidence. The practices of the defense bar are more difficult to gauge, but anecdotal evidence suggests that defense attorneys treated fingerprint evidence as an impregnable black box that did not merit the expenditure of scare resources. Fingerprint examiners themselves, by their own admission, ceased all research into the foundations of their science. The result was that the criminal justice system as a whole treated forensic fingerprint evidence as a "black box," whose outputs were scientific, unassailable, unproblematic, and error-free. Forensic fingerprint identification evolved into a practice wholly in the capture of law enforcement agencies – there were virtually no fingerprint experts who were not present or former employees of law enforcement or other government agencies, like the military – entirely lacking external oversight, regulation, quality control, or proficiency testing. Fingerprint examiners were left in the position of having their opinions treated as gospel. Perhaps the most damning anecdotal evidence that the adversarial system was not up to the task of testing fingerprint examiners' conclusions was the New York State Trooper Scandal evidence tampering scandal in which five state troopers pled guilty to fabricating fingerprint evidence in around forty cases over eight years, securing numerous criminal convictions including homicide convictions. Despite the fact that many of these fabrications were crude and easily detectable, in none of these cases, did the defense even hire a defense expert or challenge the fingerprint evidence.

When forensic fingerprint evidence was put under intensified scrutiny by the defense bar in the late 1990s, serious problems were exposed. Fingerprint evidence was implicated in several cases of false conviction. Proficiency testing revealed shockingly high error rates. And the lack of testing suggested that forensic fingerprint evidence might fail to meet the Supreme Court's standards for scientific and technical evidence. If nothing else, these developments revealed the blasé and gullible attitude that the criminal justice system had adopted toward fingerprint evidence.

As the "DNA Wars" subside, we need to guard against treating forensic DNA evidence with the same complacence with which forensic fingerprint evidence was treated for most of this century. In particular, we should be concerned about allowing law enforcement to monopolize expertise in the area of forensic DNA typing. Serious consideration should also be given to removing forensic scientists from the employ of law enforcement agencies or finding other ways of combating conscious and unconscious pro-prosecution bias; to providing for external oversight, regulation, and proficiency testing of forensic laboratories; and to providing resources to defense counsel for truly independent evaluations of forensic evidence and the scientific foundation of forensic techniques. Again, history teaches that no forensic technique is foolproof or error-free. Indeed, if ever there was a technique that claimed to be foolproof, it was fingerprint evidence. Recent history demonstrates that DNA evidence, like other types of forensic evidence, is subject to laboratory error, pro-prosecution bias, and overstatement of the scientific certainty of conclusions. Precautions should be taken to ensure that forensic DNA evidence receives ongoing scrutiny from the courts, the defense bar, and the scientific community and is not turned into a black box whose conclusions are treated as unassailable, error-free gospel.

Breadth of Databases

Perhaps the most pressing question about the future of genetic identification is how broadly databases will, or should, expand. Should genetic criminal identification databases be restricted to violent felons, felons, or to everyone convicted of a crime, no matter how petty? Should DNA samples be taken upon arrest or indictment? Should these samples be retained by the police, even in cases of dismissal, acquittal, or failure to prosecute? What about a universal database including all citizens or DNA typing at birth, as some, including most recently DNA typing's inventor, Sir Alec Jeffreys, have suggested, and as Kuwait, apparently, has begun implementing? If the innocent have nothing to fear from DNA typing, as its advocates contend, why not include everyone in the database?

History shows that each advance in the technology of criminal identification has resulted in a broadening of criminal identification databases. Anthropometric identification, relatively slow and expensive, was used primarily in prisons on felons convicted of crimes serious enough to warrant prison terms. Fingerprinting shifted the locus of identification from the prison to the police department, from conviction to arrest. It broadened criminal identification databases to include individuals guilty of petty crimes that did not ordinarily merit prison sentences. The result was that a much larger segment of the population was included in criminal identification databases, but many of those included were not violent criminals but urban poor, racial minorities, and immigrants who were particularly vulnerable to arrest for petty crimes (what today would be called "quality of life" offences) like vagrancy, public drunkenness, and prostitution.

There was an important consequence of combining expanded criminal identification databases with reformist jurisprudence, which emphasized the importance of recidivism as an indicator of propensity to crime and, therefore, future dangerousness. With expanded criminal identification databases, petty criminals, not violent felons, were most likely to accrue the requisite number of convictions to be adjudged "recidivists," not least because petty offenders, who served little, if any, prison time would be released sooner and therefore have greater opportunity for re-arrest. Thus, the weight of special punishments for recidivists tended to fall not on violent felons but on repeat petty offenders. The inevitable consequence of the combination of laws targeting recidivists – and there was a wave of such laws, called "Baumes laws," analogous to our "three-strikes-and-you're-out" laws, in the 1920s -- and comprehensive criminal identification databases was to disproportionately punish petty offenders, a result we recently experienced anew with the most recent wave of "three-strikes" laws in this country.

Essentially, the effect of fingerprint databases was to substitute recidivism, as vouched for by fingerprint records, for the definitive biological marker of criminality which criminologists had thus far failed to locate. Thus, recidivists were treated as "born criminals" and subjected to all the special measures which criminologists had spent decades concocting, in anticipation of finding that elusive marker of criminality, such as longer, sometimes indefinite, prison sentences, even for relatively minor offenses.

This perverse consequence of expanding criminal identification databases was little noticed at the time. Indeed, bigger fingerprint databases were widely viewed as a social good. As we have seen, however, in the 1930s and 40s Americans stopped short at the idea of a universal fingerprint database. Fingerprint databases were permitted to expand to include anyone convicted of a crime, no matter how petty. Civil databases, including various special groups such as civil servants, schoolteachers, military personnel, and immigrants were permitted. But an identification-free zone was created for citizens who did not fall into these special categories and managed to avoid encounters with the criminal justice system.

In the mid-century debate over the scope of databases, the resolution was to legislate away the privacy rights of those labeled "criminals" while preserving a modicum of privacy for most of those labeled "law-abiding." As we begin the debate over the extent of DNA database, the most likely outcome seems to be that a similar bargain will be struck: "tough on crime" politicians will win votes by legislating the expansion of DNA databases to include ever-larger categories of offenders. Indeed, this appears to already be occurring, although it should be noted that legislators appear to be broadening the legal scope of DNA databases far more enthusiastically than they appropriate funds to actually implement them. Given the widespread popular view of DNA as "genetic blueprint" and distrust of government, however, these same politicians will be reluctant to support a universal genetic database. Thus, DNA databases can be expected to include everyone designated "criminal" but not "law-abiding" citizens. Etzioni, in this volume, offers an example of the kind of utilitarian analysis that could support such a resolution; he argues that if an individual has been convicted of – of even suspected of -- breaking the law once, the benefits to society of storing his or her DNA in a database outweigh his or her individual privacy rights. If, however, an individual is not a convict or a suspect, the balance tips the other way.

The history of fingerprint identification teaches that our primary concern about such a "creeping" database, as Barry Steinhardt calls it, is that it threatens to inscribe race, class, or geographic inequities in arrest patterns, police practices, or criminal justice outcomes into the database. Criminal histories are not merely objective representations of individuals' antisocial behavior, nor of their potential dangerousness to society. They also reflect arrest patterns, policing practices, and biases in judicial outcomes, and as such are likely to reflect race, class, and geographic inequities. Once inscribed into the database, these inequities take on a seemingly neutral authority of their own: they appear to be pure, objective information, when in fact they may reflect the prejudices of police or judicial practitioners. In the case of fingerprinting, the criminal record, linked to the body by fingerprints, took on a life of its own appearing to convey with objective authority the degree of recidivism – and thus the degree of potential dangerousness – of the offender. The potential for such skewing of the information contained in criminal histories remains significant today, not least because of the prevalence of plea bargaining and deal-making in exchange for testimony. These practices ensure that the conviction that is officially inscribed into the criminal record is usually either lesser or greater than the offense actually committed, if indeed one was actually committed.

In short, an arrestee database, while probably the most politically, financially, and constitutionally palatable alternative, will inevitably produce a database that reflects the race, class, and geographic biases embedded in police and judicial practices. One need only look to the recent scandals over racial profiling, the appalling racial composition of our prisons, drug task forces in Texas, or the differential application of the death penalty (and presumably, therefore, of all criminal sentencing) depending on the racial dimensions of the perpetrator-victim dyad to conclude that these biases, despite decades of effort to eradicate them, remain significant. After passing through a DNA database, however, this biased information will have essentially been "laundered," and it will be treated as objective information imbued with the considerable authority of science.

There remains, however, one crucial difference between today and the late 1930s, when citizens and law enforcement first struck a bargain over biometric identification. In the 1930s, the primary application of criminal identification databases was archival: linking individual suspects to their "true" criminal histories so that they could be adjudicated with the highest degree of fairness (for them) and safety (for society). Today, the justification for both fingerprint and DNA databases has shifted more towards the investigative rather than adjudicative: the purpose of DNA databases, especially, is to solve crimes, not to determine proper punishments for those convicted. Cold searching, a novelty in the past, is becoming the principal functions of – and justification for – criminal identification databases. This brings us back to the inexorable logic of the criminal identification database in the computer age: the effectiveness of the database is wholly dependent on its size. While one might reasonably conclude that society stands to gain more by entering a serial rapist into a genetic database than a "law-abiding" citizen, one nonetheless cannot deny that a universal DNA database could solve more crimes than one restricted to convicted criminals. If criminals, especially violent ones, leave their DNA at crime scenes, why wait for them to be caught and convicted before having the ability to solve those crimes? The incredible forensic power of DNA forces us to consider much more seriously the Swiftian "modest proposal" of a universal DNA database.

Aside from enhanced crime-solving, the chief benefit of a universal database would be its equitability. David Kaye, Michael Smith, and Edward Imwinkelried suggest that a universal database would overcome the race, class, and geographic inequities inevitable in a broad convict or arrestee database. Moreover, including everyone's DNA in a single database might be the most promising way to ensure proper oversight over the database. If everyone is a potential victim of an erroneous or fabricated DNA match, of an insurance company pillaging information from a government database, or of state-sponsored eugenics, then the politicians who fund the regulatory agencies, watchdog committees, and public defenders who protect us against such events may be more likely to maintain adequate funding over the long-term, even as forensic DNA profiling inevitably ceases to be a hot issue and fades into the woodwork of police practice. It is on these grounds that Sir Alec Jeffreys, inventor of the earliest genetic identification technique, recently came out in favor of a universal database. Viewing the suspect database that is currently developing in Britain as "discriminatory," Jeffreys contends that only fair solution is to put everyone "in exactly the same boat." Then, Jeffreys argues, "the issue of discrimination disappears." Troy Duster, however, makes a compelling case that while a universal database may not perpetuate the race, class, and geographic inequities embedded in police practices and the criminal justice system, neither will it erase them. The criminal behavior of individuals stigmatized by race, class, or geography will still be documented at a higher rate than that of the privileged.

BIOMETRIC IDENTIFICATION BASED ON FINGERPRINTS

In this chapter the techniques for fingerprint identification will be explored. After explaining the theory of fingerprint verification, all current scanning technologies are described in more detail. Once it is known how these scanners identify a person by means of a fingerprint, two methods to counterfeit fingerprints are shown. All additional methods implemented by scanner manufacturers to prevent counterfeits from being successful are also described together with proposed methods how these systems could also be fooled into accepting dummy fingerprints. The consequences for systems using fingerprint verification are discussed at the end of the chapter. First, an example for fingerprint verification from practice will be given. This example also illustrates how difficult it can be to find an optimum trade-off between FAR and FRR. From a security point of view, one would want to have the FAR as small as possible. However, for acceptance of a biometry system, a large FRR is worse.

Case: Within the car industry a biometric verification system is under evaluation. Manufacturers of expensive cars are considering using fingerprint recognition as a requirement for ignition of the engine. To arm against car theft, the FAR should be as small as possible. On the other hand, suppose that the righteous owner of a car cannot use his car because his fingerprint is rejected (i.e. FRR is too high). He will consider this to be a much more serious flaw in the system than a technical failure which prevents the car from being started. This is especially true if he compares the advantages of this system with this rejection: the advantages are that the driver does not (necessarily) have to have a key to his car and a perception of higher security with respect to theft of his car. Whether indeed the security improves is questionable. Right now, we do not see car thieves trying to copy the key of your car, instead they try to by-pass the ignition mechanism where the car key is involved. Furthermore, as this article will show, it might decrease security since it is fairly easy and cheap to copy a fingerprint from a person, even without the person knowing this.

THEORY OF FINGERPRINT VERIFICATION

The skin on the inside of a finger is covered with a pattern of ridges and valleys. Already centuries ago it was studied whether these patterns were different for every individual, and indeed every person is believed to have unique fingerprints . This makes fingerprints suitable for verification of the identity of their owner. Although some fingerprint recognition systems do the comparison on the basis of actual recognition of the pattern, most systems use only specific characteristics in the pattern of ridges. These characteristics are a consequence from the fact that the papillary ridges in the fingerprint pattern are not continuous lines but lines that end, split into forks (called bifurcation), or form an island. These special points are called minutiae and, although in general a fingerprint contains about a hundred minutiae, the fingerprint area that is scanned by a sensor usually contains about 30 to 40 minutiae .

For over hundred years law enforcement agencies all over the world use minutiae to accurately identify persons . For a positive identification that stands in European courts at least 12 minutiae have to be identified in the fingerprint. The choice of 12 minutiae is often referred to as "the 12 point rule". This 12 point rule is not based on statistical calculations but is empirically defined based on the assumption that, even when a population of tens of millions of persons are considered, no two persons will have 12 coinciding minutiae in their fingerprints. Most commercially available fingerprint scanners give a positive match when 8 minutiae are found. Manufacturers claim a FAR of one in a million based on these 8 minutiae, which seems reasonable.

Fingerprint Scanning Technologies

Technologies for scanning fingerprints have evolved over the past years. The traditional method which is used by law enforcement agencies for over a hundred years now is making a copy of the print that is found at a crime scene or any other location and manually examining it to find minutiae. These minutiae are compared with prints from a database or specific ink prints, which could be taken at a later time. This method is of course based on the fact that the person who left the fingerprints is not co-operating by placing his finger on a fingerprint scanner. For systems that are commercially available (and deployed) people are required to co-operate in order to gain access to whatever is protected by the verification system.

The first generation fingerprint scanners appeared on the market in the mid eighties, so the technology is about fifteen years old. Over the past few years the technology for scanning fingerprints for commercial purposes has evolved a lot. While the first generation sensors used optical techniques to scan the finger, current generation sensors are based on a variety of techniques. The following techniques are deployed in commercial products that are currently available:

- Optical sensors with CCD or CMOS cameras
- Ultrasonic sensors
- Solid state electric field sensors
- Solid state capacitive sensors
- Solid state temperature sensors

The techniques will be described in greater detail in this section. The solid state sensors are so small that they can be built into virtually any machine. Currently a sensor is in development that will be built in a plastic card the size of a credit card, not only with respect to length and width but also with respect to thickness! It is clear that this type of sensor will give a boost to the number of applications using fingerprint technology.

Optical Sensors

With optical sensors, the finger is placed or pushed on a plate and illuminated by a LED light source. Through a prism and a system of lenses, the image is projected on a camera. This can be either a CCD camera or, its modern successor, a CMOS camera. Using frame grabber techniques, the image is stored and ready for analysis.

Ultrasonic Sensors

Ultrasonic techniques were discovered when it was noticed that there is a difference n acoustic impedance of the skin (the ridges in a fingerprint) and air (in the valleys of a fingerprint). The sensors that are used in these systems are not new, they were already being deployed for many years in the medical world for making echo's. The frequency range, which these sensors use, is from 20kHz to several GigaHertz. The top frequencies are necessary to be able to make a scan of the fingerprint with a resolution of about 500 dots per inch (dpi). This resolution is required to make recognition of minutiae possible.

Electric Field Sensors

This solid state sensor has the size of a stamp. It creates an electric field with which an array of pixels can measure variations in the electric field, caused by the ridges and valleys in the fingerprint. According to the manufacturer the variations are detected in the conductive layer of the skin, beneath the skin surface or epidermis.

Capacitive Sensors

Capacitive sensors are, just as the electric field sensors, the size of a stamp. When a finger is placed on the sensor an array of pixels measures the variation in capacity between the valleys and the ridges in the fingerprint. This method is possible since there is a difference between skin-sensor and air-sensor contact in terms of capacitive values.

Temperature Sensors

Sensors that measure the temperature of a fingerprint can be smaller than the size of a finger. Although either width or height should exceed the size of the finger, the other dimension can be fairly small since a temperature scan can be obtained by sweeping the finger over the sensor. The sensor contains an array of temperature measurement pixels which make a distinction between the temperature of the skin (the ridges) and the temperature of the air (in the valleys).

COUNTERFEITING FINGERPRINTS

The biggest problem when using biometrical identification on the basis of fingerprints is the fact that, to the knowledge of the authors, none of the fingerprint scanners that are currently available can distinguish between a finger and a well-created dummy. Note that this is contrary to what some of the producers of these scanners claim in their documentation. We will prove the statement by accurately describing two methods to create dummies that will be accepted by the scanners as true fingerprints. The two methods vary based on the cooperation of the fingerprint owner. Although there will without doubt be more ways to counterfeit fingerprints, the methods described in this article should suffice to show that all current scanners can be fooled.

Duplication With Co-operation

Duplication of a fingerprint with co-operation of its owner is of course the easiest method since it is possible to compare the dummy with the original fingerprint on all aspects and adapt it accordingly. First, a plaster cast of the finger is created. This cast is then filled with silicon rubber to create a wafer-thin silicon dummy (see also Figure 1). This dummy can be glued to anyone's finger without it being noticeable to the eye. For a thorough description of how to create such a dummy, we refer to Appendix A which describes the materials and tools that can be used. From the appendix it follows that creation of this type of dummy is possible with very liniited means within a few hours.

Fig. A wafer-thin silicon dummy of a fingerprint

Duplication Without Co-operation

For duplication of a fingerprint without co-operation of its owner it is necessary to obtain a print of the finger from for example a glass or another surface. One of the best ways to obtain such a print could be the fingerprint

Fig. A stemp type dummy of a fingeriprint

scanner itself. If the scanner is cleaned before a person will be using it, an almost perfect print is left on the scanner surface since people tend to press their finger (which is the verification finger!) firmly on the scanner. Some more expertise is required to create a dummy from such a print, but every dental technician has the skills and equipment to create one. An accurate description of how to create a dummy Of the fingerprint can be found in Appendix B. A picture of a stamp that is created using this method can be found.

ADDITIONAL TESTS OF SCANNERS

The main problem that challenges scanner manufacturers is making a distinction between dummy material that is not alive (i.e. silicone rubber) and material that is in fact not alive as well, the epidermis of a finger. Much research is done to make sure that a living finger is behind the epidermis. This research focuses on properties such as temperature, conductivity, heartbeat, blood pressure etc.. Although the methods are able to distinguish between dummies and real fingers, their operation margins have to be adjusted so radically to effectively operate indoors, outdoors, summer and winter, that a wafer-thin silicone rubber that is glued to a real finger easily passes these additional tests of scanners. For each of the possible additional tests for living fingers, a description will be given how dummies can be accepted by these systems.

Temperature

In a normal environment the temperature of the epidermis is about 8-10 degrees Celsius above the room temperature (18-20 degrees Celsius). By using the silicone rubber as described in Appendix A, the temperature transfer to the sensor decreases by at most 2 degrees if compared to a regular finger. It is clear that the difference falls with normal margins that are used on this system (at least 26-30 degrees). Sensors that are also capable of working outdoors are set to accept finger temperatures in an even broader range. Even when these sensors are compensating the fact that they are used outdoors, wafer-thin silicone rubbers won't be detected.

Conductivity

With most fingerprint scanners it is possible to add sensors which measure the conductivity of the finger. The conductivity of a regular finger is dependent on the type of skin (normal or dry). A normal conductivity value is about 200k Ohm (also dependent on the type of sensor), but the same finger will have a conductivity of several mega-Ohms during dry freezing winter weather and only several kilo-Ohms during summer when it is sweaty. Taking this into account, it is obvious that the margins are so large that putting some saliva on the silicone dummy will fool the scanners into believing it is a real finger.

Heartbeat

Several scanner manufacturers claim to detect a living finger by detecting the heartbeat in the tip of the finger. This is very well possible, although some practical problems arise from this. People actively participating in a sport can have heart rhythm of less than forty beats per minute, meaning that they should keep their finger motionless on the sensor for at least four seconds for the rhythm to be detectable. Also, the diversity in heart rhythm of a single person makes it virtually impossible to use it to take a person's heart rhythm into account when scanning the fingerprint. For example, the next day the same sportsman can have a heart rhythm of eighty beats per minute (doubled) if he decides to take the stairs instead of the elevator, just before his fingerprint is scanned. Moreover, the heartbeat of the underlying finger will be detected and accepted when a wafer-thin silicone rubber is used.

Relative Dielectric Constant

The dielectric constant of a specific material reflects the extent to which it concentrates the electrostatic lines of flux. Some manufacturers use the fact that the Relative Dielectric Constant (RDC) of human skin is different from the RDC of, for example, silicone rubber. just as with conductivity measurement in fingerprint scanners, the RDC is influenced by the humidity of the finger. To prevent an unacceptably high FRR, the margin of operation should be rather large. Putting some spirit on the silicone rubber with a wad of cotton wool before it is pressed on the fingerprint scanner fools the additional dielectric sensor. Spirit consists of 90% alcohol and 10% water. The RDCs of alcohol and water are 24 and 80 respectively, while the RDC of a normal finger is somewhere between these two values. Since the alcohol in the spirit will evaporate quicker than the water, the rate alcohol/water in the evaporating spirit will go to 0 (i.e. spirit slowly turns into water). During evaporation the RDC of the dummy will go up until it falls within the margins of the scanner and will be accepted as a real finger.

Blood Pressure

There are sensors available with which the blood pressure can be measured by using two different places on the body. They require a measurement of the heartbeat on two different places to determine the propagation speed of the heart pulse through the veins. Apart from the disadvantages that were already mentioned with the heartbeat sensors, this technique has an additional disadvantage in requiring measurement on two different places, i.e. on two hands. Similar to the heartbeat sensors, this method is not susceptible to a wafer-thin silicone rubber glued to a finger. Single point sensors are available but they must be entered directly in a vein, which obviously makes them unusable as a biometric sensor.

Detection Under Epidermis

Some systems focus on detection of the pattern of lines underneath the epidermis. The pattern of lines on this layer is identical to the pattern of lines in the fingerprint. Although this type of sensors look underneath the first layer doesn't mean that they cannot be fooled by dummy fingers, once it is known how they distinguish between the epidermis and the underlying layer.

Other Claims

Some manufacturers claim to use even more exotic detection methods and technique than the ones described above, making claims to having built a sensor which is not even known or being used in medical science. Additionally, they refuse to reveal the detection method claiming it is a company secret. Claims by these manufacturers should, without a doubt, be considered nonsense! In general, security by obscurity (trust that, by keeping specifications secret, the system will not be broken) should never be used. Although obscurity can make it more difficult for people to break the system in a brief period after introduction, most systems can be reverse engineered or worked around in ways the designers never expected.

CONSEQUENCES OF COUNTERFEIT POSSIBILITIES

The possibility to make a dummy, which will be accepted by the fingerprint scanner makes the system weak with respect to some different attacks:

1. A malicious person who wants to gain access, inconspicuously intercepts a fingerprint from someone who is granted access. With this print, a dummy is created.
2. If a righteous person is willing to co-operate, one or several dummies can be created with which this individual can give access to whomever he wishes.
3. If a righteous person handles a transaction, he can claim to be framed by a malicious person as described in point 1.

While the first two attacks on the system are possible with most verification systems, e third claim can usually be disproved since the person making the claim must have revealed something. An example is fraud with a PIN protected credit card. If the fraud is committed using the PIN code, the probability that its owner has not been careful with the PIN is much higher than the probability that the PIN system is broken. But the fingerprint verification system is very susceptible to this attack since we all leave behind fingerprints everyda, everywhere without noticing it. As long as it is still possible to use either the methodf from this article, or other methods to work around a fingerprint verification system, depl;;ient of such systems is unsuitable for virtually any application.

Case:

Suppose that a bank decides that for transactions, which exceed a certain amount of money, identification of the employee performing the transaction is required. The argumentation to use fingerprint verification instead of for example username/password combinations are that in case of fingerprint verification, the employee has to be present and cannot transfer his username/password to a colleague to perform transactions for him. Other systems that are considered, such as smart cards, can also not prevent the employee from letting other people perform a transaction. The bank trusts on the solution presented to them and decides to roll-out the fingerprint verification system throughout all offices. An employee of the bank knows that these systems can be circumvented and decides to make a dummy from a fingerprint of a colleague. The risks are small since using the fingerprint of a colleague cannot be traced back to him. To obtain the fingerprint he asks the colleague who's fingerprint he intents to use to hand him a glass or plate. This will almost certainly leave a perfect print on a clean surface, with which a dummy can be created and the fraudulent transactions can be performed. In case the malicious employee is not capable of creating a good dummy, he can always perform transactions using his own finger and claim that he is framed by a colleague the same way as described above.

Manufacturers of fingerprint scanners currently cannot deliver convincing evidence that they can make a distinction between a real, living finger and a dummy created from silicone rubber or any other material. Therefore, our advice is not to use fingerprint verification with applications where the identification serves as proof of presence. Comparing all biometric verification possibilities, fingerprint scanners are (perhaps apart from keystroke dynamics) the least secure means of verification. It is the only system where the biometrical characteristic can be stolen without the owner noticing it or reasonably being able to revent it. Even in a case where confidential computer data are protected by means of fingerprint verification we advise use of this verification only in combination with a token, for example a smart card, on which the user's template is stored. This prevents unnoticed access by someone using a dummy when the template, with which the scanned finger is compared, is stored on the computer's hard disk. The security level of the combination of fingerprint verification and smart card should be compared to username/password security. Th former can be considered more user-friendly.

With all applications that are considered to be protected by using biometric verification, techniques to compromise the system such as described in the appendices of this article should be very thoroughly examined. It should of course be taken into account that someone can break into a system if they put enough effort and resources into it (which is of course common with security issues). A problem with fingerprints is that neither the resources nor the skills to create a dummy are uncommon. Furthermore, the possibility of someone claiming to have been framed by someone else using methods that could not reasonably be prevented, must be eliminated. Otherwise the system is not suitable for the application.

4

Accuracy of Fingerprint Analysis

FINGERPRINTS: AN OVERVIEW

Latent fingerprints used in criminal investigations are often crucial pieces of evidence that can link a suspect to a crime. Latent prints are typically collected from a crime scene by specialists trained in forensic science techniques to reveal or extract fingerprints from surfaces and objects using chemical or physical methods. The fingerprint images can then be photographed, marked up for distinguishing features by latent fingerprint examiners, and used to search an automated fingerprint identification system (AFIS). An AFIS is a computer system that stores fingerprint images in an organized, searchable data structure that is widely used by criminal justice agencies to maintain databases of the fingerprints of individuals who are arrested or incarcerated.

Fingerprint databases typically contain rolled fingerprints from each finger ("tenprints") and fingerprints from the whole hand with all the fingers extended in parallel ("slaps"). If an individual whose fingerprints are in an AFIS encounters the criminal justice system again, a criminal investigator can search the AFIS to establish identity and link the individual with a particular criminal record. If a criminal investigator matches a latent print to a fingerprint in the AFIS, that individual may be linked to the crime under investigation.

An AFIS can also house repositories of latent fingerprints that remain unidentified, typically referred to as an unsolved latent file (ULF). As new fingerprints are added to the AFIS, criminal investigators can search them against the ULF collection in the hope of making a match. Matches happen regularly within one jurisdiction over time, but how are unsolved latent fingerprints collected in one jurisdiction matched against a tenprint record stored in the AFIS of another jurisdiction? Interoperability between two jurisdictions will determine whether Jurisdiction A can search the database in Jurisdiction B to find a match. Maximizing AFIS interoperability can help maximize the value of latent fingerprint evidence.

FINGERPRINT ANALYSIS

Fingerprint analysis is critical to the success of the nation's criminal justice system. In fact, fingerprints left at a crime scene — referred to as latent prints — are the most common type of forensic science evidence and have been used in criminal investigations for more than 100 years. The examination of fingerprint evidence consists of a series of steps involving the comparison of latent print to a known print (called an 'exemplar'). During this step-by-step matching process, latent print examiners must reach correct conclusions; they are also expected to produce records of the examination and, in some cases, present their conclusions — and the reasoning behind them — in court.

In recent years, the accuracy of latent print identification has been the subject of increased scrutiny, including from the National Academy of Sciences in their 2009 report, Strengthening Forensic Science in the United States: A Path Forward (pdf, 350 pages). To help address this issue, the National Institute of Justice and the U.S. National Institute of Standards and Technology (NIST) convened an expert working group to do a scientific assessment of the effects of human factors on forensic latent print analysis and to develop recommendations to reduce the risk of error.

The working group addressed issues ranging from the acquisition of impressions of friction ridge skin to courtroom testimony, and from laboratory design and equipment to emerging methods for associating latent prints with exemplars. In addition to a comprehensive discussion of how human factors relate to all aspects of latent print examinations — including communicating conclusions through reports and testimony — the report offers specific recommendations to improve the understanding and management of human-factors issues in fingerprint analysis.

CREATING A CULTURE OF OPENNESS AMONG EXAMINERS

The working group made a number of important recommendations that touch on everything from the time latent prints are submitted for examination to testimony in the courtroom. Among the recommendations is the need "to create a culture in which both management and staff understand that openness about errors is not necessarily a path to punitive sanctions but rather is part of an effective system to detect deviations from desired practices and incorrect judgments in latent print casework."

To help achieve such a culture, the panel recommended that management in forensic crime laboratories:

1. Employ a system to identify and track errors and their causes.
2. Establish policies and procedures for case review and conflict resolution, corrective action, and preventive measures.

Flowchart of Analysis, Comparison, Evaluation and Verification

One particularly helpful tool in the report is a flow chart that shows the

Analysis, Comparison, Evaluation and Verification (ACE-V) process for latent print examination that is currently used in the nation's forensic crime laboratories. This chart, developed by the working group, offers a detailed view of steps in the ACE-V process where human error risks should be minimized. This flowchart provides a valuable reminder of each step in the ACE-V process.

Research and Development Recommendations

The working group also identified a "critical need" for research into the interpretive process that is at the heart of the ACE-V process. "For example," they said, "only a handful of studies have assessed variation in the feature selection process. Generally, the findings have shown wide variation among examiners during the task of feature selection."

Other areas where research is recommended include:

- Determining the educational and cognitive abilities that should be prerequisites for training a latent print examiner.
- Developing methods that reduce the variation of the print features selected for comparisons.
- Developing a better understanding of the link between variations in feature selection and the examiner's ultimate determination.
- Addressing the utility/sufficiency of latent print analysis.
- Developing measures/metrics for latent print analysis — and using these to assess the reproducibility, reliability and validity of various interpretive stages of the process.
- Identifying key factors related to variations in the performance of latent print examiners during the interpretation process.
- Creating large, anonymous databases of exemplars and latent prints to facilitate the validation of probabilistic models and other statistical research.
- Determining the most appropriate tests of visual function for friction ridge examiners.

In particular, the working group said that the federal government should support research programs to improve automated fingerprint identification systems. Such programs could, for example:

1. Expand the algorithms used to match prints to account for the fact that the diagnostic value of minutiae depends on the region in which they are located.
2. Make fingerprint and palm print databases interoperable among local, state and federal automated identification systems.
3. Increase the compatibility between automated identification systems and other latent print software tools, including digital enhancement programs, probability calculation programs, and automated quality assessment programs.

THE HISTORY OF FINGERPRINTS

Why Fingerprint Identification?

Fingerprints offer an infallible means of personal identification. That is the essential explanation for fingerprints having replaced other methods of establishing the identities of criminals reluctant to admit previous arrests.

The science of fingerprint Identification stands out among all other forensic sciences for many reasons, including the following:

- Has served governments worldwide for over 100 years to provide accurate identification of criminals. No two fingerprints have ever been found alike in many billions of human and automated computer comparisons. Fingerprints are the very basis for criminal history foundation at every police agency on earth.
- Established the first forensic professional organization, the International Association for Identification (IAI), in 1915.
- Established the first professional certification program for forensic scientists, the IAI's Certified Latent Print Examiner (CLPE) program (in 1977), issuing certification to those meeting stringent criteria and revoking certification for serious errors such as erroneous identifications.
- Remains the most commonly used forensic evidence worldwide - in most jurisdictions fingerprint examination cases match or outnumber all other forensic examination casework combined.
- Continues to expand as the premier method for positively identifying persons, with tens of thousands of persons added to fingerprint repositories daily in America alone.
- Worldwide, fingerprints harvested from crime "scenes lead to more suspects and generate more evidence in court than all other forensic laboratory techniques combined. "

Other visible human characteristics tend to change - fingerprints are much more persistent. Barring injuries or surgery causing deep scarring, or diseases such as leprosy damaging the formative layers of friction ridge skin (injuries, scarring and diseases tend to exhibit telltale indicators of unnatural change), finger and palm print features have never been shown to move about or change their unit relationship throughout the life of a person.

In earlier civilizations, branding and even maiming were used to mark the criminal for what he or she was. The thief was deprived of the hand which committed the thievery. Ancient Romans employed the tattoo needle to identify and prevent desertion of mercenary soldiers from their ranks.

Before the mid-1800s, law enforcement officers with extraordinary visual memories, so-called "camera eyes," identified previously arrested offenders by sight. Photography lessened the burden on memory but was not the answer to the criminal identification problem. Personal appearances change.

Around 1870, French anthropologist Alphonse Bertillon devised a system to measure and record the dimensions of certain bony parts of the body. These measurements were reduced to a formula which, theoretically, would apply only to one person and would not change during his/her adult life.

The Bertillon System was generally accepted for thirty years. But it never recovered from the events of 1903, when a man named Will West was sentenced to the U.S. Penitentiary at Leavenworth, Kansas. It was discovered that there was already a prisoner at the penitentiary at the time, whose Bertillon measurements were nearly the same, and his name was William West. Upon investigation, there were indeed two men who looked exactly alike. Their names were Will and William West. Their Bertillon measurements were close enough to identify them as the same person. However, a fingerprint comparison quickly and correctly identified them as two different people. (Per prison records discovered later, the West men were apparently identical twin brothers and each had a record of correspondence with the same immediate family relatives.)

PREHISTORIC

Ancient artifacts including carvings similar to friction ridge skin have been discovered in many places throughout the world. Picture writing of a hand with ridge patterns was discovered in Nova Scotia. In ancient Babylon, fingerprints were used on clay tablets for business transactions.

BC 200s - China

Chinese records from the 221-206 BC Qin Dynasty include details about using handprints as evidence during burglary investigations. Clay seals bearing friction ridge impressions were used during both the Qin and Han Dynasties (221 BC - 220 AD).

AD 1400s – Persia

The 14th century Persian book "Jaamehol-Tawarikh" (Universal History), attributed to Khajeh Rashiduddin Fazlollah Hamadani (1247-1318), includes comments about the practice of identifying persons from their fingerprints.

1600s – Europe

In a "Philosophical Transactions of the Royal Society of London" paper in 1684, Dr. Nehemiah Grew was the first European to publish friction ridge skin observations. Dutch anatomist Govard Bidloo's 1685 book, "Anatomy of the Human Body" also described friction ridge skin (papillary ridge) details.

In 1686, Marcello Malpighi, an anatomy professor at the University of Bologna, noted fingerprint ridges, spirals and loops in his treatise. A layer of skin was named after him; "Malpighi" layer, which is approximately 1.8mm thick.

No mention of friction ridge skin uniqueness or permanence was made by Grew, Bidloo or Malpighi.

1823 - Purkinje

In 1823, John Evangelist Purkinje, anatomy professor at the University of Breslau, published his thesis discussing nine fingerprint patterns, but he too made no mention of the value of fingerprints for personal identification.

Herschel

The English first began using fingerprints in July of 1858, when Sir William James Herschel, Chief Magistrate of the Hooghly district in Jungipoor, India, first used fingerprints on native contracts. On a whim, and without thought toward personal identification, Herschel had Rajyadhar Konai, a local businessman, impress his hand print on a contract. The idea was merely "... to frighten [him] out of all thought of repudiating his signature." The native was suitably impressed, and Herschel made a habit of requiring palm prints--and later, simply the prints of the right Index and Middle fingers--on every contract made with the locals. Personal contact with the document, they believed, made the contract more binding than if they simply signed it. Thus, the first wide-scale, modern-day use of fingerprints was predicated, not upon scientific evidence, but upon superstitious beliefs. As his fingerprint collection grew, however, Herschel began to note that the inked impressions could, indeed, prove or disprove identity. While his experience with fingerprinting was admittedly limited, Sir William Herschel's private conviction that all fingerprints were unique to the individual, as well as permanent throughout that individual's life, inspired him to expand their use.

1863 - Coulier

Professor Paul-Jean Coulier, of Val-de-Grâce in Paris, published his observations that (latent) fingerprints can be developed on paper by iodine fuming, explaining how to preserve (fix) such developed impressions and mentioning the potential for identifying suspects' fingerprints by use of a magnifying glass.

Faulds 1880

During the 1870s, Dr. Henry Faulds, the British Surgeon-Superintendent of Tsukiji Hospital in Tokyo, Japan, took up the study of "skin-furrows" after noticing finger marks on specimens of "prehistoric" pottery. A learned and industrious man, Dr. Faulds not only recognized the importance of fingerprints as a means of identification, but devised a method of classification as well.

In 1880, Faulds forwarded an explanation of his classification system and a sample of the forms he had designed for recording inked impressions, to

Sir Charles Darwin. Darwin, in advanced age and ill health, informed Dr. Faulds that he could be of no assistance to him, but promised to pass the materials on to his cousin, Francis Galton.

Also in 1880, Dr. Henry Faulds published an article in the Scientific Journal, "Nature" (nature). He discussed fingerprints as a means of personal identification, and the use of printers ink as a method for obtaining such fingerprints. He is also credited with the first fingerprint identification of a greasy fingerprint left on an alcohol bottle.

1882 - Thompson

In 1882, Gilbert Thompson of the U.S. Geological Survey in New Mexico, used his own thumb print on a document to help prevent forgery. This is the first known use of fingerprints in the United States. Click the image below to see a larger image of an 1882 receipt issued by Gilbert Thompson to "Lying Bob" in the amount of 75 dollars.

1882 - Bertillon

Alphonse Bertillon, a Clerk in the Prefecture of Police of at Paris, France, devised a system of classification, known as Anthropometry or the Bertillon System, using measurements of parts of the body. Bertillon's system included measurements such as head length, head width, length of the middle finger, length of the left foot; and length of the forearm from the elbow to the tip of the middle finger. In 1888 Bertillon was made Chief of the newly created Department of Judicial Identity where he used anthropometry as the primary means of identification. He later introduced Fingerprints but relegated them to a secondary role in the category of special marks.

1883 - Mark Twain (Samuel L. Clemens)

In Mark Twain's book, "Life on the Mississippi", a murderer was identified by the use of fingerprint identification. In a later book, "Pudd'n Head Wilson", there was a dramatic court trial on fingerprint identification. A movie was made from this book in 1916 and a made-for-TV movie in 1984.

1888 - Galton

Sir Francis Galton, a British anthropologist and a cousin of Charles Darwin, began his observations of fingerprints as a means of identification in the 1880's. Juan Vucetich, an Argentine Police Official, began the first fingerprint files based on Galton pattern types. At first, Vucetich included the Bertillon System with the files.

1892 - Vucetich & Galton

Juan Vucetich made the first criminal fingerprint identification in 1892. He was able to identify Francis Rojas, a woman who murdered her two sons and cut her own throat in an attempt to place blame on another. Her bloody print was left on a door post, proving her identity as the murderer.

Sir Francis Galton published his book, "Fingerprints", establishing the individuality and permanence of fingerprints. The book included the first classification system for fingerprints. Galton's primary interest in fingerprints was as an aid in determining heredity and racial background. While he soon discovered that fingerprints offered no firm clues to an individual's intelligence or genetic history, he was able to scientifically prove what Herschel and Faulds already suspected: that fingerprints do not change over the course of an individual's lifetime, and that no two fingerprints are exactly the same. According to his calculations, the odds of two individual fingerprints being the same were 1 in 64 billion. Galton identified the characteristics by which fingerprints can be identified. A few of these same characteristics (minutia) are basically still in use today, and are sometimes referred to as Galton Details.

1897 - Haque & Bose

On 12 June 1897, the Council of the Governor General of India approved a committee report that fingerprints should be used for classification of criminal records. Later that year, the Calcutta (now Kolkata) Anthropometric Bureau became the world's first Fingerprint Bureau. Working in the Calcutta Anthropometric Bureau (before it became the Fingerprint Bureau) were Azizul Haque and Hem Chandra Bose. Haque and Bose are the two Indian fingerprint experts credited with primary development of the Henry System of fingerprint classification (named for their supervisor, Edward Richard Henry). The Henry classification system is still used in English-speaking countries (primarily as the manual filing system for accessing paper archive files that have not been scanned and computerized).

1900 - Henry

The United Kingdom Home Secretary Office conducted an inquiry into "Identification of Criminals by Measurement and Fingerprints." Mr. Edward Richard Henry (later Sir ER Henry) appeared before the inquiry committee to explain the system published in his recent book "The Classification and Use of Fingerprints." The committee recommended adoption of fingerprinting as a replacement for the relatively inaccurate Bertillon system of anthropometric measurement, which only partially relied on fingerprints for identification.

1901 - Henry

The Fingerprint Branch at New Scotland Yard (Metropolitan Police) was created in July 1901 using the Henry System of Fingerprint Classification.

- 1902: First systematic use of fingerprints in the U.S. by the New York Civil Service Commission for testing. Dr. Henry P. DeForrest pioneers U.S. fingerprinting.

- 1903: The New York State Prison system began the first systematic use of fingerprints in the U.S. for criminals.
- 1904: The use of fingerprints began in Leavenworth Federal Penitentiary in Kansas, and the St. Louis Police Department. They were assisted by a Sergeant from Scotland Yard who had been on duty at the St. Louis World's Fair Exposition guarding the British Display. Sometime after the St. Louis World's Fair, the International Association of Chiefs of Police (IACP) created America's first national fingerprint repository, called the National Bureau of Criminal Identification.
- 1905: U.S. Army begins using fingerprints.

U.S. Department of Justice forms the Bureau of Criminal Identification in Washington, DC to provide a centralized reference collection of fingerprint cards.

Two years later the U.S. Navy started, and was joined the next year by the Marine Corp. During the next 25 years more and more law enforcement agencies join in the use of fingerprints as a means of personal identification. Many of these agencies began sending copies of their fingerprint cards to the National Bureau of Criminal Identification, which was established by the International Association of Police Chiefs.

- 1907: U.S. Navy begins using fingerprints.

U.S. Department of Justice's Bureau of Criminal Identification moves to Leavenworth Federal Penitentiary where it is staffed at least partially by inmates.

- 1908: U.S. Marine Corps begins using fingerprints.
- 1915: Inspector Harry H. Caldwell of the Oakland, California Police Department's Bureau of Identification wrote numerous letters to "Criminal Identification Operators" in August 1915, requesting them to meet in Oakland for the purpose of forming an organization to further the aims of the identification profession. In October 1915, a group of twenty-two identification personnel met and initiated the "International Association for Criminal Identification" In 1918, the organization was renamed the International Association for Identification (IAI) due to the volume of non-criminal identification work performed by members. Sir Francis Galton's right index finger appears in the IAI logo. The IAI's official publication is the Journal of Forensic Identification.
- 1918: Edmond Locard wrote that if 12 points (Galton's Details) were the same between two fingerprints, it would suffice as a positive identification. Locard's 12 points seems to have been based on an unscientific "improvement" over the eleven anthropometric measurements (arm length, height, etc.) used to "identify" criminals before the adoption of fingerprints.

- 1924: In 1924, an act of congress established the Identification Division of the FBI. The IACP's National Bureau of Criminal Identification and the US Justice Department's Bureau of Criminal Identification consolidated to form the nucleus of the FBI fingerprint files.
- 1946: By 1946, the FBI had processed 100 million fingerprint cards in manually maintained files; and by 1971, 200 million cards.

With the introduction of automated fingerprint identification system (AFIS) technology, the files were split into computerized criminal files and manually maintained civil files. Many of the manual files were duplicates though, the records actually represented somewhere in the neighborhood of 25 to 30 million criminals, and an unknown number of individuals in the civil files.

- 1974: In 1974, four employees of the Hertfordshire (United Kingdom) Fingerprint Bureau contacted fingerprint experts throughout the UK and began organization of that country's first professional fingerprint organization, the National Society of Fingerprint Officers. The organization initially consisted of only UK experts, but quickly expanded to international scope and was renamed The Fingerprint Society in 1977. The initials F.F.S. behind a fingerprint expert's name indicates they are recognized as a Fellow of the Fingerprint Society. The Society hosts annual educational conferences with speakers and delegates attending from many countries.
- 1977: At New Orleans, Louisiana on 1 August 1977, delegates to the 62nd Annual Conference of the International Association for Identification (IAI) voted to establish the world's first certification program for fingerprint experts. Since 1977, the IAI's Latent Print Certification Board has proficiency tested thousands of applicants, and periodically proficiency tests all IAI Certified Latent Print Examiners (CLPEs).

Contrary to claims (in the 1990s and later) that fingerprint experts profess their body of practitioners never make erroneous identifications, the Latent Print Certification program proposed, adopted, and in-force since 1977, specifically recognizes that such mistakes do sometimes occur, and must be addressed by the Latent Print Certification Board. During the past three decades, CLPE status has become a prerequisite for journeyman fingerprint expert positions in many US state and federal government forensic laboratories. IAI CLPE status is considered by many identification professionals to be a measurement of excellence.

- 2012: The world's largest annual meeting of fingerprint experts is hosted by the IAI - this year it will be during 22-28 July in Phoenix, Arizona, USA.

- 2012: INTERPOL's Automated Fingerprint Identification System repository exceeds 150,000 sets fingerprints for important international criminal records from 190 member countries. Over 170 countries have 24 x 7 interface ability with INTERPOL expert fingerprint services.
- 2013 - America's Largest Database: The largest AFIS repository in America is operated by the Department of Homeland Security's US Visit Program, containing over 120 million persons' fingerprints, many in the form of two-finger records. The two-finger records are non-compliant with FBI and Interpol standards, but sufficient for positive identification and valuable for forensics because index fingers and thumbs are the most commonly identified crime scene fingerprints. The US Visit Program has been migrating from two flat (not rolled) fingerprints to ten flat fingerprints since 2007. "Fast capture" research will hopefully enable implementation of ten "rolled print equivalent" fingerprint recording (within 15 seconds per person fingerprinted) in future years.

The largest criminal fingerprint AFIS repository in America is the FBI's Integrated AFIS (IAFIS) in Clarksburg, WV. IAFIS has more than 60 million individual computerized fingerprint records (both criminal and civil applicant records). Old paper fingerprint cards for the civil files are still manually maintained in a warehouse facility (rented shopping center space) in Fairmont, WV, though most enlisted military service member fingerprint cards received after 1990, and all military-related fingerprint cards received after 19 May 2000, have now been computerized and can be searched internally by the FBI.

By mid-2013, the FBI plans to make civil file NGI fingerprint searches available to some US law enforcement agencies. The FBI is also planning to expand their automated identification activities to include other biometrics such as palm, face, and iris.

All US states and many large cities have their own AFIS databases, each with a subset of fingerprint records that is not stored in any other database. Many also store and search palmprints. Law enforcement fingerprint interface standards are important to enable sharing records and reciprocal searches to identify criminals.

Interpol, the European Union's Prüm Treaty, the FBI's Next Generation Identification and other initiatives seek to improve cross-jurisdiction sharing (probing and sharing/pushing) of important finger and palm print data to identify criminals.

- 2013 - World's Largest Database: As of March 2013, the Unique Identification Authority of India operates the world's largest fingerprint (multi-modal biometric) system, with over 200 million fingerprint, face and iris biometric records. UIAI plans to collect as many as 600 million multi-modal record by the end of 2014. India's

Unique Identification project is also known as Aadhaar, a word meaning "the foundation" in several Indian languages. Aadhaar is a voluntary program, with the ambitious goal of eventually providing reliable national ID documents for most of India's 1.2 billion residents.

With a database many times larger than any other in the world, Aadhaar's ability to leverage automated fingerprint and iris modalities (and potentially automated face recognition) enables rapid and reliable automated searching and identification impossible to accomplish with fingerprint technology alone, especially when searching children and elderly residents' fingerprints.

FINGERPRINT DATABASE INTEROPERABILITY

The ability of fingerprint databases in different jurisdictions to share data with each other — interoperability of the databases — can be influenced by both technology and policy. Through secure network connections, the automated fingerprint identification system (AFIS) in Jurisdiction A can be networked to the AFIS in Jurisdiction B so that either jurisdiction can search the fingerprints in the other. The two jurisdictions typically must have an AFIS manufactured by the same vendor or have a way for two different systems to communicate. The two agencies typically also must have some official agreement, such as a memorandum of understanding that defines the terms of the information sharing; otherwise the searching will be done on an ad hoc basis. The total national infrastructure of AFIS maintained by federal, state, local and tribal agencies can be thought of as the national criminal justice AFIS enterprise, and how the systems communicate with each other will depend on the network architecture and access controls.

Fingerprint searches can be done in either a vertical (e.g., local to state, state to federal) or a horizontal (e.g., local to local, state to state) manner. As a result, interoperability can be considered at different levels of geographic or jurisdictional granularity: local, regional intrastate, state, regional interstate and national. The extent to which an authorized AFIS user such as a criminal investigator can launch a latent fingerprint search from his or her computer in the national criminal justice AFIS enterprise and search for a fingerprint match in databases maintained in other jurisdictions can be thought of as the level of interoperability.

FORENSICS LAB 8.0: REVEALING LATENT FINGERPRINTS

Even someone who knows nothing else about forensics knows about fingerprints. The individuality of fingerprints had been generally accepted, as established by forensic scientists and courts, by the early 20th century, and the billions of fingerprint specimens taken since then have confirmed fingerprints as unique individual characteristics. Figure 8-1 shows a full fingerprint that we made by pressing one of our fingers to a stamp pad and then rolling it against a sheet of paper.

Fig. A typical full fingerprint taken under controlled conditions

Unfortunately, prints found at a crime scene are usually partial, broken, smeared, or otherwise inferior to the perfect specimens found on fingerprint cards, so it's often impossible to do a full comparison. Fingerprint examiners use points of comparison, also called points of identification, to compare unknown fingerprints against known specimens. A point of comparison is a particular individual feature of a particular fingerprint, such as where a ridge ends or bifurcates or the shape and number of ridges present in a whorl. If sufficient points of comparison exist between two fingerprints, it's reasonable to assume that those two prints were produced by the same finger, even if parts of the questioned print are missing, smeared, or otherwise obscured. Figure 8-2 shows a typical questioned partial fingerprint found on a questioned document. Although some ridge detail is present, it's unlikely that this print could be identified against a known print.

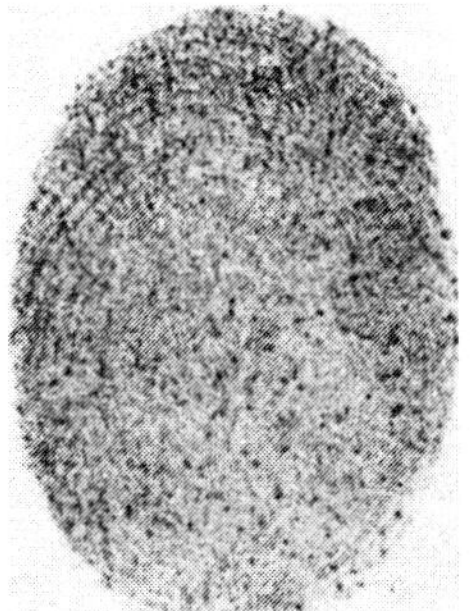

Fig. A typical questioned fingerprint

DENNIS HILLIARD COMMENTS

There are three levels of comparison:

- Level one – the overall pattern: loop, arch, whorl
- Level two – the dots, islands, bifurcations, ridge endings, etc.
- Level three – poroscopy, the presence and spatial relationship of the pores on the print ridges

With poor to moderate quality prints, no pores, smudging, or cross hatching, there needs to be more points of identification. With prints of excellent clarity and good poroscopy, identification can be made with less points of identification. The FBI has become more conscious of verification of print identification after the Brandon Mayfield mishap. There are two types of fingerprints:

Patent Fingerprints

Patent fingerprints are visible to the naked eye under ordinary light. Visible fingerprints are patent fingerprints made by fingers touching a surface after they have been in contact with ink, paint, grease, soot, blood, or some similar substance. Plastic fingerprints are patent fingerprints left on an impressionable material such as wet paint, modeling clay, tar, putty, wax, soap, and similar materials. Patent fingerprints of either type are ordinarily readily visible to crime scene investigators, and may sometimes be photographed or lifted directly. In some situations, patent fingerprints may be treated to increase their visibility or contrast against the background surface.

Latent Fingerprints

Latent fingerprints are invisible to the naked eye under ordinary light, but can be made visible by dusting, chemical development, or an alternate light source.

What's Alternate?

In forensics, the term alternate light source (or ALS) is used generically to describe any bright light source that emits light at a single wavelength or a narrow band of wavelengths. An ALS may emit light at any wavelength from far ultraviolet through the visible spectrum and into the far infrared. For example, a standard "black light" fluorescent tube is considered an ALS, as is a sodium- or mercury-vapor lamp. Beginning in the 1980s, lasers became widely used forensically as ALSs. Being expensive, bulky, and limited to a single wavelength made lasers less than idea as ALSs, so in the 1990s, lasers were gradually supplanted by portable ALSs that could be configured with filters or slits to emit narrow-band light over a wide variety of selectable wavelength ranges. Nowadays, lasers are seldom used in forensics labs. After any processing needed to reveal or enhance the print, fingerprints of either type are preserved by photographing them or by lifting them, either by carefully applying transparent lift tape to the surface that contains the print, peeling the tape from the surface, and transferring it to a card, or byelectrostatic lifting, which uses an electrostatically-charged sheet of clear plastic to attract powder applied to the fingerprints. It's at this point that the forensic scientist's job ends and the fingerprint examiner's job begin.

Dennis Hilliard comments

Lifting is generally done to allow a print to be photographed. If the object is mobile, the print is not lifted but preserved in place with lifting tape. If the surface is irregular, a lift is made for photographic purposes. Although I have never seen a fingerprint lifted by "electrostatic lifting," it is often used to lift footprints in dust from flooring. In forensic laboratories or police departments that have officers trained in fingerprint examinations, it is my experience that the prints are processed and then examined by the same analyst. In Rhode Island and in many states throughout the northeastern US, the evidence suspected of having latent prints is collected by law enforcement officers. Our examiners train these officers to partially process certain types of evidence, since time is often of the essence.

Various methods are used to reveal latent prints. Some methods are non-destructive, which means it they are tried and fail to reveal prints, other methods may be used subsequently. Other methods–notably silver nitrate development and physical developer-are destructive, either in the sense that using them precludes using alternative methods to raise the prints or that using them may preclude testing the object for other types of forensic evidence, such as blood or DNA. The particular method or methods used, and the order in which they are applied, also depends upon the nature (porous, semiporous, or nonporous) and condition (e.g., wet, dry, dirty, sticky, etc.) of the surface that contains the prints, as well as the residue that constitutes the prints, such as perspiration, blood, oil, or dust.

Visual examination is always the first step in revealing latent fingerprints. Some latent prints are patent under strong, oblique lighting. Moving small objects to different angles under a fixed light source may reveal numerous prints, as may moving the light source itself when examining larger or fixed objects. Any latent prints that are revealed under oblique lighting are photographed before any subsequent treatment is attempted. Some prints revealed by visual examination may be undetectable by any other method. Done properly, visual examination is completely non-destructive. Done improperly, the handling required for visual examination may smudge or destroy prints that are invisible visually but potentially visible using other methods.

After visual examination is complete, the usual next step is to examine the specimen by using inherent fluorescence. Various components of the fingerprint residue, including perspiration, fats, and other organic components, and foreign materials present on the fingertips when the impression was made, may fluoresce under laser, ultraviolet, or other alternate light sources. In a darkened room, the questioned surface is illuminated with the alternate light source and viewed through a filter of complementary color. For example, long-wave ultraviolet (black light) tubes emit some visible light in the deep violet part of the spectrum. Viewing a surface so illuminated through a deep yellow or orange filter blocks essentially all of the reflected incident violet light, making any inherent fluorescence emitted by the fingerprint residues in the yellow through red parts of the spectrum more clearly visible. The inherent fluorescence method is usable on any surface, including surfaces that cannot be treated with powders or chemical methods, and may reveal latent prints that are not revealed by any other method. Like visual examination, examination by inherent fluorescence is non-destructive.

After visual examination and inherent fluorescence examination are complete, other methods may be used to reveal additional latent fingerprints. Fingerprint powders, iodine fuming, and silver nitrate are considered the "classic" methods, because they have been used since the 19th century. Despite their age and the availability of newer methods, all three of these methods, with some minor improvements, remain in use today.

FINGERPRINT POWDERS

Fingerprint powders are used primarily for dusting nonporous surfaces such as glass and polished metal, most commonly to reveal latent fingerprints on immovable objects at crime scenes. Powders are often used in conjunction with super glue fuming to enable a lift to be made.

A very fine powder is applied to the area that contains the latent print. The powder adheres to the residues that make up the fingerprint. Excess powder is removed by gentle brushing or using puffs of air from a syringe. After the excess powder is removed, the fingerprint is revealed and can be photographed or lifted. Fingerprint powders are available in shades from white through black, which allows the fingerprint technician to choose a powder that contrasts with the background surface. Fluorescent fingerprint powders are useful for raising prints on printed or patterned surfaces, which might otherwise make it difficult to see the pattern of the print itself.

Magnetic fingerprint powders are used with magnetic brushes, which allow excess powder to be removed without actually touching the print. Magnetic powders are often used to raise latent fingerprints on paper surfaces, an exception to the general rule about powders being used only on non-porous surfaces.

Iodine Fuming

Iodine fuming is used to reveal prints on porous and semiporous surfaces such as paper, cardboard, and unfinished wood. The object to be treated is placed in an enclosed chamber that contains a few crystals of iodine. Gently heating the crystals causes them to sublime (go from solid phase to gas phase without passing through the liquid phase). The violet iodine vapor adheres selectively to fingerprint residues, turning them orange. These orange stains are fugitive, so they must be photographed immediately. After a period ranging from a few hours to a few days, the iodine stains disappear, leaving the specimen in its original state. The developed prints can be made semi-permanent by treating them with a starch solution, which turns the orange stains blue-black. These stains persist for weeks to months, depending on storage conditions.

Iodine Spray Reagent (ISR)

Iodine spray reagent (ISR) is a liquid analog to iodine fuming. Like iodine fuming, ISR is used to reveal prints on porous and semiporous surfaces such as paper, cardboard, and unfinished wood, but ISR can be used on specimens for which fuming is impractical. ISR is made up as two stock solutions that are combined to make the working solution. Solution A (iodine) is a 0.1% w/v solution of iodine crystals in cyclohexane. Solution B (fixer) is a 12.5% w/v solution of alpha-naphthoflavone in methylene chloride. The working solution is made up by combining A:B in a 100:2 ratio, mixing thoroughly, and filtering the working solution through a facial tissue or filter paper. The working solution is sprayed onto the questioned surface, using the finest mist possible. Latent prints develop immediately and should be photographed as soon as possible.

Roll Your Own ISR

We didn't have any alpha-naphthoflavone on hand (or any cyclohexane, for that matter), so we decided to see what we could accomplish with what we did have on-hand. Iodine has such a high affinity for the fats present in fingerprints that we thought almost any iodine solution should work, at least after a fashion. As it turned out, we were right.

We transferred a gram or so of iodine to a small spray bottle and added a few mL of lighter fluid, which formed a beautiful violet solution. Not all of the iodine dissolved, and we'd used the last of our lighter fluid, so we topped off the bottle with 70% ethanol. Iodine in ethanolic solution is brown, so we weren't surprised to see the solution turn a deep purple-brown color. One of us then pressed his fingers to a sheet of copy paper. We sprayed that area of the paper (in the sink; spraying iodine is very messy) and used a hair dryer to evaporate the solvent. Figure 8-3 shows the results. Not ideal, certainly, but much better than what we expected.

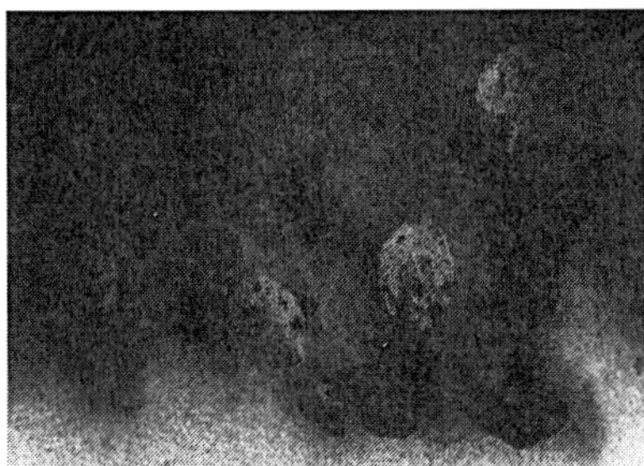

Fig. Fingerprints revealed by spraying with an iodine solution

Silver Nitrate

Silver nitrate is also used to reveal prints on paper and similar surfaces. The surface is treated with a dilute solution of silver nitrate by spraying or immersion. The soluble silver nitrate reacts with the sodium chloride (salt) present in sweat to produce insoluble silver chloride. The surface may or may not be rinsed with water after treatment to remove excess silver nitrate. In either case, the treated surface is exposed to sunlight or an ultraviolet light source, which reduces the silver chloride to metallic silver, revealing the prints as gray-black stains. Careful observation is required to make sure the prints are not overdeveloped, particularly if the surface was not rinsed after treatment. In extreme cases of over-development, the entire surface may turn black. Silver nitrate development is destructive, and so is used only after iodine fuming and other development methods. Three variants of silver nitrate solution are used, a 1% w/v aqueous solution, a 3% w/v aqueous solution, and a 3% w/v ethanolic solution. The alcohol solution is used on surfaces such as wax paper, coated cardboard, and polystyrene foam that repel water and so cause the aqueous solutions to bead.

Silver nitrate is used last, if it is used at all, because using it precludes subsequently using any other development method. Silver nitrate may succeed where other development methods fail, because silver nitrate reacts with the non-volatile sodium chloride present in fingerprint residues. Very old fingerprints may have lost all of their volatile residues, but the sodium chloride residue remains. Silver nitrate has been used successfully to develop latent prints that are years, decades, even centuries old.

Ninhydrin

Ninhydrin was introduced in 1954 as the first of the modern fingerprint development methods. In 1910, the English organic chemist Siegfried Ruhemann synthesized ninhydrin (triketohydrindene hydrate), and reported that it reacts with amino acids to form a violet dye that was subsequently named Ruhemann's Purple (RP). Forensic scientists must have been asleep at the switch, because it wasn't until 44 years later that S. Oden and B. von Hoffsten reported in the March 6, 1954 issue of Nature that ninhydrin could be used to develop latent fingerprints. Although amino acids are present in only tiny amounts in fingerprint residues, RP is so intensely colored that ninhydrin development produces stark visible images of latent prints.

Like iodine fuming and silver nitrate development, ninhydrin development is most useful for prints on porous surfaces. The questioned surface is simply sprayed with or dipped in a dilute solution of ninhydrin. After a period ranging from a few minutes to several hours, the prints self-develop as purple stains. In some cases, allowing development to continue for 24 to 48 hours reveals additional latent prints. Processing can be sped up by heating and humidifying the treated surface with a steam iron. After development with ninhydrin, prints may be sprayed with a 5% w/v solution of zinc chloride in a 25:1 mixture of MTBE (methyl tert-butyl ether) and anhydrous ethanol. This reagent causes a color shift from purple to yellow-orange and makes the developed prints fluoresce under an ALS. Two variants of ninhydrin solution are used, depending on the surface to be treated. The standard formulation is a 0.5% w/v solution of ninhydrin in a 3:4:93 mixture of methanol:isopropanol:petroleum ether. The alternate formulation is a 0.6% w/v solution of ninhydrin in acetone.

DFO

DFO, also known by its chemical name of 1,8-diazafluoren-9-one, operates by the same mechanism as ninhydrin, reacting with amino acids in fingerprint residues to form visible stains. DFO stains are much fainter than those produced by ninhydrin, but DFO stains fluoresce directly, without after-treatment. DFO was popularized by UK police forces, and is still more widely used in British Commonwealth countries than elsewhere. DFO is somewhat controversial. Many experts maintain that DFO is more sensitive than ninhydrin and provides superior detail. Other experts have questioned the reliability of DFO, and prefer to use ninhydrin alone. If DFO is used, it must be used before ninhydrin, PD, or silver nitrate is applied. DFO reagent is a 0.05% solution of DFO crystals in a solution made up of methanol:ethyl acetate:acetic acid:petroleum ether in a 20:20:4:164 ratio. DFO is applied by spraying or immersion, followed by drying, retreating the surface, drying again, heating the treated surface to 50 Â°C to 100 Â°C for 10 to 20 minutes, and finally by viewing under an alternate light source at 495 nm to 550 nm. Raised prints are photographed through an orange filter. Because DFO reagent is expensive and the required procedure is complex and time-consuming, DFO treatment is used less often than it might otherwise be.

Sudan Black

Sudan black is a dye that reacts with the sebaceous perspiration component of fingerprints to form a blue-black stain. Sudan black is used primarily for wet surfaces, including those contaminated with beverages, oil, grease, or foods, and is also useful for post-processing of cyanoacrylate-fumed prints, particularly those on the interior of latex or rubber gloves. The Sudan black reagent is simply a 1% w/v solution of Sudan black in 60% to 70% ethanol, although it is usually made up by dissolving the solid dye in 95% ethanol and then adding distilled water to reduce the ethanol concentration to 60% to 70%. In use, the questioned surface is immersed in the Sudan black solution for about two minutes and then rinsed gently with water. Raised prints are visible as blue-black stains.

Blood Reagents

Blood reagents are used to develop latent prints and enhance visible prints that include blood in the fingerprint residues. These agents are also commonly used to reveal latent bloodstained footprints, hand marks, and so on. Amido black, the oldest of these reagents, is a dye that stains proteins in blood residues blue-black. Several variants of amido black reagent are used. The most common is a 0.2% w/v solution of amido black in a 9:1 mixture of methanol:acetic acid. The questioned surface is sprayed with or immersed in this solution and allowed to soak for 30 seconds to one minute, after which it is rinsed with a 9:1 methanol:acetic acid solution. The dye/rinse procedure can be repeated to increase contrast. After a final rinse with water, the specimen is dried and photographed. Surfaces that are likely to be damaged by methanol can be treated with an alternate formulation that contains 0.3% amido black w/v, 0.3% sodium carbonate w/v, and 2.0% 5-sulfosalicylic acid w/v in an 89:5:5:1.25 mixture of water:acetic acid:formic acid:Kodak Photo-Flo 600. Questioned surfaces are sprayed with or immersed in this solution for three to five minutes and then rinsed with water. Again, the treatment can be repeated to increase contrast. An alternate water-based amido black reagent can be made up that is 0.2% w/v with respect to amido black and 1.9% w/v with respect to citric acid in a 998:2 mixture of water:Kodak Photo-Flo 600. The questioned surface is sprayed with or immersed in this solution and allowed to soak for 30 seconds to one minute, after which it is rinsed with water. Repeated treatments may be used to increase contrast.

Leucocrystal violet (LCV) is an alternative to amido black, and provides similar results. LCV reagent is a solution in 3% hydrogen peroxide that is 0.2% w/v with respect to LCV, 2.0% w/v with respect to 5-sulfosalicylic acid, and 0.74% w/v with respect to sodium acetate. When this solution is sprayed on a questioned surface, the prints develop in about 30 seconds, after which the surface is blotted dry and photographed. Coomassie brilliant blue is another alternative to amido black that provides similar results. Coomassie brilliant blue reagent is a 0.1% solution of Coomassie brilliant blue R dye in a 2:9:9 mixture of acetic acid:methanol:water. When this solution is sprayed on a questioned surface, the prints develop in 30 to 90 seconds, after which the surface is rinsed with a 2:9:9 solution of acetic acid:methanol:water. Repeated treatments may be used to increase contrast. After the final treatment, the surface is rinsed with distilled water, dried, and photographed.

Crowle's double stain is still another alternative to amido black that also provides similar results. As you might expect, the developer solution uses two dyes. It contains 0.015% w/v Coomassie brilliant blue R and 0.25% crocein scarlet 7B in a 3:5:92 mixture of trichloroacetic acid:acetic acid:water. The questioned surface is sprayed with or immersed in this solution, allowed to soak for 30 to 90 seconds, and then rinsed with a 3:97 mixture of acetic acid:water. After the final treatment, the surface is rinsed with water, dried, and photographed. DAB, also known by its chemical name of 3,3'-diaminobenzidine tetrahydrochloride, is the newest of the blood reagents. The DAB method is relatively complicated and expensive, but it sometimes provides usable results where no other method works. The DAB process requires four reagents. Solution A, the fixer, is a 2% w/v solution of 5-sulfosalicylic acid in distilled water. Solution B, the buffer, is a 1:8 mixture of 1 M pH 7.4 phosphate buffer solution in distilled water. Solution C is a 1% w/v solution of DAB in distilled water. Developer is made up by combining 180 parts by volume of Solution B with 20 parts Solution C and one part 30% hydrogen peroxide.

Prints may be developed by the DAB submersion method or the DAB tissue method, depending on which is better suited for the specimen. For the submersion method, the specimen is soaked for 3 to 5 minutes in a tray of Solution A to fix the prints, followed by rinsing for 30 seconds to one minute in a tray of distilled water. After the first rinse, the specimen is soaked for up to five minutes in a tray of developer, until maximum contrast is achieved. After a final rinse in distilled water, the specimen is air-dried or dried with a hair dryer, after which it is photographed. For the tissue method, the surface is covered with an unscented facial tissue or a thin paper towel, which is then sprayed with Solution A and allowed to soak for three to five minutes. The tissue is removed and the area is rinsed for 30 seconds to one minute with distilled water. A new tissue is placed over the subject area, saturated with developer, and allowed to soak for up to five minutes. When development is complete, the area is rinsed thoroughly with distilled water, dried, and photographed.

Small Particle Reagent (SPR)

Small particle reagent (SPR) is a liquid suspension of solid particles of dark gray molybdenum disulfide, applied to the questioned surface by spraying or dipping. SPR works in the same way as fingerprint powders-by physical adhesion of particles to fatty fingerprint residues-but unlike dry fingerprint powders, SPR can be used for processing wet surfaces, including surfaces soaked in liquid accelerants and other organic solvents. SPR is also used on glossy nonporous surfaces such as glass and plastic, coated glossy papers, and surfaces covered with glossy paint. Prints raised by SPR are extremely fragile, and should be photographed before any attempt is made to lift them. Modified versions of SPR are available in white, gray, and fluorescent forms, all of which are based on chemicals other than molybdenum disulfide.

Roll Your Own SPR

Although SPR is readily available from forensic suppliers, you may want to try making your own. To do so, add about 5 g of molybdenum disulfide (dry powder lubricant sold as Moly Lube and similar trade names) to about 100 mL of water to which you've added 1 mL of liquid dish washing detergent. Agitate this suspension thoroughly before use, and apply it by spraying or dipping. Allow the SPR to act for one minute, and then rinse the surface gently with water. We tried this ad hoc method, and it actually kind of worked.

Super Glue Fuming

Super Glue fuming, also called cyanoacrylate fuming from the primary component of Super Glue, was discovered by accident in 1976 when Masao Soba noticed white fingerprints on the surface of a super glue container. Frank Kendall improved the process and adapted it to latent fingerprint development, reporting his findings in a 1980 paper. Since that time, Super Glue fuming has become one of the most frequently-used latent print development processes.

Super Glue fuming is used to develop latent prints on nonporous glossy surfaces such as glass, plastic, and polished metal. Like dusting, cyanoacrylate fuming is a physical process. Cyanoacrylate vapor is selectively attracted to fingerprint residues, where it builds up as a crystalline white deposit. The developed latent prints may be photographed as is, or may be dusted or treated with various dyes that enhance the visibility and contrast of the prints. Although the exact mechanism remains unknown, it's suspected that the super glue fumes are catalyzed by the tiny amount of moisture attracted by the sodium chloride residues in latent fingerprints.

The standard method for Super Glue fuming is to place the object to be fumed in an enclosed chamber (aquariums are often used) that contains a small electric heater. An aluminum weighing boat is placed on the heater, and the temperature set to high. When the boat is hot, a few mL of cyanoacrylate are added to the boat. Fuming commences immediately, and is ordinarily complete after 30 seconds to 10 or 15 minutes. Alternatively, a cotton ball can be soaked in 0.5 M sodium hydroxide and allowed to dry. Once dry, the cotton ball is placed in the chamber and moistened with a few drops of cyanoacrylate. Fuming begins within a few seconds and is allowed to continue until the latent prints become visible.

Super Glue fumed prints can be photographed directly, or treated with dyes to increase the visibility and contrast of the prints and make them easier to see against a patterned background surface. With the exception of Sudan black, most of these dyes are fluorescent, with absorption wavelengths that match to the emission wavelengths of commonly-used forensic alternate light sources. For example, rhodamine 6G may be excited with a light source from 495 nm (blue) to 540 nm (green-yellow), with maximum absorption at 525 nm (green). Rhodamine 6G fluoresces in the range of 555 nm (yellow-green) to 585 nm (orange), with maximum emission at 566 nm (yellow). By viewing the treated surface through a filter that blocks wavelengths below about 555 nm but passes longer wavelengths, the treated fingerprints are visible as a yellow glow against a dark background.

Various fluorescent dyes are used alone or in combination to post-process super glue fumed prints. Standard individual dyes include rhodamine 6G,Ardrox, 7-(p-methoxybenzylamino)-4-nitrobenz-2-oxa-1,3-diazole (MBD for short),basic yellow 40, safranin O, and thenoyl europium chelate. The most commonly used combination is RAM, a mixture of rhodamine 6G, Ardrox P133D, and MBD. Other mixtures are also used, including RAY (rhodamine 6G, Ardrox, and basic yellow 40), and MRM 10 (MBD, rhodamine 6G, and basic yellow 40).

Physical Developer (PD)

Physical developer (PD) is useful for developing latent fingerprints on most porous surfaces and some nonporous surfaces. It is particularly useful for revealing latent prints on paper currency, paper bags, and porous surfaces that have been wet. PD is a destructive process, and so is always used last if at all. PD is an alternative to the silver nitrate method. You can use one or the other, but not both. Whichever you use must be the last method you apply. PD is normally used after DFO and/or ninhydrin, and often reveals latent prints that neither of these methods revealed.

The "physical" part of the name is a misnomer. PD is not a physical process (like dusting), but a chemical one. It depends on a redox reaction that reduces silver ions to metallic silver, which stain the latent fingerprints a gray-black color. The PD working solution is unstable, in the sense that it must be used immediately after it is made up, but it is this very instability that allows PD to work as well as it does. PD is expensive, complex, finicky, destructive, and requires a great deal of experience to get good results. Despite these criticisms, PD is used because it often gets results when no other method works. For this reason, many forensics labs routinely use PD as the final step in processing latent prints.

The PD process requires four solutions, with a fifth solution optional. Solution A is a 2.5% w/v solution of maleic acid in distilled water. Solution B (redox solution) is an aqueous solution that is 3% w/v with respect to ferric nitrate, 8% w/v with respect to ferrous ammonium sulfate, and 2% w/v with respect to citric acid. Solution C (detergent) is an aqueous solution that is 0.3% w/v with respect to n-dodecylamine acetate and 0.4% w/v with respect to Synperonic-N. Solution D is a 20% w/v solution of silver nitrate. Solution E (bleach) is a 1:1 mixture of standard chlorine laundry bleach with water.

The specimen to be treated is first placed in a tray of solution A and agitated for five minutes or until any bubbling has ceased, whichever is longer. The specimen is then transferred to a second tray that contains the working PD redox solution, made by combining solutions B:C:D in a 100:4:5

ratio and mixing thoroughly. The specimen is soaked in the working PD redox solution for 5 to 15 minutes, with constant agitation, after which it is rinsed thoroughly with water, air-dried or dried with a hair dryer, and photographed.

Bleach solution may be applied at the operator's discretion as a final step. This solution darkens the developed prints, lightens the background, and removes ninhydrin stains, but may also eliminate detail that is visible before bleaching. The specimen is simply dipped in the bleach solution for about 15 seconds and then rinsed, dried, and photographed. It's important that the final rinse be thorough, because otherwise the specimen may degrade very quickly.

Adhesive Surface Techniques

The FBI uses the phrase adhesive surface techniques to describe four processing methods used to raise latent prints on sticky surfaces such as the adhesive side of sticky tapes, sticky labels, peel-and-stick plastics, and so on. Surprisingly, three of the four methods–alternate black powder, ash gray powder, and sticky-side powder–are powder based. (One might think that these powders would adhere equally well to the adhesive and the fingerprint residues, but they do not.) All three powders are applied in the same way–made into a thin paste, brushed onto the questioned surface, and rinsed off with cold water–and provide similar results. The primary differences among these three powders are their colors, chosen to provide contrast against different surface colors. The final method, a 0.1% w/v solution of gentian violet(also called crystal violet) in water. The questioned surface is sprayed with or immersed in the gentian violet solution for one to two minutes, after which it is removed and rinsed with cold water. The gentian violet stains the latent prints, which can then be viewed and photographed under ordinary light.

Vacuum Metal Deposition (VMD)

Vacuum metal deposition (VMD) is similar conceptually to cyanoacrylate fuming, but substitutes metal vapor for cyanoacrylate vapor. The questioned surface is placed in a chamber from which the air is evacuated. The chamber also contains small pieces of gold and zinc that can be heated electrically until they vaporize. The specimen is exposed first to gold vapor, and then to zinc vapor. The metal vapors adhere selectively to fingerprint residues, revealing latent prints as metal-plated traces on a pristine substrate. Because VMD requires expensive equipment and materials, its use is limited to well-equipped and well-funded forensics labs.

The FBI categorizes these processes as Standard (used routinely) or Optional (used only in special situations or as supplements to Standard processes), as follows:

Standard Processes

Adhesive surface techniques (alternate black powder, ash gray powder, gentian violet, and sticky-side powder), amido black (methanol base), amido black (water base – Fischer 98), DAB, DFO, fingerprint powders, iodine fuming, ISR, LCV, ninhydrin (petroleum ether base), PD, RAM, silver nitrate, Sudan black, Super Glue fuming, and VMD.

Optional Processes

Amido black (water base), Coomassie brilliant blue, Crowle's double stain, fluorescent super glue dyes (Ardrox, MBD, MRM 10, rhodamine 6G, safranin O, and thenoyl europium chelate), Liqui-Drox, and ninhydrin (acetone base).

Obviously, from this plethora of techniques–and these are only the most popular of a larger group–we needed to choose a subset for the lab sessions in this section. We settled on iodine fuming, silver nitrate, ninhydrin, and Super Glue fuming, which happen to be the Big Four in real forensics labs and also have the advantage of being inexpensive and use materials that are relatively easy to acquire. We'll also use gentian violet to develop prints on cellophane tape, and dusting to develop prints on glass. Finally, we'll use two liquids found in most homes to reveal latent fingerprints on brass cartridge cases.

DNA FINGER PRINTING AND CRIMINAL JUSTICE

Crime is as old as human civilization. The day, human beings civilized criminal behaviours were defined. Social norms were settled for identifying what is in the interest of society and what is harmful for societal well-being. For attaining this, it was needed that the people who do not observe the societal norm of behaviour should be identified and punished by which, society be made crime free. This gave birth to the processes of detection, investigation and administration of criminal justice. In Twenty first century crime is posing greater challenge for the societal existence and law enforcement agencies are unable in tackling this problem by resorting traditional methods of investigation. Crime, particularly mass killing by terrorists, Naxalites and other professional criminals are occurring without any check. Law enforcement agencies are now finding it self unable to detect criminals, even if detected the rate of conviction is very poor due to lack of evidence. Criminal law is losing its deterrent effect and common citizenery is losing faith in law. Now citizens are much security conscious, always fearing of victimization resulting into either surrendering to criminal elements or taking law into their hands for revenge or security and thus furthering the problem of criminality.

The traditional evidences against accused have been eye witnesses, confessions and statement of approvers. Eye witness has now become a rare species, the reason being due to the technological development modus operandi of committing crime has changed and also now crimes is committed in well planned manner. Even if the eye witness is available he changes his version of statement day by day because the fear of criminal elements or corruption.

Now a days the criminal elements are not feeling any kind of deterrence from the prevailing criminal justice system and result is crime waves. This situation in criminal justice is needed to be coped with immediately:

> "The days of the 'perfect crime' are numbered. New Technological break through in crime detection may soon render current police method obsolete. A careful criminal cover up obvious clues by wiping away his fingerprints or wearing shoes that make no distinctive marks. But nearly everyone leaves a small perhaps microscopic-part of themselves behind, especially following a crime of violence. And crime solving techniques now on drawing board or in the experimental phase hold good...."

Evidentiary clues are always available at the site of every crime and science has progressed so much that it can identify, compare and link even tiniest connecting clue found. Correct identification of culprits, victims and mutilated, putrefied corpses has always been a legal, social and emotional problem before the public, police and courts. The discovery of DNA fingerprinting in the end of twentieth century has brought a sea change in the identification scenario. DNA fingerprinting helps law enforcement agencies or courts to determine an individual's involvement in a crime and settle dispute over paternity. It was developed by Sir Alec Jeffreys, a Professor at the University of Leicester in United Kingdom in 1984.

DNA FINGERPRINTING:

DNA (Deoxyribo Nucleic Acid) is a chemical structure that forms chromosomes. A piece of chromosomes that dictates a particular trait is called gene. Structurally DNA is a double helix two strands of genetic material spiraled around each other. Each strand contains a sequence of bases also called nucleotides. A base is one of four chemicals :- adenine, guanine, cytosine and thymine. The two strands of DNA are connected at each base. The chemical structure of every person's DNA is same. The only difference between people is the order of base pairs. There are so many millions of base pairs in each person's DNA that every person has a different sequence. Using these sequences every person could be identified solely by the sequence of their base pairs. To identify individuals, forensic scientists scan 13 DNA regions that vary from person to person and use the data to create a DNA profile of that person. This DNA profile is commonly called DNA fingerprint. There is extremely small chance that another person has the same DNA profile for a particular set of regions. The DNA profile has now become an important forensic method to identify criminal and victims. DNA fingerprinting is prepared to identify criminal from hair, blood, semen or other biological materials found at the scene of a violent crime. It depends on the fact that no two persons have exactly same DNA sequence. The DNA samples of the culprit can be obtained from the scene of crime itself. For example blood samples from a scene of murder or samples of seminal fluids deposited on the clothes or on the surface or in the body of victim of rape can be used to acquire DNA fingerprint of the culprit. This is compared with those taken and prepared from the possible suspect in the case.

DNA FINGERPRINTING TECHNOLOGIES USED IN FORENSIC INVESTIGATIONS:

Restriction Fragment Length Polymorphism (RFLP)

RFLP is a technique for analyzing the variable length of DNA fragments that results from digesting a DNA sample with special kind of enzyme. This enzyme, a restriction endonuclease cuts DNA at a specific sequence pattern known as restriction endonuclease recognition sites in a DNA sample generates variable lengths of DNA fragments, which are separated by using gel electrophoresis. For gel electrophoresis the DNA fragments are poured into a gel such as agarose and an electric charge is applied to the gel, with the positive charge at bottom and negative charge at the top. Because DNA has a slightly negative charge, the pieces of DNA will be attracted towards bottom of the gel, the smaller pieces, however, will be able to move more quickly and thus further towards the bottom than the larger pieces. The different sized pieces of DNA will therefore be separated by size, with the smaller pieces towards bottom and larger pieces towards top. They are then hybridized with DNA probes that bind to a complimentary DNA sequence in the sample and after that X-Ray image is taken. This allows to identify, in a particular person's DNA, the occurrence and frequency of the particular genetic pattern.

RFLP was one of the first applications of DNA analysis to forensic investigation. In RFLP relatively large amounts of DNA is required. In addition, samples degraded by environmental factors such as dirt or mold, do not work well with RFLP. Therefore this method is not usually used now a days.

Polymerase Chain Reaction Anlalysis (PCR)

Polymerase chain reaction is used to make millions of exact copies of DNA from a biological sample. DNA amplification with PCR allows DNA analysis on biological samples as small as few skin cells. The ability of PCR to amplify such tiny quantities of DNA enables even highly degraded samples to be analysed. Great care, however, is needed to be taken to prevent contamination with other biological materials during identification, collection and preservation of a sample.

Short Tandem Repeat Analysis (STR)

Every strand of DNA has pieces that contains genetic information which informs an organism's development (exons) and pieces that, apparently, supply no relevant genetic information at all (introns). Although the introns may seem useless, it has been found that they contain repeated sequence of base pairs. These sequences are called Variable Number Tandem Repeats (VNTRs). A given person's VNTRs come from the genetic information donated by his parents. Because VNTRs patterns are inherited generally, a given person's VNTR pattern is unique. For precise and clear identification by DNA fingerprinting base sequence is studied in specific regions of DNA strand. This technology is called short Tandem Repeat (STR) technology. STR technology is used to evaluate specific regions within nuclear DNA. Variability in STR regions can be used to distinguish one DNA profile from another. Usually a standard set of 13 specific STR regions are used in identifying an individual.

Mitochondrial DNA Analysis (mt. DNA)

mt. DNA analysis uses DNA extracted from a cellular organelle called mitochondria. Older biological samples which lack nucleated cellular material such as hair, bones and teeth cannot be analysed with nuclear DNA, they can be analysed with mt DNA and this DNA material is analysed and DNA fingerprint is prepared by RFLP, PCR or STR analysis method.

DNA FINGERPRINTING IN CRIME INVESTIGATION

Crime has become a major challenge before the society. In the era of scientific development modus operandi of criminals have become very sophisticated due to which traditional evidences are rarely available against the criminal, even it is difficult to identify the criminal, resulting into failure of criminal justice system. For prevention of crime and deterring criminal elements from committing crime it is not only enough to define activity as an offence and prescribe severe punishment but deterrence situation should be created for criminals and prospective criminals by making efficient and prompt investigation and ultimately by conviction of criminals. DNA fingerprinting is now becoming very potent means by which criminal may be identified and convicted. DNA fingerprinting has excellent features in identification of criminal. It provides correct individualization of each individual on earth as each person has unique DNA configuration and sequencing of base pairs. Therefore, if biological evidentiary clue is available the criminal may be identified without any fault. On crime scene always some biological evidentiary clues are available particularly skin, blood stains, semen, swabs and hair.

Skin is the clue material in cases where skin has been abraded in assault. Beating with Lathi or any hard and blunt instrument abraded and picked up skin. Where a person has been dragged up, the skin may be abraded and used surface may carry skin fragments. Where a person has been hanged, the ligature will carry the skin fragments. In sexual assault also skin fragment is found stick in the nail of victim and accused. By using skin fragments DNA fingerprint may be prepared and criminal may be identified.

On crime scene most usual biological evidentiary clues found is hairs. A person sheds hairs everywhere. It is natural phenomenon, even very vigilant criminal has no control over this phenomenon, at the crime scene investigating officer can collect this evidence by which DNA profile may be prepared and perpetrator can be produced before the court for conviction. There is cross transference of hairs in close contacts: rape, scuffles etc. They are also picked up by the weapon of offence. DNA fingerprint prepared from the hair roots identifies the person beyond reasonable doubts.

Bitten off, broken or paired nails provide body cells. Nail scrapings carry scratched body materials. Such scrapings are common in rape and scuffles where the victim has resisted the onslaught of the culprit. Nail scrapings have great potential for identification of the victim and criminal.

At the site of crime body fluids are usually found and DNA fingerprint from these materials gives right direction to investigating agency in solving the case and ultimately the criminal is convicted. Transference or cross transference of blood is common in the offences against person. Blood is most frequently utilized body material in DNA fingerprinting. In a criminal case blood available may be liquid blood, blood clot, blood stains, blood spatters, crust, blood powder or washed blood on all sorts of objects and surface. It may be found at the scene, with the victim, with the culprit, on the weapon of offence, on the discarded garments or anywhere where the culprit has washed his hands or other blood smeared articles or discarded and thrown blood smeared articles.

Prior to use of DNA fingerprinting, matters involving sexual assault could be solved primarily by circumstantial evidences only. These are such type of offences which are committed in solitary and solitude condition therefore direct evidences are not available. It was very difficult for the victim of rape to prove the offence in the absence of circumstantial evidences or an eyewitness, which is very rare. The advent of DNA profiling has created a great hope in tackling sexual assault cases. Semen, swabs and saliva are important biological evidentiary clues in sexual assault cases from which DNA fingerprint can be prepared and criminal may be identified. Saliva is rich in body cells, hence it is amenable to DNA fingerprinting. It is found in a variety of offences like scuffles, murders, trespass and sexual assaults viz. bite marks by the victim or by culprit, cigarette stubs, beer bottles, tumblers and tea cup etc. usually carry saliva stains of suspects. In rape cases DNA fingerprinting adduces is great help in investigation and prosecution of criminal. Now the rapist may be promptly apprehended by getting DNA fingerprint from the semen, swabs and saliva found at crime scene. Even if the semen is contaminated with any material, vaginal fluid, rectal matter in sodomy cases, saliva in oral sex or with the semen of other rapists in gang rapes, it can be fixed unmistakably and real or every culprit can be individualized and identified.

In the 21st century crime and criminality is creating great menace for societal existence. Investigating agencies are now feeling handicapped in investigating offences with the help of traditional methods of investigation. In this situation portryal of the infalliability of DNA profile in individualization and identification of criminal and thus its unrivalled ability to solve crime have beamed a ray of hope to criminal justice system for tackling the problem of criminality. This is the reason that whole through world governments are investing money in doing research in this area and some countries have taken steps for preparation of National DNA Data base (NDNAD) by which, culprits should be promptly identified. Carole Mc Cartney observed that :-

> "Portrayals of the infallibility of DNA and its unrivalled ability to solve crime have led to determined effort and financial investment in significantly increasing the forensic use of DNA. Indeed, one of the National DNA Database (NDNAD) 'strategic objectives' is stated as being to demonstrate value for money from the database, especially in crime detection and through this crime reduction."

Various countries are now making striving for establishment of NDNAD as in modern era when criminals, modus operandi have become very shophisticated and crimes are committed in very planned and organized manner, in this situation criminal may be detected and criminality can be tackled only by using modern forensic methods; but up to now Indian law enforcement agencies are adhering to traditional methods of investigation and proving the case against the accused. This situation is worstly affecting the criminal justice in India and criminal activities are increasing unchecked.

In India NDNAD has not been established, which is much needed for proper tackling of criminality and ultimately for prevention of crime. Lord Brown in case of Marper observed about the deterrence effect of NDNAD that :-

> "... The more complete the database, the better the chance of detecting criminals, both those guilty of crimes past and those whose crimes are yet to be committed. The better chance too of deterring from future crime those whose profiles are already on the database. And these, of course, are not only benefits. The larger the database, the less call there will be to round up the usual suspects. Instead those amongst the usual suspects who are innocent will at once be exonerated."

DNA fingerprinting particularly establishment of NDNAD has immensely changed the policing and play very vital role in prevention of crime. Watson observed that :-

> "It is claimed that science and technology play a vital role in modern policing, with the NDNAD revolutionizing crime detection. DNA has not merely enhanced existing police capacity, but has even begun to replace the slow, tedious and expensive traditional investigative methods of police interviews."

When from the crime scene a full DNA profile has been obtained, but there is no match on the NDNAD, then this DNA profile may be used for 'familial searching'. The profile comparison is made for close matches e.g. where subject could be a parent, child or sibling of someone whose profile is on the NDNAD. If no similar profile is available in NDNAD, then all the suspected person's DNA is prepared and compared with DNA fingerprint prepared from biological material found at crime scene. By this way actual criminal may be identified from suspected persons.

LEGAL MEASURES IN INDIA

Indian legislature has shown its great determination in facing the challenges of crime due to changed modus operandi of scientifically advanced criminal elements, by enacting law to permit law enforcement agencies to adopt modern technologically developed methods of investigation. DNA fingerprinting is one of such method of criminal investigation. Particularly in 2005 Indian Parliament has ingrained the provisions in Criminal Procedure Code permitting investigation to use DNA fingerprinting.

Medical examination of accused person

During investigation if police officer arrests on charge of an offence of such nature and committed in such circumstances that police officer is believing that medical examination will afford evidences of his involvement in the crime, police officer get him medically examined by a registered medical practitioner. Term 'nature and circumstances' used in Sec.53 of Cr.P.C. indicates when police officer has proceeded the crime scene and recovered biological clues there, he may take resort of this provision and for comparing the biological clue with his body materials he may be medically examined. In medical examination now a days DNA fingerprinting is most important means for determining identification of criminal. Provision contained in Sec.53 of Cr.P.C. specifically enumerates this method for identification of criminal. This situation shows adaptability of Indian Law for modern scientific developments. :

Explanation : In this section and in Section 53-A and 54. :

- "Examination" shall include the examination of blood, blood stains, semen, swabs in case of sexual offences, sputum and sweat, hair samples and finger nail clippings by the use of modern and scientific techniques including DNA profiling and such other tests which the registered medical practitioner thinks necessary in a particular case......"

Medical examination of accused and victim in sexual assault cases

Offences of sexual assault are committed in solitary and solitude condition, due to this circumstantial evidence and testimony of victim are only available for identification and prosecution of accused. These evidences are not satisfactory for booking and conviction of criminal and ultimately for coping the increasing incidents of sexual assaults. Advent of DNA fingerprinting has now changed the scenario. In sexual assault always biological clues are available at the crime scene and clothing of accused and victim, from them DNA fingerprint may be prepared and criminal is identified. During the trial circumstantial evidence and victims testimony would be corroborated by DNA fingerprint. By this criminal will be penalized and problem of increasing sexual assault may be coped effectively. For tackling problem of sexual assault Criminal Procedure Code gives specific direction to investigating officer for getting DNA fingerprint from biological material found at crime scene, with accused and victim. Sec. 53-A of criminal procedure code directs registered medical practitioner to take material from arrested person on accusation of committing rape for preparation of DNA profile :

"53-A (2) The registered medical practitioner conducting such examination shall, without delay, examine such person and prepare a report of his examination giving the following, particulars, namely :-

(i) .....
(ii) .....
(iii) Marks of injury, if any, on the person of the accused.
(iv) The description of material taken from the person of the accused for DNA profiling, and ....."

In sexual assaults biological matters such as semen, blood, blood stains and hairs of accused are usually found in or on the person of victim. If these biological material is recovered and DNA profile is prepared and it is compared with DNA profile of suspected persons, the offender will be identified. Circumstantial evidences and testimony of victim are corrborated by DNA fingerprinting resulting into conviction of offender ultimately resulting into curbing the problem of sex crimes. Criminal Procedure Code has enabled investigating agency for getting biological clues by medical examination of victim of sex crime for the purpose of DNA fingerprinting.

Evidentiary value of DNA fingerprinting

DNA fingerprint is not considered as primary evidence till date. Under Indian Evidence Act DNA fingerprint report is considered as expert evidence.

Sec.45 of Evidence Act makes provision for admissibility of expert opinion in a trial:-

> "When court has to form an opinion upon a point of foreign law or of science or art, or as to identity of handwriting or finger impressions, the opinions upon that point of persons specially skilled in such foreign law, science or art or in questions as to identity of handwriting or finger impressions are relevant facts. Such persons are called experts."

DNA fingerprint is admissible evidence, it is considered as expert opinion about the DNA profile of a person, therefore made admissible u/s. 45 Evidence Act.

CONCLUDING REMARKS

For societal wellbeing and proper justice administration it is essential that a criminal case should be properly investigated and criminal should be identified and convicted. Initially criminal justice system depended on the testimony of eye witness and circumstantial evidences for investigation, prosecution and conviction. Crime investigators usually resorted third degree methods for crime investigation. Due to cultural change, now third degree methods are considered inhuman. Eye witnesses have no willingness to come forth and crime is committed in well planned manner by using technologically developed gazettes furthering the problem of investigator in obtaining circumstantial clues. In this situation for tackling the problem of criminality, for doing proper, prompt and effective investigation modern scientific techniques beams a ray of hope to criminal justice system. DNA fingerprinting is one of the modern investigative measures which may provide effective and prompt solution to tackle the problem of criminality.

DNA fingerprinting has now emerged as very potent means of identifying criminal. By use of DNA fingerprint culprit may be identified and ultimately he may be penalized for his criminal behaviour. Major problem with the use of this measure of investigation is that the police officers are lacking willingness in using this scientifically developed measure and at the same time they are not having required skill in collection of biological clues found at crime scene. Collection of samples at scene of crime requires some skill and observance of basic rule of hygene. DNA samples are extremely susceptible to contamination. Proper training should be imported to police officer for collection of biological materials found at crime scene and coordination between forensic experts and police officer should be strengthened. DNA fingerprinting technology can be better utilized only when National DNA Database is established. Government should take necessary steps for establishment of National DNA Data base. For more evidentiary value evidence law should also be amended after considering unique and infallible identifying quality of DNA fingerprinting.

5

How Fingerprinting Works

A woman has been murdered. When the detectives arrive on the scene, the house is in shambles. Clothes are strewn about the floor, lamps are ov¬erturned and there's no sign of the assailant. Then, one of the detectives picks up a glass. On its side is a smudged, bloodythumbprint. He takes it down to the lab, where it's analyzed and matched to a recorded set of prints. The detectives catch their killer.

Forensics Image Gallery

This scen¬e has been replayed in one crime drama after another. Ever since scientists discovered that every person's fingerprints are unique, and police officers realized this singularity could help them catch criminals, fingerprints have been an integral part of the law enforcement process. Today, fingerprints are also used to prevent forged signatures, identify ¬accident victims, verify job applicants and provide personalized access to everything from ATMs to computer networks.

But fingerprinting has come a long way from the days when police officers lifted prints from a crime scene and checked them manually against their files. Modern fingerprinting techniques can not only check millions of criminal records simultaneously, but can also match faces, backgrounds and other identifiable characteristics to each perpetrator. What are the basic characteristics of a fingerprint? How long have people been using prints as a form of identification? Find out in the next section.

A minor scrape, scratch or even burn won't affect the structure of the ridges in your fingerprints -- new skin reforms in its original pattern as it grows over the wound. But each ridge is also connected to the inner skin by small projections called papillae. If these papillae are damaged, the ridges are wiped out and the fingerprint destroyed.

Some criminals have tried to evade capture by tampering with their own fingerprints. Chicago bankrobber John Dillinger reportedly burned his fingertips with acid in the 1930s. Recently, a man in Lawrence, Mass., tried to hide his identity by cutting and stitching up all ten of his fingertips (fortunately, a police officer recognized his face). But as fingerprint technology becomes a common form of authentication from bank vaults to luxury cars, law enforcement officials worry that would-be criminals might try to steal entire fingers for the prints. In one case, robbers in Malaysia cut off a man's fingers so they could steal his Mercedes. Companies that make biometrics security equipment realize the potential dangers of this system, and are now creating scanners that detect blood flow to make sure the finger is still alive.

WHAT ARE FINGERPRINTS?

Fingerprints are the tiny ridges, whorls and valley patterns on the tip of each finger. They form from pressure on a baby's tiny, developing fingers in the womb. No two people have been found to have the same fingerprints -- they are totally unique. There's a one in 64 billion chance that your fingerprint will match up exactly with someone else's. Fingerprints are even more unique than DNA, the genetic material in each of our cells. Although identical twins can share the same DNA -- or at least most of it -- they can't have the same fingerprints.

Fingerprinting is one form ofbiometrics, a science that uses people's physical characteristics to identify them. Fingerprints are ideal for this purpose because they're inexpensive to collect and analyze, and they never change, even as people age. Although hands and feet have many ridged areas ¬that could be used for identification, fingerprints became a popular form of biometrics because they are easy to classify and sort. They're also accessible.

Fingerprints are made of an arrangement of ridges, called friction ridges. Each ridge contains pores, which are attached to sweat glands under the skin. You leave fingerprints on glasses, tables and just about anything else you touch because of this sweat. All of the ridges of fingerprints form patterns called loops, whorls or arches:

- Loops begin on one side of the finger, curve around or upward, and exit the other side. There are two types of loops: Radial loops slope toward the thumb, while ulnar loops slope toward the little finger.
- Whorls form a circular or spiral pattern.
- Arches slope upward and then down, like very narrow mountains.

Scientists look at the arrangement, shape, size and number of lines in these fingerprint patterns to distinguish one from another. They also analyze very tiny characteristics called minutiae, which can't be seen with the naked eye. If fingerprints are so unique and subtle, how are they recorded accurately? In the next section, we'll learn about dactyloscopy, or the art of fingerprinting.

FINGERPRINT BASICS

Fingerprints are one of those bizarre twists of nature. Human beings happen to have built-in, easily accessible identity cards. You have a unique design, which represents you alone, literally at your fingertips. How did this happen?

People have tiny ridges of skin on their fingers because this particular adaptation was extremely advantageous to the ancestors of the human species. The pattern of ridges and "valleys" on fingers make it easier for the hands to grip things, in the same way a rubber tread pattern helps a tire grip the road.

The other function of fingerprints is a total coincidence. Like everything in the human body, these ridges form through a combination of genetic and environmental factors. The genetic code in DNA gives general orders on the way skin should form in a developing fetus, but the specific way it forms is a result of random events. The exact position of the fetus in the womb at a particular moment and the exact composition and density of surrounding amniotic fluid decides how every individual ridge will form.

So, in addition to the countless things that go into deciding your genetic make-up in the first place, there are innumerable environmental factors influencing the formation of the fingers. Just like the weather conditions that form clouds or the coastline of a beach, the entire development process is so chaotic that, in the entire course of human history, there is virtually no chance of the same exact pattern forming twice.

Consequently, fingerprints are a unique marker for a person, even an identical twin. And while two prints may look basically the same at a glance, a trained investigator or an advanced piece of software can pick out clear, defined differences. This is the basic idea of fingerprint analysis, in both crime investigation and security. A fingerprint scanner's job is to take the place of a human analyst by collecting a print sample and comparing it to other samples on record. In the next few sections, we'll find out how scanners do this.

OPTICAL SCANNER

A fingerprint scanner system has two basic jobs -- it needs to get an image of your finger, and it needs to determine whether the pattern of ridges and valleys in this image matches the pattern of ridges and valleys in pre-scanned images. There are a number of different ways to get an image of somebody's finger. The most common methods today are optical scanning and capacitance scanning. Both types come up with the same sort of image, but they go about it in completely different ways.

The heart of an optical scanner is a charge coupled device (CCD), the same light sensor system used indigital cameras and camcorders. A CCD is simply an array of light-sensitive diodes called photosites, which generate an electrical signal in response to light photons. Each photosite records a pixel, a tiny dot representing the light that hit that spot. Collectively, the light and dark pixels form an image of the scanned scene (a finger, for example). Typically, an analog-to-digital converter in the scanner system processes the analog electrical signal to generate a digital representation of this image. See How Digital Cameras Workfor details on CCDs and digital conversion.

The scanning process starts when you place your finger on a glass plate, and a CCD camera takes a picture. The scanner has its own light source, typically an array of light-emitting diodes, to illuminate the ridges of the finger. The CCD system actually generates an inverted image of the finger, with darker areas representing more reflected light (the ridges of the finger) and lighter areas representing less reflected light (the valleys between the ridges).

Before comparing the print to stored data, the scanner processor makes sure the CCD has captured a clear image. It checks the average pixel darkness, or the overall values in a small sample, and rejects the scan if the overall image is too dark or too light. If the image is rejected, the scanner adjusts the exposure time to let in more or less light, and then tries the scan again.

If the darkness level is adequate, the scanner system goes on to check the image definition (how sharp the fingerprint scan is). The processor looks at several straight lines moving horizontally and vertically across the image. If the fingerprint image has good definition, a line running perpendicular to the ridges will be made up of alternating sections of very dark pixels and very light pixels. If the processor finds that the image is crisp and properly exposed, it proceeds to comparing the captured fingerprint with fingerprints on file. We'll look at this process in a minute, but first we'll examine the other major scanning technology, the capacitive scanner.

Capacitance Scanner

Like optical scanners, capacitive fingerprint scanners generate an image of the ridges and valleys that make up a fingerprint. But instead of sensing the print using light, the capacitors use electrical current.

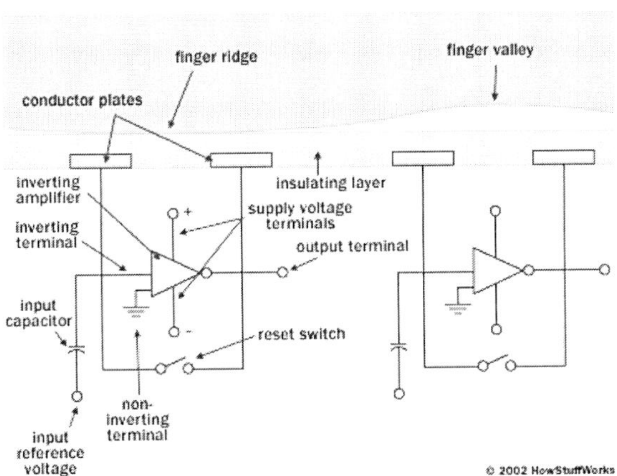

The diagram below shows a simple capacitive sensor. The sensor is made up of one or moresemiconductor chips containing an array of tiny cells. Each cell includes two conductor plates, covered with an insulating layer. The cells are tiny -- smaller than the width of one ridge on a finger. The sensor is connected to an integrator, an electrical circuit built around an inverting operational amplifier. The inverting amplifier is a complex semiconductor device, made up of a number of transistors, resistors and capacitors. The details of its operation would fill an entire article by itself, but here we can get a general sense of what it does in a capacitance scanner. (Check out this page on operational amplifiers for a technical overview.)

Like any amplifier, an inverting amplifier alters one current based on fluctuations in another current (see How Amplifiers Work for more information). Specifically, the inverting amplifier alters a supply voltage. The alteration is based on the relative voltage of two inputs, called the inverting terminal and the non-inverting terminal. In this case, the non-inverting terminal is connected to ground, and the inverting terminal is connected to a reference voltage supply and a feedback loop. The feedback loop, which is also connected to the amplifier output, includes the two conductor plates.

As you may have recognized, the two conductor plates form a basic capacitor, an electrical component that can store up charge (see How Capacitors Work for details). The surface of the finger acts as a third capacitor plate, separated by the insulating layers in the cell structure and, in the case of the fingerprint valleys, a pocket of air. Varying the distance between the capacitor plates (by moving the finger closer or farther away from the conducting plates) changes the total capacitance (ability to store charge) of the capacitor. Because of this quality, the capacitor in a cell under a ridge will have a greater capacitance than the capacitor in a cell under a valley.

To scan the finger, the processor first closes the reset switch for each cell, which shorts each amplifier's input and output to "balance" the integrator circuit. When the switch is opened again, and the processor applies a fixed charge to the integrator circuit, the capacitors charge up. The capacitance of the feedback loop's capacitor affects the voltage at the amplifier's input, which affects the amplifier's output. Since the distance to the finger alters capacitance, a finger ridge will result in a different voltage output than a finger valley.

The scanner processor reads this voltage output and determines whether it is characteristic of a ridge or a valley. By reading every cell in the sensor array, the processor can put together an overall picture of the fingerprint, similar to the image captured by an optical scanner.

The main advantage of a capacitive scanner is that it requires a real fingerprint-type shape, rather than the pattern of light and dark that makes up the visual impression of a fingerprint. This makes the system harder to trick. Additionally, since they use a semiconductor chip rather than a CCD unit, capacitive scanners tend to be more compact that optical devices.

Analysis

In movies and TV shows, automated fingerprint analyzers typically overlay various fingerprint images to find a match. In actuality, this isn't a particularly practical way to compare fingerprints. Smudging can make two images of the same print look pretty different, so you're rarely going to get a perfect image overlay. Additionally, using the entire fingerprint image in comparative analysis uses a lot of processing power, and it also makes it easier for somebody to steal the print data.

Instead, most fingerprint scanner systems compare specific features of the fingerprint, generally known asminutiae. Typically, human and computer investigators concentrate on points where ridge lines end or where one ridge splits into two (bifurcations). Collectively, these and other distinctive features are sometimes called typica. The scanner system software uses highly complex algorithms to recognize and analyze these minutiae. The basic idea is to measure the relative positions of minutiae, in the same sort of way you might recognize a part of the sky by the relative positions of stars. A simple way to think of it is to consider the shapes that various minutia form when you draw straight lines between them. If two prints have three ridge endings and two bifurcations, forming the same shape with the same dimensions, there's a high likelihood they're from the same print.

To get a match, the scanner system doesn't have to find the entire pattern of minutiae both in the sample and in the print on record, it simply has to find a sufficient number of minutiae patterns that the two prints have in common. The exact number varies according to the scanner programming.

Pros and Cons

There are several ways a security system can verify that somebody is an authorized user. Most systems are looking for one or more of the following:

- What you have
- What you know
- Who you are

To get past a "what you have" system, you need some sort of "token," such as an identity card with a magnetic strip. A "what you know" system requires you to enter a password or PIN number. A "who you are" system is actually looking for physical evidence that you are who you say you are -- a specific fingerprint, voice or iris pattern.

"Who you are" systems like fingerprint scanners have a number of advantages over other systems. To name few:

- Physical attributes are much harder to fake than identity cards.
- You can't guess a fingerprint pattern like you can guess a password.
- You can't misplace your fingerprints, irises or voice like you can misplace an access card.
- You can't forget your fingerprints like you can forget a password.

But, as effective as they are, they certainly aren't infallible, and they do have major disadvantages. Optical scanners can't always distinguish between a picture of a finger and the finger itself, and capacitive scanners can sometimes be fooled by a mold of a person's finger. If somebody did gain access to an authorized user's prints, the person could trick the scanner. In a worst-case scenario, a criminal could even cut off somebody's finger to get past a scanner security system. Some scanners have additional pulse and heat sensors to verify that the finger is alive, rather than a mold or dismembered digit, but even these systems can be fooled by a gelatin print mold over a real finger. (This site explains various ways somebody might trick a scanner.) To make these security systems more reliable, it's a good idea to combine the biometric analysis with a conventional means of identification, such as a password (in the same way an ATM requires a bank card and a PIN code).

The real problem with biometric security systems is the extent of the damage when somebody does manage to steal the identity information. If you lose your credit card or accidentally tell somebody your secret PIN number, you can always get a new card or change your code. But if somebody steals your fingerprints, you're pretty much out of luck for the rest of your life. You wouldn't be able to use your prints as a form of identification until you were absolutely sure all copies had been destroyed. There's no way to get new prints.

But even with this significant drawback, fingerprint scanners and biometric systems are an excellent means of identification. In the future, they'll most likely become an integral part of most peoples' everyday life, just like keys, ATM cards and passwords are today.

THE FINGERPRINTING PROCESS

The technique of fingerprinting is known as dactyloscopy. Until the advent of digital scanningtechnologies, fingerprinting was done using ink and a card.

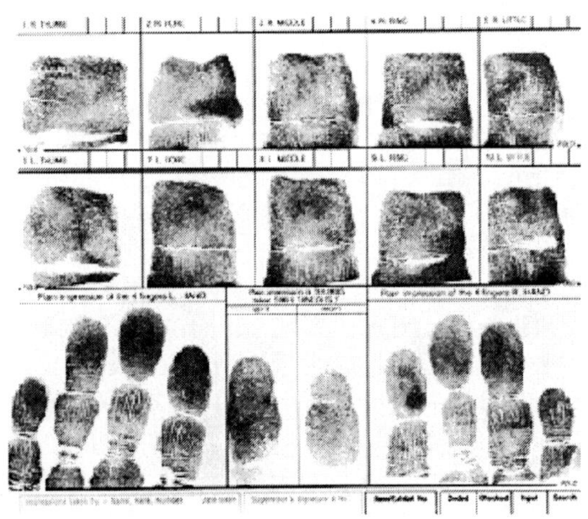

To create an ink fingerprint, the person's finger is first cleaned with alcohol to remove any sweat and dried thoroughly. The person rolls his or her fingertips in ink to cover the entire fingerprint area. Then, each finger is rolled onto prepared cards from one side of the fingernail to the other. These are called rolled fingerprints. Finally, all fingers of each hand are placed down on the bottom of the card at a 45-degree angle to produce a set of plain (or flat) impressions. These are used to verify the accuracy of the rolled impressions. Today, digital scanners capture an image of the fingerprint. To create a digital fingerprint, a person places his or her finger on an optical or silicon reader surface and holds it there for a few seconds. The reader converts the information from the scan into digital data patterns. The computer then maps points on the fingerprints and uses those points to search for similar patterns in the database.

Law enforcement agents can analyze fingerprints they find at the scene of a crime. There are two different types of prints:

- Visible prints are made on a type of surface that creates an impression, like blood, dirt or clay.
- Latent prints are made when sweat, oil and other substances on the skin reproduce the ridge structure of the fingerprints on a glass, murder weapon or any other surface the perpetrator has touched. These prints can't be seen with the naked eye, but they can be made visible using dark powder, lasers or other light sources. Police officers can "lift" these prints with tape or take special photographs of them.

When did this basic form of identification become a law enforcement staple? How did Babylonians and the ancient Chinese use fingerprints? Go to the next section to find out.

HISTORY OF FINGERPRINTING

There are records of fingerprints being taken many centuries ago, although they weren't nearly as sophisticated as they are today. The ancient Babylonians pressed the tips of their fingertips into clay to record business transactions. The Chinese used ink-on-paper finger impressions for business and to help identify their children.

However, fingerprints weren't used as a method for identifying criminals until the 19th century. In 1858, an Englishman named Sir William Herschel was working as the Chief Magistrate of the Hooghly district in Jungipoor, India. In order to reduce fraud, he had the residents record their fingerprints when signing business documents.

A few years later, Scottish doctor Henry Faulds was working in Japan when he discovered fingerprints left by artists on ancient pieces of clay. This finding inspired him to begin investigating fingerprints. In 1880, Faulds wrote to his cousin, the famed naturalist Charles Darwin, and asked for help with developing a fingerprint classification system. Darwin declined, but forwarded the letter to his cousin, Sir Francis Galton.

Galton was a eugenicist who collected measurements on people around the world to determine how traits were inherited from one generation to the next. He began collecting fingerprints and eventually gathered some 8,000 different samples to analyze. In 1892, he published a book called "Fingerprints," in which he outlined a fingerprint classification system -- the first in existence. The system was based on patterns of arches, loops and whorls.

Meanwhile, a French law enforcement official named Alphonse Bertillon was developing his own system for identifying criminals. Bertillonage (or anthropometry) was a method of measuring heads, feet and other distinguishing body parts. These "spoken portraits" enabled police in different locations to apprehend suspects based on specific physical characteristics. The British Indian police adopted this system in the 1890s.

Around the same time, Juan Vucetich, a police officer in Buenos Aires, Argentina, was developing his own variation of a fingerprinting system. In 1892, Vucetich was called in to assist with the investigation of two boys murdered in Necochea, a village near Buenos Aires. Suspicion had fallen initially on a man named Velasquez, a love interest of the boys' mother, Francisca Rojas. But when Vucetich compared fingerprints found at the murder scene to those of both Velasquez and Rojas, they matched Rojas' exactly. She confessed to the crime. This was the first time fingerprints had been used in a criminal investigation. Vucetich called his system comparative dactyloscopy. It's still used in many Spanish-speaking countries.

Sir Edward Henry, commissioner of the Metropolitan Police of London, soon became interested in using fingerprints to nab criminals. In 1896, he added to Galton's technique, creating his own classification system based on the direction, flow, pattern and other characteristics of the friction ridges in fingerprints. Examiners would turn these characteristics into equations and classifications that could distinguish one person's print from another's. The Henry Classification System replaced the Bertillonage system as the primary method of fingerprint classification throughout most of the world.

In 1901, Scotland Yard established its first Fingerprint Bureau. The following year, fingerprints were presented as evidence for the first time in

English courts. In 1903, the New York state prisons adopted the use of fingerprints, followed later by the FBI.

But how has fingerprinting changed since the 19th century? In the next section, we'll find out about modern fingerprinting techniques.

MODERN FINGERPRINTING TECHNIQUES

The Henry system finally enabled law enforcement officials to classify and identify individual fing¬erprints. Unfortunately, the system was very cumbersome. When fingerprints came in, detectives would have to compare them manually with the fingerprints on file for a specific criminal (that's if the person even had a record). The process would take hours or even days and didn't always produce a match. By the 1970s, computers were in existence, and the FBI knew it had to automate the process of classifying, searching for and matching fingerprints. The Japanese National Police Agency paved the way for this automation, establishing the first electronic fingerprint matching system in the 1980s. Their Automated Fingerprint Identification Systems (AFIS), eventually enabled law enforcement officials around the world to cross-check a print with millions of fingerprint records almost instantaneously.

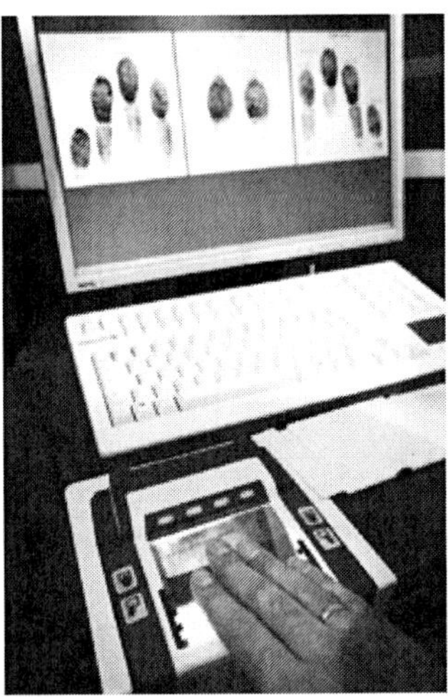

Fig.A background and identity check fingerprint capture machine is displayed in London. Modern technologies have made fingerprinting a much more effective means of identification.

AFIS collects digital fingerprints with sensors. Computer software then looks for patterns and minutiae points (based on Sir Edward Henry's system) to find the best match in its database. The first AFIS system in the U.S. was speedier than previous manual systems. However, there was no coordination between different agencies. Because many local, state and federal law enforcement departments weren't connected to the same AFIS system, they couldn't share information. That meant that if a man was arrested in Phoenix,

Ariz. and his prints were on file at a police station in Duluth, Minn., there might have been no way for the Arizona police officers to find the fingerprint record. That changed in 1999, with the introduction of Integrated AFIS (IAFIS). This system is maintained by the FBI's Criminal Justice Information Services Division. It can categorize, search and retrieve fingerprints from virtually anywhere in the country in as little as 30 minutes. It also includes mug shots and criminal histories on some 47 million people. IAFIS allows local, state and federal law enforcement agencies to have access to the same huge database of information. The IAFIS system operates 24 hours a day, 365 days a year.

But IAFIS isn't just used for criminal checks. It also collects fingerprints for employment, licenses and social services programs (such as homeless shelters). When all of these uses are taken together, about one out of every six people in this country has a fingerprint record on IAFIS.

OTHER BIOMETRICS

Fingerprinting isn't the only way to catch a criminal, or perform one of the many other biometrics-driven technologies now available. Eyescans, voice fingerprints and evenDNA are now providing means of identification, as well as access to everything from ATMs to cars.

Here are just a few of thebiometrics you might be using in the near future:

- Eye scans: Both the retina (the layer of tissue in the back of the eye that converts light into nervesignals) and the iris (the colored part of the eye) have unique characteristics that make them highly accurate biometrics. For a retinal scan, a person holds his or her eye close to the scanning device for 10 to 15 seconds while a low-intensity lightand sensor analyze distinct patterns. Although retinal scans are used in very high-security institutions like power plants and the military areas, they are currently too expensive to be practical for widespread use. Irises have more than 200 different unique identifying characteristics (about six times more than fingerprints) ranging from rings to freckles. Iris identification systems take only about two seconds to scan the iris and look for patterns. They're used in some prisons and a few airports.
- Ear scans: Ears are unique in size, shape and structure. Scientists use these traits to develop biometric scans of the ear. In ear scans, a camera creates an image of the ear that is analyzed for identifying characteristics.
- Voice fingerprints: Every time a new Osama bin Laden tape comes out, the FBI Audio Lab in Quantico,Va. runs it through a voice analyzer, which captures the frequency, intensity and other measurements to determine whether the tape is authentic. These so-called "voice fingerprints" aren't as definitive as fingerprints or DNA, but they can help distinguish one person from another.

- DNA fingerprints: Every individual has unique DNA. While you can change your appearance, you can't change your DNA. Because of this, scientists are starting to use DNA analysis to link suspects to blood, hair, skin and other evidence left at crime scenes. DNA fingerprinting is done by isolating the DNA from human tissues. The DNA is cut using special enzymes, sorted and passed through a gel. It's then transferred to a nylon sheet, where radioactive probes are added to produce a pattern -- the DNA fingerprint.

Some of these technologies are still in development, so it isn't yet known which is the most effective form of identification. And of course, some types of biometrics are better suited for specific tasks than others. For example, voice fingerprints are most appropriate for phone financial transactions.

THE BASICS OF FINGERPRINT IDENTIFICATION

The skin on the inside surfaces of our hands, fingers, feet, and toes is "ridged" or covered with concentric raised patterns. These ridges are called friction ridges and they serve the useful function of making it easier to grasp and hold onto objects and surfaces without slippage. It is the many differences in the way friction ridges are patterned, broken, and forked which make ridged skin areas, including fingerprints, unique.

GLOBAL VERSUS LOCAL FEATURES

We make use of two types of fingerprint characteristics for use in identification of individuals: Global Features and Local Features. Global Features are those characteristics that you can see with the naked eye. Global Features include:

- Basic Ridge Patterns
- Pattern Area
- Core Area
- Delta
- Type Lines
- Ridge Count

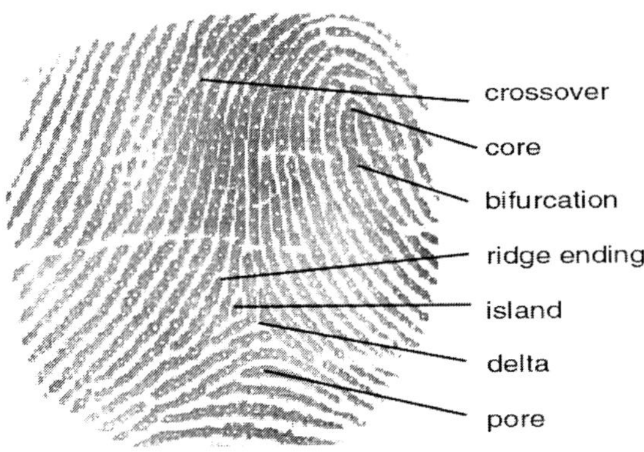

The Local Features are also known as Minutia Points. They are the tiny, unique characteristics of fingerprint ridges that are used for positive identification. It is possible for two or more individuals to have identical global features but still have different and unique fingerprints because they have local features - minutia points - that are different from those of others.

Fingerprint Scanning

Fingerprint scanning is the acquisition and recognition of a person's fingerprint characteristics for identification purposes. This allows the recognition of a person through quantifiable physiological characteristics that verify the identity of an individual. There are basically two different types of finger-scanning technology that make this possible. One is an optical method, which starts with a visual image of a finger. The other was a semiconductor-generated electric field to image a finger.

There are different ways to identify fingerprints. They include traditional police methods of matching minutiae, straight pattern matching, moiré fringe patterns and ultra sonic.

Fingerprint Matching

Fingerprint matching techniques can be placed into two categories: minutiae-based and correlation based. Minutiae-based techniques first find minutiae points and then map their relative placement on the finger. However, there are some difficulties when using this approach. It is difficult to extract the minutiae points accurately when the fingerprint is of low quality. Also this method does not take into account the global pattern of ridges and furrows. The correlation-based method is able to overcome some of the difficulties of the minutiae-based approach. However, it has some of its own shortcomings. Correlation-based techniques require the precise location of a registration point and are affected by image translation and rotation.

Fingerprint Classification

Large volumes of fingerprints are collected and stored everyday in a wide range of applications including forensics, access control, and driver license registration. An automatic recognition of people based on fingerprints requires that the input fingerprint be matched with a large number of fingerprints in a database To reduce the search time and computational complexity, it is desirable to classify these fingerprints in an accurate and consistent manner so that the input fingerprint is required to be matched only with a subset of the fingerprints in the database.

Fingerprint classification is a technique to assign a fingerprint into one of the several pre-specified types already established in the literature which can provide an indexing mechanism. Fingerprint classification can be viewed as a coarse level matching of the fingerprints. An input fingerprint is first matched at a coarse level to one of the pre-specified types and then, at a finer level, it is compared to the subset of the database containing that type of fingerprints only. We have developed an algorithm to classify fingerprints into five classes, namely, whorl, right loop, left loop, arch, and tented arch.

Image capture

There are two approaches for capturing the fingerprint image for matching: minutia matching and global pattern matching. Minutia matching is a more microscopic approach that analyzes the features of the fingerprint, such as the location and direction of the ridges, for matching. The only problem with this approach is that it is difficult to extract the minutiae points accurately if the fingerprint is in some way distorted. The more macroscopic approach is global pattern matching where the flow of the ridges is compared at all locations between a pair of fingerprint images; however, this can be affected by the direction that the image is rotated.

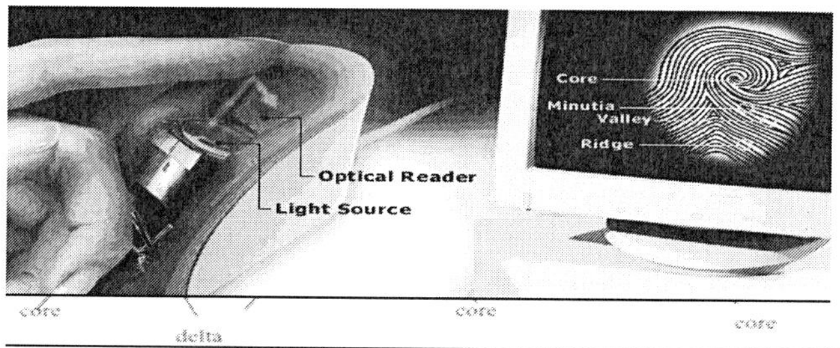

core delta core core

Types of scanners

- Optical Scanner - captures a fingerprint image using a light source refracted through a prism
- Thermal Scanner - very small sensor that produces a larger image of the finger and is contrast-independent
- Capacitive Scanner - uses light to illuminate a finger placed on a glass surface and records the reflection of this light with a solid-state camera

Each of these devices use light to measure the ridges and non-ridges, take an original fingerprint image, capture the minutia points and create an identifying template from the minutia points.

Image Processing

Following the image capture, image processing is performed to achieve a definitive match on the individual. At this stage, image features are detected and enhanced for verification against the stored minutia file. Image enhancement is used to reduce any distortion of the fingerprint caused by dirt, cuts, scars, sweat and dry skin.

Image Verification

At the verification stage, the image of the fingerprint is compared against the authorized user's minutia file to determine a match and grant access to the individual.

Potential Issues

Although fingerprint recognition is the cheapest form of biometric security available and is widely accepted by public and law enforcement communities as reliable identification, its disadvantage is that it requires close physical contact with scanning device. Some other issues includes

- Privacy
- False Rejection
- False Acceptance
- Accuracy
- Privacy: Comparison and storage of unique biological traits makes some individuals feel that their privacy is being invaded. Many associate fingerprint scanning with the fingerprinting of alleged criminals and are therefore hesitant to accept this technology.
- False Rejection Rate (FRR): False rejection occurs when a registered user does not gain access to the system. This person has then been falsely rejected from access.
- False Acceptance Rate (FAR): False acceptance is when an unauthorized user gains access to a biometrically protected system.
- Accuracy: Although fingerprints are unique to an individual, there are instances where a fingerprint may become distorted and authorization will not be granted to the user. As discussed above in image processing, dirt, cuts, scars, sweat and dry skin can cause fingerprint distortion.

IRIS RECOGNITION

Iris recognition combines computer vision, pattern recognition, statistics, and the human-machine interface. The purpose is real-time, high confidence recognition of a person's identity by mathematical analysis of the random patterns that are visible within the iris of an eye from some distance. Because the iris is a protected internal organ whose random texture is stable throughout life, it can serve as a kind of living passport or a living password that one need not remember but one always carries along. The iris's random patterns are unique to each individual — a human "bar code" or living passport. No two irises are alike. Each person has a distinct pattern of filaments, pits and striations in the colored rings surrounding the pupil of each eye. This pattern is stable throughout life. Unlike fingerprints, iris "prints" are not subject to environmental damage. An internal organ, the iris is protected by the cornea. Because these structures are transparent, the iris can be easily identified with a high degree of certainty up to three feet away.

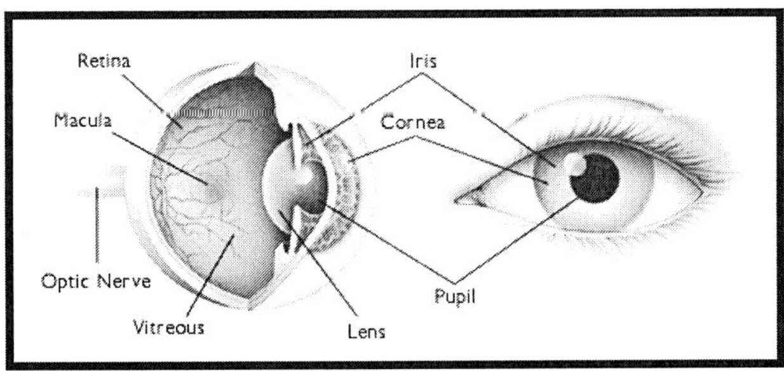

Iris Acquisition

The first step is location of the iris by a dedicated camera no more than 3 feet from the eye. The monochrome camera uses both visible and infrared light, the latter of which is located in the 700-900nm range.

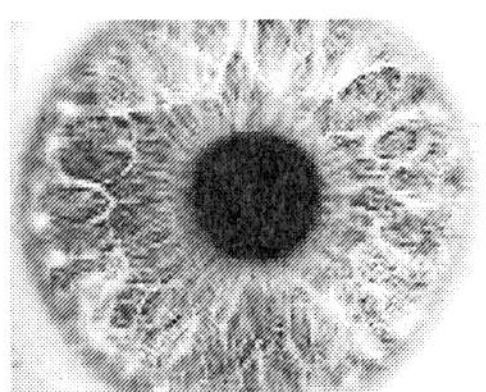

Iris

The retina, a thin nerve (1/50th of an inch) on the back of the eye, is the part of the eye which senses light and transmits impulses through the optic nerve to the brain - the equivalent of film in a camera. Blood vessels used for biometric identification are located along the neural retina, the outermost of retina's four cell layers.

Iris scan

These systems take advantage of random variations in the visible features the iris, the colored ring around the pupil. After a iris has been scanned once, a unique file is placed in a database. Iris scanning takes advantage of random variations in the visible features the iris, the colored part of the eye. The iris consists largely of a system of muscle that expands and contract the pupil in response to changing lighting conditions. The details of each iris are unique, that is, no two are exactly alike, not even among twins, not even in your own two eyes. The structure of the iris develops in the embryo, assuming its lifelong character by the seventh or eighth month. Some color changes can occur in the first months of life, which explains why some babies who are born with blue eyes may end up with brown or some other color.

After taking a picture of the eye, the system samples the radial and angular variations of each individual iris to form an Iris Code, a digital file that serves as a reference in database. At 512 bytes, the file is quite small, since it is a hexadecimal code reference rather than an actual iris image. Research confirms an extraordinarily high level of statistical reliability for the system. A person using the system simply looks into a camera. The computer program then locates the iris. Next, the system locates the iris' outer and inner edges. The monochrome camera uses both visible and infrared (700-900nm) light. The program maps segments of the iris into hundreds of vectors. Position, orientation and spatial frequency provide the basis for calculation of the Iris Code. The system also manages to take into account normal changes in the eye. For example, the system compensates for papillary expansion and contraction. It can also detect reflections from the cornea.

Drawbacks

Iris Scanning offers numerous benefits over other biometrics, although there are a few definite issues surrounding this technology. First and foremost, iris scanning is fairly expensive. Iris scanning produces an accurate image with any number of distractions, but it still depends on user cooperation; a user must know he is being scanned and agree to it for the scan to work. This is because a person must sit quite still – even just momentarily – to get an accurate scan.

How Is the Performance of a Biometric Measured

Biometric performance is most commonly measured in two ways: False Rejection Rate (FRR), and False Acceptance Rate (FAR). The FRR is the probability that are not authenticated to access your account. A strict definition states that the FRR is the probability that a mated comparison (i.e. 2 biometric samples of the same eye) incorrectly determines that there is no match. The FAR is the chance that someone other than you is granted access to your account, in other words, the probability that a non-mated comparison (i.e. two biometric samples of different eye) match. FAR and FRR numbers are generally expressed in terms of probability. When comparing biometric systems, a low false acceptance rate is most important when security is the priority. Whereas, a low false rejection rate is most important when convenience is the priority. All biometric implementations balance these two criteria. Some systems use very high FAR's such as 1 in 300. This means that the likelihood that the system will accept someone other than the enrolled user is 1 in 300. However, the likelihood that the system will reject the enrolled user (its FRR) is very low, giving them ease of use, but with low security. Most iris scan systems should be able to run with FAR's of 1 in 1,000,000or better. Another factor that must be taken into account when determining the necessary FAR and FRR for the organization is the actual quality of the database in the user population. Experience with several thousand users, and the experience of its customers, indicates that a percentage of the population do not have database of sufficient quality to allow for authentication of the individual. In the general office worker population approximately 2.5% of employees fall into this group. For these users, a smartcard token with password authentication is recommended. Statistical improvements in false rejection rates can also be achieved by requiring the user to use more than one eye to authenticate. Such techniques are referred to as flexible verification. Within the industry, FAR and FRR numbers are often quoted by competing vendors. However, the environments and testing methodologies, which are used to arrive at these statistics, vary greatly. Therefore, the reported FAR and FRR cannot be relied upon as a definitive measure of performance. Most companies when publishing their FRR only publish the 2 or 3 test attempt statistic. Furthermore, it is continually enhancing the hardware and software to improve the security and usability of the products, which mean the numbers are constantly improving.

Applications

Biometrics has come in a large way in India, though it is yet to be implemented physically for various security systems. The prices of biometric products and systems are falling as demand for the technology grows and more vendors enter the market. Fraud, security breaches, and human administrative error are driving the rapid expansion of biometric technology. Existing methods of electronic identification--ID cards, card keys, and PIN numbers--are inadequate. These can be stolen or exchanged for fraudulent purposes. It isn't surprising, then, that the new field of biometrics appeals strongly to banks, credit card firms, government welfare offices, departments of correction, and law enforcement organizations.

Security in defense establishments has always been a sore point, stressed upon by one and all. In addition to physical security where in the physical access to the high security areas is controlled and restricted, security to the data and information in the large number of computers needs to be addressed urgently and effectively. For this, biometrics-fingerprint recognition is by far the most cost effective, suitable solution available in the market today. This may be used singly or in conjunction with some other biometric technology like Retina scan or Iris recognition or even with non-biometric means like smart cards.

6

Fingerprint Systems

FINGERPRINT SYSTEMS

The principle of fingerprint systems is schematically. In an enrollment process, the system captures finger data from an enrollee with sensing devices, extracts features from the finger data, and then record them as template with a personal information, e.g. a personal identification number (PIN), of the enrollee into a database. We are using the word "finger data" to mean not only features of the fingerprint but also other features of the finger, such as "live and well" features. In an identification (or verification) process, the system captures finger data from a finger with sensing devices, extracts features, identifies (or verifies) the features by comparing with templates in the database, and then outputs a result as "Acceptance" only when the features correspond to one of the templates.

Most of fingerprint systems utilize optical or capacitive sensors for capturing fingerprints. These sensors detect difference between ridges and valleys of fingerprints. Optical sensors detect difference in reflection. Capacitive sensors, by contrast, detect difference in capacitance. Some systems utilize other types of sensors, such as thermal sensors, ultrasound sensors. In this paper we examine fingerprint systems which utilize optical or capacitive sensors.

A RISK ANALYSIS FOR FINGERPRINT SYSTEMS

Generally, fingerprint systems capture images of fingerprints, extract features from the images, encrypt the features, transmit them on communication lines, and then store them as templates into their database. Some systems encrypt templates with a secure cryptographic scheme and manage not whole original images but compressed images. Therefore, it is said to be difficult to reproduce valid fingerprints by using the templates. Some systems are secured against a so-called replay attack in which an attacker copies a data stream from a fingerprint scanner to a server and later replays it, with an one time protocol or a random challenge response device. We are here not concerned with the security for the communication and database of fingerprint systems, assuming that they are secure enough by being protected with some encryption schemes. In this section, we would like to address security of fingerprint scanners.

When a legitimate user has registered her/his live finger with a fingerprint system, there would be several ways to deceive the system. In order to deceive the fingerprint system, an attacker may present the following things to its fingerprint scanner.

1. The registered finger: The highest risk is being forced to press the live finger against the scanner by an armed criminal, or under duress. Another risk is being compelled the legitimate user to fall asleep with a sleeping drug, in order to make free use of her/his live finger. There are some deterrent techniques against these crimes. Combination with another authentication method, such as by PINs, passwords, or ID cards, would be helpful to deter the crimes. Furthermore, a duress control enables the users to alarm "as under duress" with a secret code or manner, which is different with a PIN or usual manner respectively. Combining a duress control with a fingerprint system would provide a helpful measure to apply to someone for protection. Similarly, a two persons control, namely a two persons rule, where the authentication process requires two persons properties, i.e., fingerprints, or would be helpful to deter the crimes.
2. An unregistered finger (an imposter's finger): An attack against authentication systems by an imposter with his own biometrics is referred to as non-effort forgery. Commonly, accuracy of authentication of fingerprint systems are evaluated by the false rejection rate (FRR) and the false acceptance rate (FAR). The FAR is important indicator for the security against such a method as with an unregistered finger. Moreover, fingerprints are usually categorized as "loops," "whorls," "arches," and others. If an attacker knows what category of the registered finger is, an unregistered finger of which pattern is similar to the registered one would be presented to the scanner. In this case, the probability of the acceptance may be different from the ordinary FAR. From this point of view, the accuracy of authentication for the system should be evaluated not only, as usual, for fingers throughout the categories of fingerprints but also for fingers within each category. Another attacker may modify his fingerprint by painting, cutting, or injuring his own fingertip. However, it is thought to be very difficult to deceive the fingerprint system with such a modified finger. The reason for this is that fingerprints are so random the attacker cannot identify which patterns should be modified. Ordinarily, ten-odd

features, such as ridge's and valley's distinctive patterns are used for the authentication.

3. A severed fingertip from the registered finger: A horrible attack may be performed with the finger which is severed from the legitimate user's hand. Even if the finger severed from the user's half-decomposed corpse, the attacker may use, for the wrong purpose, a scientific crime detection technique to clarify its fingerprint. In the same way as the above-mentioned registered finger, combination with another authentication method, or a duress/two-persons control would be helpful to deter these crimes. The detection whether the finger is alive or not would be helpful as well.
4. A genetic clone of the registered finger: In general, it is said that identical twins do not have the same fingerprint, and neither would clones. The reason is that fingerprints are not entirely genetically determined, and rather determined in part by its pattern of nerve growth into the skin. As a result, this is not exactly the same even in identical twins. However, it is also said that fingerprints are different in identical twins, but only slightly different. If the genetic clone's fingerprint is similar to the registered finger's, an attacker may try to deceive fingerprint systems by using it. Now is the time when we must keep a close watch on such possibility with genetic engineering,
5. An artificial clone of the registered finger: More casual attacks against fingerprint systems may use an artificial finger. An artificial finger can be produced as a printed fingerprint with a copier or a desk top publishing (DTP) technique as well as forged documents. If an attacker can make a mold of the registered finger directly modeling it, s/he can produce an artificial finger with some materials. Even if not, s/he may make a mold of the registered finger modeling its residual fingerprint at second hand, so as to produce an artificial finger. And, if an attacker can make an artificial finger which can deceive a fingerprint system, one of countermeasures against the attack obviously is the detection whether or not the finger is alive. Again, combination with another authentication method, or two-persons control would be also helpful to deter the crimes.
6. The others: In some fingerprint systems, an error in authentication may be caused by making noise or flashing a light against the fingerprint scanner, or by heating up, cooling down, humidifying, impacting on, or vibrating the scanner outside its environmental tolerances. Some attackers may use the error to deceive the system. This method is well-known as a "fault based attack," and may be carried out with above-mentioned attacks. Furthermore, a

7fingerprint image may be stood out in strong relief against the scanner surface, if we spray some materials on the surface. The image would be the residual fingerprint of a registered finger. In this case, a bald thing or finger, regardless of alive or not, which are pressed on the surface, may be accepted by the fingerprint system.

As fingerprint systems come into wide use, they are still more exposed themselves to a risk of casual attacks. Apart from duress or other crimes, our great concern, in this paper, is the possibilities of attacks with artificial fingers.

TYPES OF PRINTS

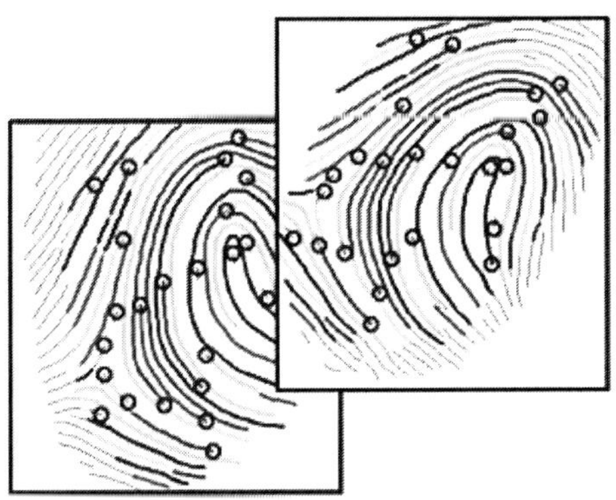

In general, the purpose of collecting fingerprints is to identify the fingerprint-leaver. This person may be the suspect, a victim, or even someone else who happened to be at the place where the print was found. There are three types of fingerprints that can be found: latent, patent, and plastic. Latent fingerprints are made of the sweat and oil on the skin's surface. This type of fingerprint is invisible to the naked eye and requires additional processing in order to be seen. This processing can include basic powder techniques or the use of chemicals. Patent fingerprints can be made by blood, grease, ink, or dirt. This type of fingerprint is easily visible to the human eye. Plastic fingerprints are three-dimensional impressions and can be made by pressing your fingers in fresh paint, wax, soap, or tar. Like patent fingerprints, plastic fingerprints are easily seen by the human eye and do not require additional processing for visibility purposes.

SURFACE CHARACTERISTICS AND COLLECTION METHODS

Characteristics of the surface in which the print is found are important in deciding which collection methods should be employed on scene. The general characteristics of the surface are: porous, non-porous smooth and non-porous rough. The distinction between porous and non-porous surfaces is their ability to absorb liquids. Liquids sink in when dropped onto a porous surface, while they sit on top of a non-porous surface. Porous surfaces include paper, cardboard, and untreated wood. Non-porous smooth surfaces include varnished or painted surfaces, plastics, and glass. Non-porous rough surfaces include vinyl, leather, and other textured surfaces. For porous surfaces, scientists sprinkle chemicals like ninhydrin over the prints and then take photographs of the developing fingerprints. For non-porous smooth surfaces, experts use powder-and-brush techniques, followed by lifting tape. For rough surfaces, the same powdering process should be used, but instead of using regular lifting tape for these prints, scientists use something that will get into the grooves of the surface. A gel-lifter or Mikrosil (a silicone casting material) can be used effectively to lift a print on these surfaces.

Analysis of Collected Prints

Once someone collects the print, the analysis can begin. During analysis, examiners determine whether there is enough information present in the print to be used for identification. This includes determining class and individual characteristics for the unknown print. Class characteristics are the characteristics that narrow the print down to a group but not an individual. The three fingerprint class types are arches, loops, and whorls. Arches are the least common type of fingerprint, occurring only about 5% of the time. This pattern is characterized by ridges that enter on one side of the print, go up, and exit on the opposite side. Loops are the most common, occurring 60-65% of the time. This pattern is characterized by ridges that enter on one side of the print, loop around, and then exit on the same side. Whorls present a circular type of ridge flow and occur 30-35% of the time. Individual characteristics are those characteristics that are unique to an individual. They are tiny irregularities that appear within the friction ridges and are referred to as Galton's details. The most common types of Galton's details are bifurcation, ridge endings, and dots or islands.

Comparison of Prints

After an examiner completes the analysis, they compare the unknown print side by side with a known print. The unknown print is the print found at the crime scene, and the known print is the print of a possible suspect. First, the class characteristics are compared. If the class characteristics of the two prints are not in agreement, then the first print is automatically eliminated. If this is the case, another known print may be compared to the unknown print. If the class characteristics appear to match, the examiner then focuses on the individual characteristics. They look at each individual characteristic point by point until they have found a possible match.

Evaluation of Comparison

After the examiner completes the comparison, they can make a proper evaluation. If there are any unexplainable differences between the unknown and known fingerprints, then they can exclude the known fingerprint as the source. This means that if the class characteristics are in disagreement, then the conclusion would be exclusion. However, if the class characteristics as well as the individual characteristics are in agreement and if there are no unexplainable differences between the prints, the conclusion would be identification. In some cases, neither of these conclusions is possible. There may not be a sufficient quality or quantity of ridge detail to effectively make a comparison, making it impossible to determine whether or not the two prints came from the same source. In these instances, no conclusion can be made and the report will read "inconclusive." The three possible results that can be made from a fingerprint examination are therefore exclusion, identification, or inconclusive.

Verification of the Evaluation

After the first examiner reaches one of the three conclusions, another examiner must verify the results. During this verification process, the entire exam is repeated. The second examiner does the repeated exam independently from the first exam, and for an identification conclusion, both examiners must agree. If they agree, the fingerprint evidence becomes a much stronger piece of evidence if and when it goes to court.

Databases such as AFIS (Automated Fingerprint Identification System) have been created as ways of assisting the fingerprint examiners during these examinations. These databases help provide a quicker way to sort through unlikely matches. This leads to quicker identification of unknown prints and allows fingerprints to be as widely used as they are in criminal investigations.

DISHONEST ACTS WITH ARTIFLCIAL FINGERS

The assumption that artificial fingers can be accepted by fingerprint systems, we discuss dishonest acts against the system with the artificial fingers. Several patterns of dishonest acts, with artificial fingers, in a fingerprint system are shown in Table 2.1. In this table, L(X) and L(Y) denote live fingers of persons X and Y respectively. A(X) and A(Y) denote artificial fingers which are molded after L(X) and L(Y) respectively. A(Z) denotes artificial fingers which are created artificially without being molded after live fingers. There may be at the total of 25 possible combinations because we can enroll and

Table. Several patterns of dishonest acts with artifical fingers in a fingerprint system

	Verification / Identification				
Enrollment	L(X)	A(X)	L(Y)	A(Y)	A(Z)
L(X)	(1)	(2)	- *	-	-
A(Y)	-	-	(3)	(4)	-
A(Z)	-	-	-	-	(5)

verify each of L(X), L(Y), A(X), A(Y) and A(Z) in the system. However, we show 15 in 25 combinations and focus on 5 combinations, from (1) to (5), in the table on the following assumptions;

- Only X can enroll fingers in the system,
- but X cannot enroll Y's live finger L(Y),
- the fingers must be enrolled with a genuine X's PIN,
- and X can obtain and enroll A(Y) and A(Z) in the system.

The case (1) is the proper way in the system. The case (2) or (5) is also the proper way in some systems which positively utilize artificial fingers. However, we discuss dishonest acts of artificial fingers regarding the cases from (2) to (5) as dishonest ways. The possible dishonest acts are:

- Some other persons than X can pretend, in the system, to be X by presenting artificial fingers in all the cases of (2), (4) and (5), and
- Y can pretend, in the system, to be X by presenting her/his own fingers L(Y) in the case (3).

In addition, X can deny the participation showing by means of evidence that her/his own live fingers cannot be accepted by the system, when the dishonest act was detected in the cases (3), (4) and (5).

Most discussion on dishonest acts with artificial fingers have mainly focused on the case (2). However, it should be noted that the dishonest acts, which correspond to the cases (4) and (5), can be done if artificial fingers are accepted by the system. The cases (3), (4) and (5) can be probably prevented by a practical measure that enrollment is closely watched to prevent using artificial fingers. Moreover, dishonest acts which correspond to the cases, from (2) to (5), never occur if the system can successfully reject artificial fingers.

Accuracy of authentication of fingerprint systems are commonly evaluated by the false rejection rate (FRR) and the false acceptance rate (FAR). In Table 2.1, the case (1) and the case indicated by the sign '*' correspond to those which are commonly evaluated by the FRR and FAR in its ordinary performance test for live finger samples, respectively. It is important that the acceptance rate for the cases from (2) to (5) should be evaluated if the system cannot perfectly reject artificial fingers.

EXPERIMENTS

HOW TO MAKE ARTIFICIAL FINGERS

There are several major ways to make an artificial finger of a given live finger. First of all, an impression must be obtained from the live finger. The fingerprint image of the impression is mostly the lateral reversal, i.e., transposed from left to right as in a mirror reflection, of the original. This impression may be used as an artificial finger to deceive the system, if an attacker can make an impression without being lateral reversal. However, most of the impressions have a lateral reversal fingerprint image. When a live finger was pressed on an impression material, a fingerprint image on the impression may be used as a mold of an artificial finger. If an impression was captured with a digital camera as an electronic fingerprint image, it may be set right side left by an image processing software, and then printed on a material to make an artificial finger. This electronic fingerprint image also can be used to make a mold, as it is. The impression may be a residual or inked fingerprint. Even if the fingerprint image is latent, some techniques for scientific crime detection can enhance it with some material, e.g. aluminum powder, a ninhydrin solution, a cyanoacrylate adhesive, then producing the artificial finger with them. Besides these, we can, without original live fingers, create an artificial finger with a fictitious fingerprint, such as, using a so-called fingerprint generator.

In our experiments, we make artificial fingers by the following two ways, which are; (1) we make an impression directly pressing a live finger to plastic material, and then mold an artificial finger with it, and (2) we wapture an fingerprint image from a residual fingerprint with a digital microscope, and then make a mold to produce an artificial finger. The material for artificial fingers is gelatin which is obtained from bone, skin, etc., of animals. In this paper we are using the word "gummy" to refer not a sol but a gel of gelatinous materials. Accordingly, we use the terms "gummy" fingers to refer artificial fingers which are made of gelatin, since its toughness are nearly equal to gummies which are one kind of sweets, and also made of gelatin with some additives such as sugar and/or fruit juice. Here, the origin of the word "gummy" is "gummi" in the German language.

Table. Fingerprint Types of Experiments.

Experiment	Enrollment	Verification
Type 1	Live Finger	Live Finger
Type 2	Live Finger	Gummy Finger
Type 3	Gummy Finger	Live Finger
Type 4	Gummy Finger	Gummy Finger

- e.g., Live Fingers, Generators,
- Experiment Enrollment Verification
- Impression Type 1 Live Finger Live Finger
- e c7., Moids, Pesiduaf F'Oluerprints, Type 2 Live Finger Gummy Finger
- Type 3 Gummy Finger Live Finger
- ID Type 4 iGummy Finger Gummy Finger

Experimental Procedures

The goal of experiments which we conducted is to examine whether fingerprint systems, which are commercially available, accept the artificial fingers or not. Accordingly, we examined the acceptance rates of fingerprint systems by using the artificial fingers, i.e.. gummy fingers, and live fingers. The following describes procedures of the experiments.

1. Types of artificial fingers: Two types of artificial fingers were examined. One is produced by cloning with a plastic mold. The other is produced by cloning from a residual fingerprint. These are detailed in APPENDIX A.
2. Types of experiments: Four types of experiments (shown in Table 3.1) were conducted. Difference in the types are follows:
   - Type 1: A subject presents her/his live finger to verify with a template which was made by enrolling the live finger.
   - Type 2: A subject presents her/his live finger to verify with a template which was made by enrolling her/his live finger.
   - Type 3: A subject presents her/his live ringer to verify with a template which was made by enrolling her/his gummy finger.
   - Type 4: A subject presents her/his gummy finger to verify with a template which was rnade by enrolling the artificial finger.
3. Rules in experiments: We conducted the experiments under the extra rules as follows:
   - We allowed a tester as deputy for the subject to present or enroll the subject's artificial finger.
   - In Type 4 experiment, we also allowed that an artificial finger which is presented is not always the same as one which was used in enrollment.
   - The subject or tester must intentionally present the live/gummy finger to fit the center of it to that of a scanning area of the fingerprint devise.
   - The gummy fingers can be modified their shape to fit the scanning area.
4. The acceptance rates: Only one live/gummy finger must be enrolled as a template while we allowed to retry in enrollment. We attempted one-to-one verification 100 times in each type of experiment for each fingerprint system counting the number of times that it accepts a finger presented. As a result, we measured the acceptance rates in verification for the fingerprint systems.
5. Subjects: The subjects are five persons whose ages are from 20's to 40's, in the experiment for the gummy fingers cloned with a plastic mold, whereas the subject is only one person in the experiment for the gummy fingers cloned from a residual fingerprint.
6. Artificial Fingers: Gummy fingers of each subject were made in the two ways which we explained in APPENDIX A and used in the experiments.
7. Fingerprint systems: We tested 11 types of fingerprint systems which are shown in APPENDIX B.1. Each of them consists of a finger device and a software for verification. We set a threshold value for the highest security level if the fingerprint system allows to adjust

or select threshold values in verification. All of fingerprint devices can be connected with a personal computer (PC), and used for access control. The procedures for fingerprint systems are detailed in APPENDIX B.2.

EXPERIMENTAL RESULTS AND DISCUSSIONS

CLONING WITH A PLASTIC MOLD

Artificial Fingers

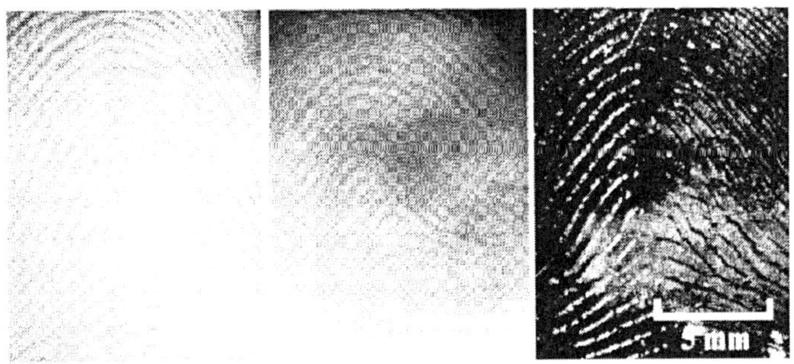

(a) Live Finger (b) Silicone Finger (c) Gummy Finger

Fig. The photomicrographs of a live finger and its artificial fingers.

The photomicrographs of a live finger and its artificial fingers. The gummy finger which is cloned by using a plastic mold with an impression of the live finger. The molded gummy fingers are rather transparent and amber while having ridges and valleys similar to those of the live finger, in terms of the outside appearance. Fingerprint images of a live finger, a silicone finger and a gummy finger, which were displayed by the system with Device C (equipped with an optical scanner). Here, the silicone finger is the artificial finger made of silicone rubber, and presented so as to compare with the others.

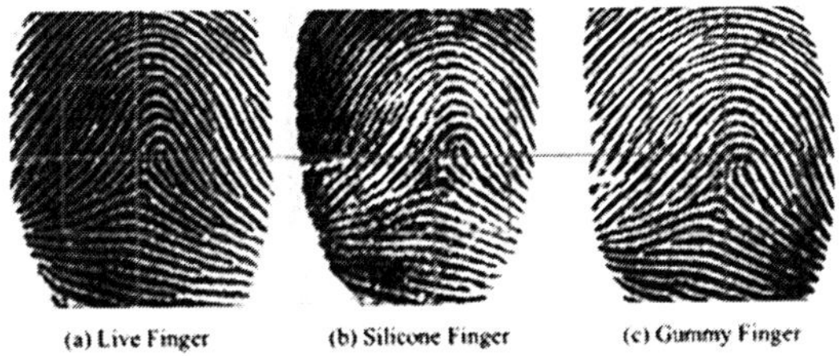

Fig. Fingerprint images of a live finger, a silicene finger and a gummy finger, which were displayed by the system with Device C equipped with an optical scanner.

The captured image of the gummy finger is very similar to that of the live fingerprint images of a live finger and a gummy finger, which were displayed by the system with Device H (equipped with a capacitive sensor).

While some defects are observed in the image (right side) of a gummy finger, both of the images are similar to each other. Here, the reason why we do not present the image of a silicone finger is that it cannot be accepted by the system with Device H.

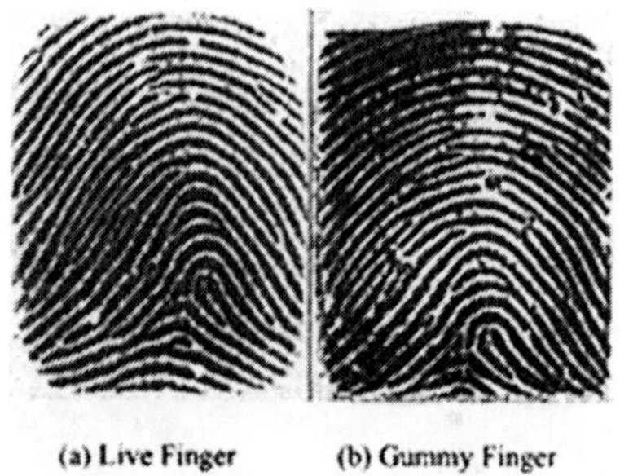

Fig. The sample images of fingers which derive from the same finger and were displayed on the screen when being scanned by the capactive sensor of Device H.

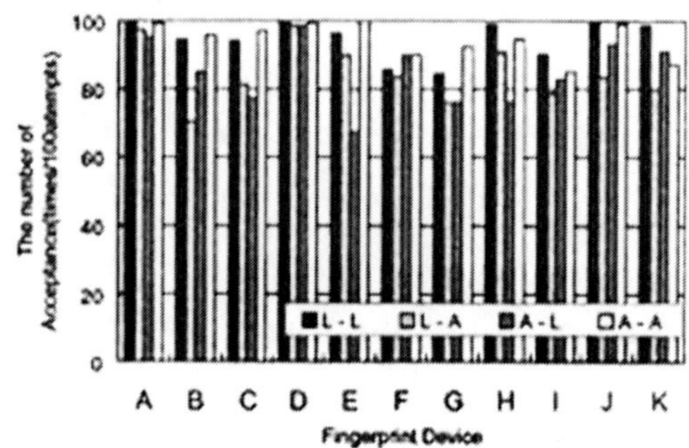

Fig. Average number of acceptance for each device, in terms of gummy fingers which wee cloned with plastic molds. Here, the number of the subjects is five.

Acceptance Rates of the Artificial Fingers

The results of the experiments for the artificial fingers, which were cloned with molds. It was found through the experiments that we could enroll the gummy fingers in all of the 11 types of fingerprint systems. It was also found that all of the fingerprint systems accepted thegummy fingers in their verification procedures with the probability of 68-100%.

CLONING FROM A RESIDUAL FINGERPRINT

Artificial Fingers

The outside appearance of the mold which we used in our experiments. A photograph of the gummy finger which was produced from a residual fingerprint on a glass plate, enhancing it with a cyanoacrylate adhesive. We applied a technique for processing printed circuit boards to the production of the molds for cloning the gummy fingers. The fingerprint image of the gummy finger, which was displayed by the system with Device H (equipped with a capacitive sensor).

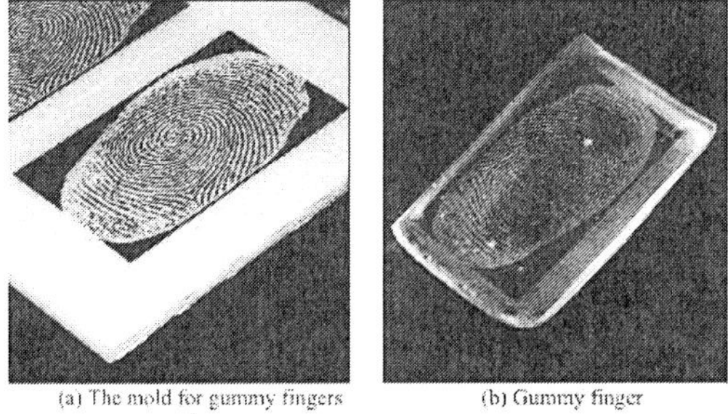

Fig. Photographs of the outside appearance of the mold and a gummy finger. The gummy finger was produce from a residual fingerprint on a glass plate, enchancing it with a cyanoacrylate adhesive

Fig. The Fingerprint image of the gummy finger, which was displayed by the system with Device H (equipped with a capacitive sensor.)

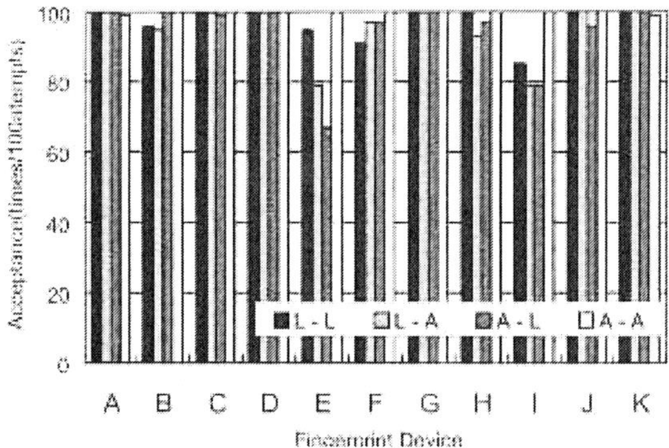

Fig. Average number of acceptance for each device, in terms of gummy fingers which were cloned from residual fingerprints. Here, the subject is one person.

Acceptance Rates of the Artificial Fingers

The results of the experiments for an artificial finger, which was cloned from a residual fingerprint. As a result, we could enroll the gummy finger in all of the 11 types of fingerprint systems. It was found that all of the fingerprint systems accepted the gummy finger in their verification procedures with the probability of more than 67%.

Discussions

The number of samples in the experiments is so small that we cannot compare performance of the fingerprint systems. However, the number of samples is enough for us to see evidence that the gummy fingers could be accepted by commercial fingerprint systems. Based on our analysis, these variations may be caused by deformation of gummy fingers. We found that some of artificial fingers were damaged while being heated by the fingerprint sensors in the experiments. Some of the sensors frequently heated up when repeating verification in a short period. We think that the number of acceptance will increase if we pause for cooling every time after verification. The gummy fingers can be modified their shape to fit the scanning area. Accordingly, we cut the gummy fingers to fit the sensing area for some devices. This might cause decrease of errors in positioning the gummy fingers. Another reason why the number of acceptance varies is that we allowed retrying in enrollment. Finally, all of gummy fingers could be enrolled in the end in our experiments, even if the systems employ some protection. Also, the number of acceptance of live fingers is greater than that of gummy fingers for some systems that may employ so-called live and well detection.

We have investigated the difference in characteristics of live, gummy and silicone fingers. While silicone fingers were impossible to measure, the moisture and electric resistance of the gummy fingers could be measured. We used a moisture meter, and a digital multimeter (range: from 0 to 40 Mohms). According to this comparison, gummy fingers are more similar to live fingers in their characteristics than silicone fingers. The compliance was also examined for live and gummy fingers. Here, the compliance indicated by the change in resonance frequency (i.e., tactile sensor output) as the function of the pressure (i.e., pressure sensor output). In brief, **Fig.**4.8 shows that the live finger is softer than the gummy finger. We found that these fingers are clearly different in compliance.

CONCLUSIONS

We illustrated a risk analysis for fingerprint systems. The risk analysis revealed that there are many attack ways to deceive the systems, even if their templates and communication are protected by a secure measure. Conventional arguments tend to focus on a question how to detect use of artificial fingers, which derive from live fingers of legitimate users. However, as we pointed out, there can be various dishonest acts using artificial fingers against the systems. We also pointed out that artificial fingers can be made not only of silicone but also of gelatin, and examined 11 types of fingerprint systems whether or not they accept the gummy fingers. Consequently, all of these systems accepted the gummy fingers all in their enrollment procedures and also with the rather higher probability in theirverification procedures. The results are enough for us to see evidence that artificial fingers can be accepted by commercial fingerprint systems. The objection will no doubt be raised that it is very difficult to take an impression of the live finger from a legitimate user without the cooperation of her/him. Therefore, we demonstrated that the gummy fingers made from residual fingerprints can be accepted by all of the 11 systems.

After we started this study, we come to know by the published book, that Dutch researchers reported that an artificial finger, which was made of silicone rubber, putting saliva on its surface can be accepted by fingerprint systems with capacitive devices. Their study and ours share certain similarities in that both intend to encourage the suppliers and users of fingerprint systems to reconsider security of their systems. While their study is seen to be of use in designing fingerprint systems, unfortunately, details of the experimental conditions have not been described.

As we mentioned, a user of biometric systems cannot frequently replace or change her/his biometric data because of limits of biometric data intrinsic to her/himself. For example, gelatin, i.e., an ingredient of gummies, and soft plastic materials are easy to obtain at grocery stores and hobby shops, respectively. The fact that gummy fingers, which are easy to make with cheap, easily obtainable tools and materials, can be accepted suggests review not only of fingerprint systems but also of biometric systems. Manufacturers and vendors of biometric systems should carefully examine the security of their systems against artificial clones. Also, they should make public results of their examination, which lead users of their system to a deeper understanding of the security. The experimental study on the gummy fingers will have considerable impact on security assessment of fingerprint systems.

Declaration: We would like to stress that this study intends to encourage the suppliers and users of fingerprint systems to reconsider security of their systems, not to criticize libelously fingerprint systems for their security, and not to compare their performance. For this purpose, we have collected, for our experiments, as many fingerprint systems commercially vailable as we could, and have detailed their specification and the experimental conditions.

AUTOMATED FINGERPRINT IDENTIFICATION SYSTEM

Prior to the industrial revolution and the mass migrations to the cities, populations lived mostly in rural communities where everyone knew everyone else and there was little need for identification. Indeed, there were no police forces, no penitentiaries, and very few courts. As cities became crowded, crime rates soared and criminals flourished within a sea of anonymity. Newspapers feasted on stories of lawlessness, legislatures quickly responded with more laws and harsher penalties (especially for repeat offenders), and police departments were charged with identifying and arresting the miscreants. Identification systems—rogues' galleries, anthropometry, Bertillon's "portrait parlé", and the Henry system—emerged and quickly spread worldwide at the end of the 19th and beginning of the 20th century.

The late 1960s and early 1970s witnessed another era of civil turmoil and an unprecedented rise in crime rates, but this era happened to coincide with the development of the silicon chip. The challenges inherent in identification systems seemed ready-made for the solutions of auto¬matic data processing, and AFIS—Automated Fingerprint Identification System—was born.

During this same period, The RAND Corporation, working under a national grant, published The Criminal Investigative Process (Greenwood et al., 1975), a comprehensive study and critique of the process by which crimes get solved—or do not. Generally critical of traditional methods used by detectives, the study placed any hopes for improvement on physical evidence in general and latent prints in particular. In a companion study, Joan Petersilia concluded that:

No matter how competent the evidence techni¬cian is at performing his job, the gathering of physical evidence at a crime scene will be futile unless such evidence can be properly processed and analyzed. Since fingerprints are by far the most frequently retrieved physical evidence, mak¬ing the system of analyzing such prints effective will contribute the most toward greater success in identifying criminal offenders through the use of physical evidence. (Petersilia, 1975, p 12)

Though new technology was already in development at the Federal Bureau of Investigation (FBI), it would be a popular movement at the local and state levels that would truly test Petersilia's theory.

Need For Automation

In 1924, the FBI's Identification Division was established by authority of the United States congressional budget ap¬propriation bill for the Department of Justice. The identifica¬tion division was created to provide a central repository of criminal identification data for law enforcement agencies throughout the United States. The original collection of fingerprint records contained 810,188 records. After its cre¬ation, hundreds of thousands of new records were added to this collection yearly, and by the early 1960s the FBI's criminal file had grown to about 15 million individuals. This was in addition to the 63 million records in the civilian file, much of which was the result of military additions from World War II and the Korean conflict.

Almost all of the criminal file's 15 million individuals contained 10 rolled fingerprints per card for a total of 150 million single fingerprints. Incoming records were manually classified and searched against this file using the FBI's modified Henry system of classification. Approxi¬mately 30,000 cards were searched daily. The time and human resources to accomplish this daily workload continued to grow. As a card entered the system, a preliminary gross pattern classification was assigned to each fingerprint by technicians. The technicians could complete approximately 100 fingerprint cards per hour. Complete classification and searching against the massive files could only be accomplished at an average rate of 3.3 cards per employee per hour. Obviously, as the size of the criminal file and the daily workload increased, the amount of resources required continued to grow. Eventually, classification extensions were added to reduce the portion of the criminal file that needed to be searched against each card. Nonetheless, the manual system used for searching and matching fingerprints was approaching the point of being unable to handle the daily workload.

Although punch card sorters could reduce the number of fingerprint cards required to be examined based on pattern classification and other parameters, it was still necessary for human examiners to scrutinize each fingerprint card on the candidate list. A new paradigm was necessary to stop the increasing amount of human resources required to process search requests. A new automated approach was needed to (1) extract each fingerprint image from a tenprint card, (2) process each of these images to produce a reduced-size template of characteristic information, and (3) search a database to automatically produce a highly reduced list of probable candidate matches (Cole, 2001, pp 251–252).

Early AFIS Development

In the early 1960s, the FBI in the United States, the Home Office in the United Kingdom, Paris Police in France, and the Japanese National Police initiated projects to develop automated fingerprint identification systems. The thrust of this research was to use emerging electronic digital com¬puters to assist or replace the labor-intensive processes of classifying, searching, and matching tenprint cards used for personal identification.

FBI AFIS Initiative

By 1963, Special Agent Carl Voelker of the FBI's Identifi¬cation Division realized that the manual searching of the criminal file would not remain feasible for much longer. In an attempt to resolve this problem, he sought the help of en¬gineers Raymond Moore and Joe Wegstein of the National Institute of Standards and Technology (NIST).1 After describ¬ing his problem, he asked for assistance in automating the FBI's fingerprint identification process.

The NIST engineers first studied the manual methods used by human fingerprint technicians to make identifications. These methods were based on comparing the minutiae (i.e., ridge endings and ridge bifurcations) on fingerprint ridg¬es. If the minutiae from two fingerprints were determined to be topologically equivalent, the two fingerprints were declared to be identical—that is, having been recorded from the same finger of the same person. After this review, and after studying additional problems inherent with the inking process, they believed that a computerized solution to auto¬matically match and pair minutiae could be developed that would operate in a manner similar to the techniques used by human examiners to make fingerprint identifications. But to achieve this goal, three major tasks would have to be accomplished. First, a scanner had to be developed that could automatically read and electronically capture the inked fingerprint image. Second, it was necessary to accurately and consistently detect and identify minutiae existing in the captured image. Finally, a method had to be developed to compare two lists of minutiae descriptors to determine whether they both most likely came from the same finger of the same individual.

The Identification Division of the FBI decided that the approach suggested by Moore and Wegstein should be followed. To address the first two of the three tasks, on December 16, 1966, the FBI issued a Request for Quotation (RFQ) "for developing, demonstrating, and testing a device for reading certain fingerprint minutiae" (FBI, 1966). This con¬tract was for a device to automatically locate and determine the relative position and orientation of the specified minutiae in individual fingerprints on standard fingerprint cards to be used for testing by the FBI. The requirements stated that the reader must be able to measure and locate minutiae in units of not more than 0.1 mm and that the direction of each minutiae must be measured and presented as output in units of not more than 11.25 degrees (1/32 of a full circle). The initial requirements called for a prototype model to process 10,000 single fingerprints (1,000 cards). Contractors were also instructed to develop a proposal for a subsequent contract to process 10 times that number of fingerprints.

The 14 proposals received in response to this RFQ were divided into 5 broad technical approaches. At the conclusion of the proposal evaluation, two separate proposals were funded to provide a basic model for reading fingerprint im¬ages and extracting minutiae. Both proposed to use a "flying spot scanner" for capturing the image. But each offered a different approach for processing the captured image data, and both seemed promising. One contract was awarded to Cornell Aeronautical Labs, Inc., which proposed using a general-purpose digital computer to process binary pixels and develop programs for detecting and providing measure¬ment parameters for each identified minutiae. The second contract was awarded to North American Aviation, Inc., Autonetics Division, which proposed using a special-purpose digital process to compare fixed logical marks to the image for identifying, detecting, and encoding each minutia.

While the devices for fingerprint scanning and minutiae detection were being developed, the third task of comparing two minutiae lists to determine a candidate match was ad¬dressed by Joe Wegstein (Wegstein, 1969a, 1970, 1972a/b, 1982; Wegstein and Rafferty, 1978, 1979; Wegstein et al., 1968). He developed the initial algorithms for determining fingerprint matches based on the processing and compari¬son of two lists describing minutiae location and orientation. For the next 15 years, he continued to develop more reliable fingerprint matching software that became increasingly more complex in order to account for such things as plastic distortion and skin elasticity. Algorithms he developed were embedded in AFISs that were eventually placed in operation at the FBI and other law enforcement agencies.

By 1969, both Autonetics and Cornell had made significant progress on their feasibility demonstration models. In 1970, a Request for Proposal (RFP) was issued for the construc¬tion of a prototype fingerprint reader to reflect the experi¬ence gained from the original demonstration models with an additional requirement for speed and accuracy. Cornell was awarded the contract to deliver the prototype reader to the FBI in 1972. After a year's experience with the prototype system, the FBI issued a new RFP containing additional re¬quirements such as a high-speed card-handling subsystem. In 1974, Rockwell International, Inc., was awarded a contract to build five production model automatic fingerprint reader systems. This revolutionary system was called Finder. These readers were delivered to the FBI in 1975 and 1976. The next 3 years were devoted to using these readers in the conver¬sion of 15 million criminal fingerprint cards (Moore, 1991, pp 164–175). As it became apparent that the FBI's efforts to automate the fingerprint matching process would be successful, state and local law enforcement agencies began to evaluate this new technology for their own applications. The Minneapolis–St. Paul system in Minnesota was one of the first automat¬ed fingerprint matching systems (after the FBI's) to be in¬stalled in the United States. Further, while the United States was developing its AFIS technology in the 1960s, France, the United Kingdom, and Japan were also doing research into automatic fingerprint image processing and matching.

French AFIS Initiative

In 1969, M. R. Thiebault, Prefecture of Police in Paris, re¬ported on the French efforts. (Descriptions of work done by Thiebault can be found in the entries listed in the Additional Information section of this chapter.) France's focus was on the solution to the latent fingerprint problem rather than the general identification problem that was the concern in the United States. The French approach incorporated a vidicon (a video camera tube) to scan photographic film transparencies of fingerprints. Scanning was done at 400 pixels per inch (ppi), which was less than an optimal scan rate for latent work. This minutiae matching approach was based on special-purpose, high-speed hardware that used an array of logical circuits. The French also were interested in resolving the problem of poor fingerprint image quality. In order to acquire a high-contrast image that would be easy to photograph and process, a technique was devel¬oped to record live fingerprint images photographically using a principle of "frustrated total internal reflection" (FTIR). Although not put into large-scale production at that time, 20 years later FTIR became the cornerstone for the development of the modern-day livescan fingerprint scan¬ners. These are making the use of ink and cards obsolete for nonforensic identification purposes today.

By the early 1970s, the personnel responsible for develop¬ment of France's fingerprint automation technology had changed. As a result, there was little interest in pursuing automated fingerprint identification research for the next several years. In the late 1970s, a computer engineering subsidiary of France's largest financial institution responded to a request by the French Ministry of Interior to work on automated fingerprint processing for the French National Police. Later, this company joined with the Morphologic Mathematics Laboratory at the Paris School of Mines to form a subsidiary called Morpho Systems that went on to develop a functioning. Currently, Morpho Systems is part of Sagem (also known as Group SAFRAN).

United Kingdom AFIS Initiative

During the same period of time, the United Kingdom's Home Office was doing research into automatic fingerprint identification. Two of the main individuals responsible for the United Kingdom's AFIS were Dr. Barry Blain and Ken Millard. (Papers produced by Millard are listed in the Addi¬tional Information section of this chapter). Like the French, their main focus was latent print work. By 1974, research was being done in-house with contractor assistance from Ferranti, Ltd. The Home Office developed a reader to detect minutiae, record position and orientation, and determine ridge counts to the five nearest neighbors to the right of each minutia. This was the first use of ridge count information by an AFIS vendor (Moore, 1991).

Japanese AFIS Initiative

Like France and the United Kingdom, Japan's motivation for a fingerprint identification system was directed toward the matching of latent images against a master file of rolled fingerprints. Japan's researchers believed that an accurate latent system would naturally lead to the development of an accurate tenprint system.

By 1966, the Osaka Prefecture Police department housed almost 4 million single fingerprints. An early automation effort by this agency was the development of a pattern classification matching system based on a 17- to 20-digit number encoded manually (Kiji, 2002, p 9). Although this approach improved the efficiency of the totally manual method enormously, it had inherent problems. It required a great deal of human precision and time to classify the latents and single fingerprints; was not fully suitable for latent matching; and produced a long list of candidates, resulting in expensive verifications.

Within a few years, the fingerprint automation focus of Japanese researchers had changed. By 1969, the Iden¬tification Section of the Criminal Investigation Bureau, National Police Agency of Japan (NPA), approached NEC to develop a system for the computerization of fingerprint identification. NEC determined that it could build an auto¬mated fingerprint identification system employing a similar minutiae-based approach to that being used in the FBI system under development. At that time, it was thought that a fully automated system for searching fingerprints would not be realized for 5 to 10 years. In 1969, NEC and NPA representatives visited the FBI and began to learn about the current state of the art for the FBI's AFIS plans. During the same period, NPA representatives also col¬laborated with Moore and Wegstein from NIST. Additional AFIS sites were visited where information was acquired regarding useful and worthless approaches that had been attempted. All of this information was evaluated and used in the development of the NEC system.

For the next 10 years, NEC worked to develop its AFIS. In addition to minutiae location and orientation, this sys¬tem also incorporated ridge-count information present in the local four surrounding quadrants of each minutiae under consideration for pairing. By 1982, NEC had suc¬cessfully installed its system in the NPA and started the card conversion process. Within a year, latent inquiry searches began. In 1980, NEC received a U.S. patent for automatic minutiae detection. It began marketing its automated fingerprint iden¬tification systems to the United States a few years later.

The Politicization of Fingerprints and the San Francisco Experiment

Early development and implementation of automated fingerprint systems was limited to national police agencies in Europe, North America, and Japan. But the problems associated with huge national databases and the newborn status of computer technology in the 1970s limited the utility of these systems. Government investment in AFIS was justified largely on the promise of efficiency in the processing of incoming tenprint records. But funding these expensive systems on the local level would demand some creativity (Wayman, 2004, pp 50–52).

Following the success of the FBI's Finder, Rockwell took its system to market in the mid-1970s. Rockwell organized a users group for its Printrak system and sponsored an annual conference for customers and would-be custom¬ers. Starting with a beta-site in San Jose, California, more than a dozen installations were completed in quick succes¬sion. Peggy James of the Houston Police Department, Joe Corcoran from Saint Paul, Donna Jewett from San Jose, and others devoted their energies toward educating the international fingerprint community on the miracle of the minutiae-based Printrak system.

Each system that came online trumpeted the solution of otherwise unsolvable crimes and the identity of arrested criminals. A users group newsletter was published and distributed that highlighted some of the best cases and listed the search statistics of member agencies. Ken Moses of the San Francisco Police Department had at¬tended several of those Printrak conferences and became a staunch crusader for fingerprint automation. In three successive years, he persuaded the Chief of Police to include a Printrak system in the city budget, but each time it was vetoed by the mayor. After the third mayoral veto, a ballot proposition was organized by other politicians. The proposition asked citizens to vote on whether they wanted an automated fingerprint system. In 1982, Proposition E passed with an 80% plurality.

The mayor refused to approve a sole-source purchase from Rockwell, even though it was the only system in the world being marketed. She insisted on a competitive bid with strict evaluation criteria and testing. While on a trade mission to Japan, the mayor learned that the Japanese National Police were working with NEC to install a finger¬print system, but NEC stated that the system was being developed as a public service and the company had no plans to market it. After meeting with key Japanese of¬ficials, NEC changed its mind and agreed to bid on the San Francisco AFIS. When the bids were opened, not only had Printrak and NEC submitted proposals, but a dark horse named Logica had also entered the fray. Logica had been working with the British Home Office to develop a system for New Scotland Yard.

San Francisco retained systems consultant Tim Ruggles to assist in constructing the first head-to-head benchmark tests of competing in-use fingerprint systems. The test was most heavily weighted toward latent print accuracy, and a set of 50 latent prints graded from poor to good from actual past cases was searched against a prescribed ten¬print database. All tests were conducted at the respective vendor's home site.2 NEC was awarded the contract and installation was completed in December 1983.

Besides being the first competitive bid on 1980s technol¬ogy, what differentiated the San Francisco system from those that had gone before was organizational design. AFIS was viewed as a true system encompassing all as¬pects of friction ridge identification—from the crime scene to the courtroom. The AFIS budget included laboratory and crime scene equipment, training in all phases of forensic evidence, and even the purchase of vehicles. In 1983, a new crime scene unit was organized specifically with the new system as its centerpiece. Significant organizational changes were put into effect:

1. All latents that met minimum criteria would be searched in AFIS.
2. A new unit called Crime Scene Investigations was created and staffed on a 24/7 schedule.
3. Department policies were changed to mandate that patrol officers notify crime scene investigators of all felonies with a potential for latent prints.

4. All crime scene investigators who processed the crime scenes were trained in the use of the system and en¬couraged to search their own cases.
5. Performance statistics were kept from the beginning, and AFIS cases were tracked through the criminal jus¬tice system to the courts.

The result of the San Francisco experiment was a dramatic 10-fold increase in latent print identifications in 1984. The district attorney demanded and got five new positions to prosecute the AFIS cases. The conviction rate in AFIS-generated burglary cases was three times higher than in burglary cases without this type of evidence (Figure 6–1; Bruton, 1989).

AFIS FUNCTIONS AND CAPABILITIES

Identification bureaus are legislatively mandated to main¬tain criminal history records. Historically, this meant huge file storage requirements and cadres of clerks to maintain and search them. Demographic-based criminal history computers were established well ahead of AFIS, first as IBM card sort systems and then as all-digital information systems with terminals throughout the state and, via the National Crime Information Center (NCIC) network and the National Law Enforcement Teletype System (Nlets), throughout the nation. These automated criminal his¬tory systems became even more labor-intensive than the paper record systems they supposedly replaced. In many systems, more paper was generated and placed into the history jackets along with the fingerprint cards, mug shots, warrants, and other required documents.

AFIS revolutionized state identification bureaus because it removed from the paper files the last document type that could not previously be digitized—the fingerprint card. State identification bureaus could now bring to their legislatures cost–benefit analyses that easily justified the purchase of an automated fingerprint system through the reduction of clerical personnel.

Local and county jurisdictions did not usually enjoy the eco¬nomic benefits of state systems. Pre-AFIS personnel levels were often lower and controlled more by the demands of the booking process than by file maintenance. AFIS generally increased staffing demands on the latent and crime-scene-processing side because it made crime scene processing dramatically more productive. Local and county AFIS purchases were usually justified on the basis of their crime-solving potential.

Technical Functions. Law enforcement AFISs are composed of two interdependent subsystems: the tenprint (i.e., criminal identification) subsystem and the latent (i.e., criminal investigation) subsystem. Each subsystem oper¬ates with a considerable amount of autonomy, and both are vital to public safety. The tenprint subsystem is tasked with identifying sets of inked or livescan fingerprints incident to an arrest or citation or as part of an application process to determine whether a person has an existing record.

In many systems, identification personnel are also charged with maintaining the integrity of the fingerprint and criminal history databases.

Identification bureau staffs are generally composed of fingerprint technicians and supporting clerical personnel.

An automated tenprint inquiry normally requires a minu¬tiae search of only the thumbs or index fingers. Submitted fingerprints commonly have sufficient clarity and detail to make searching of more than two fingers unnecessary. Today's AFIS can often return a search of a million records in under a minute. As databases have expanded across the world, some AFIS engineers have expanded to searching four fingers or more in an effort to increase accuracy. The latent print or criminal identification subsystem is tasked with solving crimes though the identification of latent prints developed from crime scenes and physical evidence. Terminals used within the latent subsystem are often specialized to accommodate the capture and digital enhancement of individual latent prints. The latent subsys¬tem may be staffed by latent print examiners, crime scene investigators, or laboratory or clerical personnel. The staff of the latent subsystem is frequently under a different command structure than the tenprint subsystem and is often associated with the crime laboratory.

The search of a latent print is more tedious and time-consuming than a tenprint search. Latent prints are often fragmentary and of poor image quality. Minutiae features are normally reviewed one-by-one before the search be¬gins. Depending on the portion of the database selected to be searched and the system's search load, the response may take from a few minutes to several hours to return.

Most law enforcement AFIS installations have the ability to perform the following functions:

- Search a set of known fingerprints (tenprints) against an existing tenprint database (TP–TP) and return with results that are better than 99% accurate.3
- Search a latent print from a crime scene or evidence against a tenprint database (LP–TP).
- Search a latent from a crime scene against latents on file from other crime scenes (LP–LP).
- Search a new tenprint addition to the database against all unsolved latent prints in file (TP–LP).

Enhancements have been developed to allow other func¬tions that expand AFIS capabilities, including:

- Addition of palmprint records to the database to allow the search of latent palmprints from crime scenes.
- Interfacing of AFIS with other criminal justice informa¬tion systems for added efficiency and "lights out"4 operation.
- Interfacing of AFIS with digital mug shot systems and livescan fingerprint capture devices.
- Addition of hand-held portable devices for use in identity queries from the field. The query is initiated by scanning one or more of

the subject's fingers, extracting the mi¬nutiae within the device, and transmitting to AFIS, which then returns a hit or no-hit (red light, green light) result. Hit notification may be accompanied by the thumbnail image of the subject's mug shot.

- Multimodal identification systems, including fingerprint, palmprint, iris, and facial recognition, are now available.

System Accuracy

Most dedicated government computer systems are based on demographic data such as name, address, date of birth, and other information derived from letters and numbers. For example, to search for a record within the motor vehicle database, one would enter a license number or operator data. The success of the search will be dependent on the accuracy with which the letters and numbers were originally perceived and entered. The inquiry is straightforward and highly accurate at finding the desired record.

Automated fingerprint systems are based on data extract¬ed from images. Although there is only one correct spelling for a name in a motor vehicle database, a fingerprint image can be scanned in an almost infinite number of ways. Success in searching fingerprints depends on the clarity of the images and the degree of correspondence between the search print and the database print (compression and algorithms are two other factors that can affect accuracy). In the case of searching a new tenprint card against the tenprint database, there is usually more than enough im¬age information present to find its mate 99.9% of the time in systems with operators on hand to check respondent lists (rather than true "lights out" operations).

A latent print usually consists of a fragmentary portion of a single finger or piece of palm, though the quality of some latent impressions can exceed their corresponding images of record. The amount of information present in the image is usually of lesser quality and often is contaminated with background interference. Entering latents into the com¬puter has a subjective element that is based on the experi¬ence of the operator. Based on latent print acceptance test requirements commonly found in AFIS proposals and contracts, the chances of a latent print finding its mate in the database is about 70 to 80%. Naturally, the better the latent image, the higher the chances of success. Inversely, the chance of missing an identification, even when the mate is in the database, is 25%. Especially in latent print searches, failure to produce an identification or a hit does not mean the subject is not in the database. Other factors beyond the knowledge and control of the operator, such as poor-quality database prints, will adversely affect the chances of a match.

Because of the variability of the images and the subjectiv¬ity of the terminal's operator, success is often improved by conducting multiple searches while varying the image, changing operators, or searching other systems that may contain different copies of the subject's prints. It is com¬mon that success comes only on multiple attempts.

Fingerprint Standards

Law enforcement agencies around the world have had standards for the local exchange of inked fingerprints for decades. In 1995, Interpol held a meeting to address the transfer of ink-and-paper fingerprint cards (also known as forms) between countries. The local standards naturally had different text fields, had different layouts of text fields, were in different languages, and were on many different sizes of paper. Before that effort could lead to an interna¬tionally accepted fingerprint form, Interpol moved to the electronic exchange of fingerprints. In the ink-and-paper era, the standards included fiber con¬tent and thickness of the paper, durability of the ink, size of the "finger boxes", and so forth. With the move in the early 1990s toward near real-time responses to criminal finger¬print submittals, there came a new set of standards.

The only way to submit, search, and determine the status of fingerprints in a few hours from a remote site is through electronic submittal and electronic responses. The source can still be ink-and-paper, but the images need to be digi¬tized and submitted electronically to address the growing demand for rapid turnaround of fingerprint transactions. The FBI was the first agency to move to large-scale electronic submission of fingerprints from remote sites. As part of the development of IAFIS, the FBI worked very closely with NIST to develop appropriate standards for the electronic transmission of fingerprint images.

Starting in 1991, NIST held a series of workshops with forensic experts, fingerprint repository managers, industry representatives, and consultants to develop a standard, under the ANSI guidelines, for the exchange of fingerprint images. It was approved in November 1993, and the formal title was "Data Format for the Interchange of Fingerprint Information (ANSI NIST CSL 1 1993)". This standard was based on the 1986 ANSI/National Bureau of Standards minutiae-based standard and ANSI/NBS ICST 1 1986, a standard that did not address image files. This 1993 NIST standard (and the later revisions) became known in the fingerprint technology world simply as the "ANSI/NIST standard". If implemented correctly (i.e., in full compliance with the standard and the FBI's implementa¬tion), it would permit fingerprints collected on a compliant livescan from any vendor to be read by any other compliant AFIS and the FBI's yet-to-be-built (at that time) IAFIS.

The standard was deliberately open to permit communities of users (also known as domains of interest) to customize it to meet their needs. Some of the customizable areas were image density (8-bit gray scale or binary) and text fields associated with a transaction (e.g., name, crime). The idea was that different communities of users would write their own implementation plans. The mandatory parts of the ANSI/NIST standard were the definitions of the record types, the binary formats for fingerprint and signature images and, within certain record types, the definition of "header" fields such as image compression type. Record Types. For a transaction to be considered ANSI/NIST compliant, the data must be sent in a structured fashion with a series of records that align with ANSI/NIST record types as implemented in a specific user domain (e.g., Interpol).

- All transmissions (also known as transactions) have to start with a type 1 record that is basically a table of contents for the transmission, the transaction type field (e.g., CAR for "criminal tenprint submission—answer required"), and the identity of both the sending and receiving agencies.
- Type 2 records can contain user-defined information associated with the subject of the fingerprint transmis¬sion (such as name, date of birth, etc.) and the purpose of the transaction (arrest cycle, applicant background check, etc.). These fields are defined in the domain-of-interest implementation standard (e.g., the FBI's EFTS). Note that type 2 records are also used for responses from AFISs. They fall into two sets: error messages and search results. The actual use is defined in the domain specification.
- Types 3 (low-resolution gray scale), 4 (high-resolution gray scale), 5 (low-resolution binary), and 6 (high-resolution binary) were set up for the transmission of fingerprint images at different standards (500 ppi for high resolution and 256 ppi for low resolution) and im¬age density (8 bits per pixel for grayscale) or binary (1 bit per pixel for black and white). Note that all images for records type 3 through 6 are to be acquired at a minimum of 500 ppi; however, low-resolution images are down-sampled to 256 ppi for transmission. There are few, if any, ANSI/NIST implementations that support type 3, 5, or 6 images (see explanation below). None of these three record types are recommended for use by latent examiners and fingerprint technicians.
- Type 7 was established for user-defined images (e.g., latent images, faces) and, until the update of the ANSI/NIST standard in 2000, it was the record type for exchanging latent images. This record type can be used to send scanned copies of identity documents, and so forth. Again, the domain specification determines the legitimate uses of the type 7 record.
- Type 8 was defined for signatures (of the subject or per¬son taking the fingerprints), and it is not used in many domains.
- Type 9 was defined for a minimal set of minutiae that could be sent to any AFIS that was ANSI/NIST-compliant.

The first such implementation plan was the FBI's EFTS issued in 1994. The EFTS limited what record types, of the nine defined in the ANSI/NIST standard, the FBI would use, and defined the type 2 data fields. The key decision the FBI made was that it would only accept 500-ppi gray-scale im¬ages or, in ANSI/NIST parlance, type 4 images. As a result of that decision, all law enforcement systems since then have specified type 4 images and do not accept types 3, 5, or 6, which as a result have fallen into disuse for these applications in the United States. The type 4 records start out with header information in front of the image. The headers tell the computer which finger the image is from, whether it is from a livescan or an inked card, the image size in the number of pixels of width and height, and whether the image is from a rolled impression or a flat or plain impression.

Image Quality. Both the ANSI/NIST standard and the EFTS lacked any metrics or standards for image qual¬ity. The FBI then appended the EFTS with an image quality standard (IQS) known as Appendix F. (Later, a reduced set of image quality specifications were added as Appendix G because the industry was not uniformly ready to meet Appendix F standards.) The IQS defines minimal accept¬able standards for the equipment used to capture the fingerprints. There are six engineering terms specified in the IQS. They are:

1. Geometric image accuracy—the ability of the scanner to keep relative distances between points on an object (e.g., two minutiae) the same relative distances apart in the output image.
2. Modulation transfer function (MTF)—the ability of the scanning device to capture both low-frequency (ridges themselves) and high-frequency (ridge edge details) information in a fingerprint at minimum standards.
3. Signal-to-noise ratio—the ability of the scanning device to digitize the information without introducing too much electronic noise (that is, with the pure white image parts appearing pure white and the totally black image parts appearing totally black).
4. Gray-scale range of image data—avoiding excessively low-contrast images by ensuring that the image data are spread across a minimal number of shades of gray.
5. Gray-scale linearity—as the level of gray changes in a fingerprint capture, the digital image reflects a corre¬sponding ratio of gray level across all shades of gray.
6. Output gray-level uniformity—the ability of the scan¬ning device to create an image with a continuous gray scale across an area on the input image (tested using a special test image) that has a single gray level.

Interestingly, only two of these six image quality stan¬dards apply to latent scanning devices: geometric image accuracy and MTF. In fact, the FBI does not certify (see below for a discussion of certified products) scanners for latent use but recommends that latent examiners purchase equipment they are comfortable with using from an image-quality perspective. But EFTS Appendix F does mandate that latent images be captured at 1,000 ppi.

There are no standards for the quality of the actual finger¬print, but livescan and AFIS vendors have rated fingerprint quality for years. They know that fingerprint quality is possibly the strongest factor in the reliability of an AFIS's successfully matching a fingerprint to one in the repository. These ratings are often factored into the AFIS algorithms. In a paper titled "The Role of Data Quality in Biometric Systems" (Hicklin and Khanna, 2006), the authors wrote the following:

Note that this definition of data quality goes be¬yond most discussions of biometric quality, which focus on the concept of sample quality. Sample quality deals with the capture fidelity of the subject's physical characteristics and the intrinsic data content of those characteristics. However, an equally important issue for any operational system is metadata quality: databases need to be con-cerned with erroneous relationships between data elements, which generally come from administra¬tive rather than biometric-specific causes.

Although no standard exists for fingerprint image quality, NIST has researched the relationship between calculated image quality (using algorithms similar to those employed by AFIS vendors) and successful match rates in automated fingerprint identification systems. This led NIST to develop and publish a software utility to measure fingerprint image quality.

The software is entitled NIST Fingerprint Image Software 2. It was developed by NIST's image group for the FBI and the U.S. Department of Homeland Security and is available free to U.S. law enforcement agencies as well as to bio¬metrics manufacturers and researchers. The CD contains source code for 56 utilities and a user's guide.

The following summary is from the NIST Web site in 2007: New to this release is a tool that evaluates the quality of a fingerprint scan at the time it is made. Problems such as dry skin, the size of the fingers and the quality and condition of the equip¬ment used can affect the quality of a print and its ability to be matched with other prints. The tool rates each scan on a scale from 1 for a high-quality print to 5 for an unusable one. "Although most commercial fingerprint systems already include proprietary image quality software, the NIST soft¬ware will for the first time allow users to directly compare fingerprint image quality from scanners made by different manufacturers," the agency said.

Certified Products List. To assist the forensic community to purchase IQS-compliant equipment, the FBI established a certification program. The vendors can self-test their equipment and submit the results to the FBI where, with the technical assistance of Mitretek, the results are evaluated. If the results are acceptable, a letter of certification is sent to the vendor. It is important to know that, for capture devices, it is a combination of the optics (scanner), image processing software, and the operating system that is tested. Therefore, letters of certification are not issued for a scanner but for a scanner and PC configu¬ration that includes a specific scanner model, connected to a PC running a specific operating system, and any image-enhancement scanner drivers used. At the rate at which manufacturers upgrade scanners, it can be hard to purchase previously certified pieces of equipment. A complete list of all certified equipment is maintained on the FBI's Web site under the CJIS section.

Compression. About the same time as the writing of the EFTS, the FBI decided on the compression stan¬dard for ANSI/NIST transmissions. Given that the data rate (bandwidth) of telecommunications systems was very low in 1993 compared to today's rates and that the cost of disk storage was quite high, the FBI elected to compress fingerprint images using a technique called wavelet scalar quantization (WSQ).

The initial plan was for tenprint transmissions to be com¬pressed with WSQ at 20:1 and for latent images to remain uncompressed. An FBI fingerprint card in the early 1990s had a surface area for fingerprints that was 8 inches wide and 5 inches high for a total of 40 square inches. Scanning at 500 ppi in both the 8-inch direction (X) and the 5-inch di¬rection (Y) yielded a total of 10 million bytes of information (10 MB). Compression at 20:1 would produce a half (0.5) MB file that was much easier to transmit and store.

At the 1993 IAI Annual Training Conference in Orlando, Fl, the IAI Board of Directors expressed its concerns to the IAFIS program director about the proposed compression rate of 20:1. The FBI agreed to support an independent as¬sessment of the impact of compression on the science of fingerprint identification by the IAI AFIS committee, under the Chairmanship of Mike Fitzpatrick of Illinois (IAI AFIS Committee, 1994). As a result of the study, the FBI agreed to reduce the average compression to 15:1 (Higgins, 1995, pp 409–418).5

As other domains of interest adopted the ANSI/NIST stan¬dard around the world (early adopters included the Royal Canadian Mounted Police and the United Kingdom Home Office), they all used the EFTS as a model and all incor¬porated the IQS standard by reference. With one or two exceptions, they also adopted WSQ compression at 15:1.

With the move to higher scan rates for tenprint transac¬tions, the compression technology of choice is JPEG 2000, which is a wavelet-based compression technique. Currently (as of 2007), there are at least five 1000-ppi tenprint, image-based automated fingerprint identification systems using JPEG 2000. Both Cogent and Motorola have delivered 1000-ppi systems. It is anticipated that the other vendors will deliver such systems as the demand increases. Given that older livescan systems operating at 500 ppi can submit transactions to these new automated fingerprint identification systems, it is important that they be capable of working in a mixed-density (500-ppi and 1000-ppi) environment.

All four major AFIS vendors demonstrated the capabil¬ity to acquire, store, and process 1000-ppi tenprints and palmprints during the 2005 Royal Canadian Mounted Police AFIS Benchmark. It is important to note that these systems acquire the known tenprint and palm images at 1000 ppi for archiving but down-sample them to 500 ppi for searching and creating an image to be used in AFIS. Currently, 1000-ppi images are used primarily for display at latent examiner workstations. As automated fingerprint identification systems move to using third-level features, it is assumed that the higher resolution images will play a role in the algorithms.

THE AUTOMATIC FINGERPRINT IDENTIFICATION SYSTEM

The following sections describe the components and algorithms that make up a typical fingerprint recognition system. While individual systems will not necessarily do things the way they are described here, the basic principles are described and examples are given wherever possible. Note that this is not always possible, due to the proprietary nature of the technology and algorithms involved. Also, there are two main system methodologies for fingerprint matching, those that utilize minutiae matching, and those that use pattern matching on the overall ridge structure.

CAPTURE DEVICES

The following sections describe the function of each component of a fingerprint capture device.

Scanning

For fingerprints there are a number of alternative data capture devices (Fig 0 1). These include optical scanners, silicon based capacitance and thermal sensors, pressure based sensor and ultrasound devices. The main objective of a fingerprint scanner, regardless of the method it uses, is to provide the system with an image of the fingerprint that is as accurate as possible. For most applications, the image is produced at a resolution of 500 dpi using an 8-bit gray-scale. The following sections describe how each of the different kinds of fingerprint scanners works.

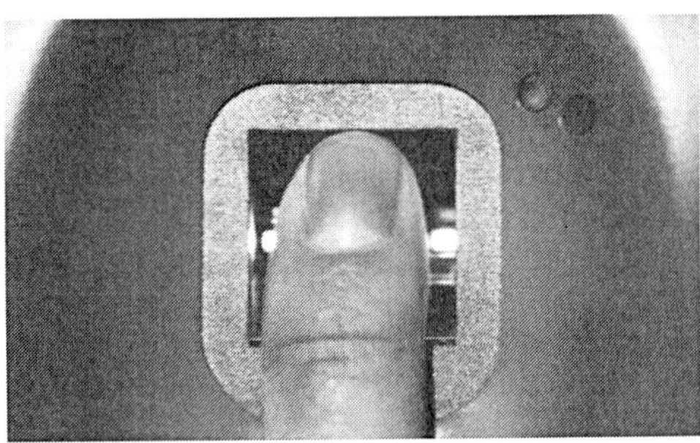

Fig. Fingerprint scanner

Optical

Optical devices are one of the more common fingerprint scanning devices. They are based on the reflection changes that occur when a light source interacts with the ridge lines of a fingerprint. This is most commonly achieved through the use of Frustrated Total Internal Reflection (FTIR). The light source shines onto a special reflection surface, which reflects the light differently depending on the pressure applied to it. A light sensor is used to capture the 'image' of the fingerprint. The general layout of a typical optical scanner. Due to the involvement of pressure, this type of scanner returns different quality images depending on the pressure applied to the reflective surface (by the fingerprint). If too little pressure is applied, the sensors may not be able to create an image at all. Alternatively, if too much pressure is applied, the skin in between the fingerprint ridges will also be in contact with the reflective surface, causing a lack of definition between ridges and valleys. Optical scanners are also affected by dirty fingerprints, which can also result in unusable images.

Due to the required placements of the light source, reflective surface and light sensors, optical scanners are typically physically large components. However some companies have developed improved methods to reduce the size of the component. An example of this is the Surface Enhanced Irregular Reflection (SEIR) technology patented by SecuGen Corporation. Another weakness of optical scanners is that they must be cleaned regularly, to avoid dust; dirt and oil build up on the reflective surface. Also, optical scanners are the most vulnerable to physical replay attacks. This is where the latent print of the previous user is used to gain access to the system. This can theoretically be achieved in some scanners simply by shining an external light at the correct angle. Optical models generally claim that they cannot be fooled by a 2-dimensional image of a fingerprint; however a simple 3D model (e.g. latex rubber or similar product) is often sufficient.

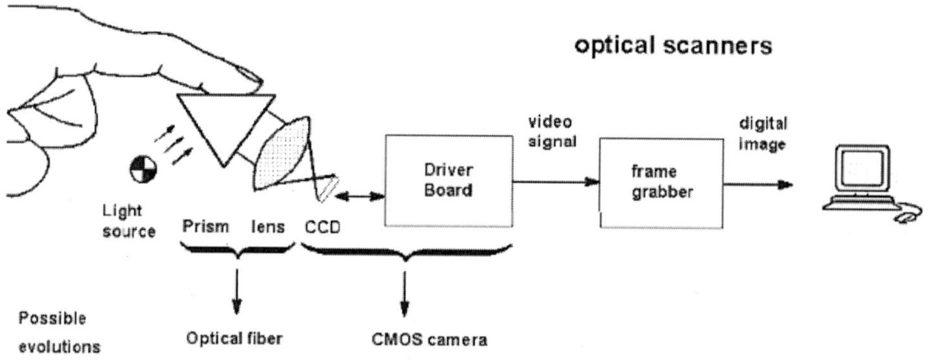

Fig. General layout of an optical fingerprint scanner, reproduced from [Atmel Corporation 2001]

Capacitance

Capacitive fingerprint scanners create images of a fingerprint through the use of rectangular arrays of capacitors. The capacitors are located under a very thin protective layer that must be thick enough to protect the capacitors, but also thin enough so it does not obscure the readings. When a finger is placed on the scanner, the capacitors measure the capacitance difference generated by the different distances between the capacitors and the ridge lines and furrows. This concept is depicted in Fig 0 3. One disadvantage of using this method is that the capacitance difference is significantly affected by moisture. Hence, fingers that are too wet, oily or dry will generate low quality or unusable images. Also, the capacitors themselves are vulnerable to Electro-Static Discharge (ESD) and external electric fields. As a result the grounding of capacitance scanners is very important to prevent damage to the capacitors. One major advantage of capacitance scanners is their reduced size. Scanners using this method have been produced small enough to fit on PCM/CIA cards, and hence are convenient for use in small portable devices such as laptop computers and mobile phones.

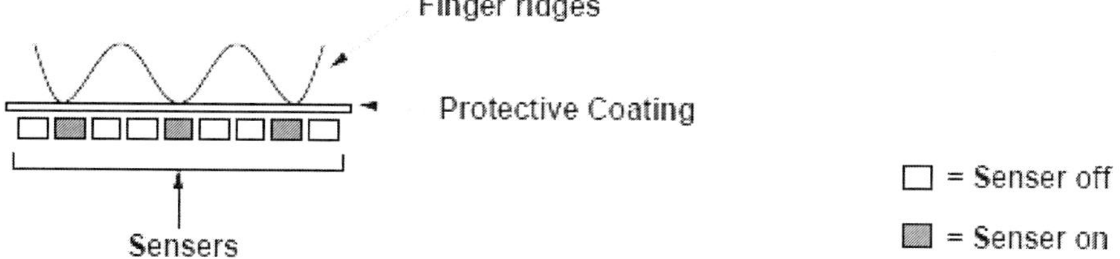

Fig. Depiction of a capacitance scanner

Thermal

Thermal fingerprint scanners work in a manner similar to capacitance scanners. However, instead of measuring the difference in temperature between the ridges and valleys of the fingerprint (which are too small to measure effectively), the fingerprint image is generated by converting the temperature differential on a sensor. For example, before a fingerprint is placed on the scanner, all the thermal sensors are at equilibrium with the air temperature around the scanner. Then when a fingerprint is placed on the sensor, those points where a ridge is in contact will cause a change in temperature in the sensor for that point. However, those points where there is a valley will still be measuring the local air temperature; hence those points have no temperature differential, and are not activated.

The major drawback of this method is that the temperature difference disappears after less than a tenth of a second as the finger and the sensors reach equilibrium. As a result, the image is only available for a short period of time. This means that a user has a very short amount of time to position his/her finger appropriately, before the image is gone. In order to counter this, Atmel Corporation, a major producer of thermal fingerprint scanners has developed and patented a sweeping technique. This technique works by sweeping a finger over the sensor array, which takes the images from different times over the sweep (in a manner that is not affected by the sweeping speed) and reconstructs the total fingerprint image from these images using a proprietary and private algorithm. This allows the scanner to be reduced to approximately 1/5 the size of a normal square array scanner and eliminates the problem of the temperature equalization, as the sweeping motion ensures that any one sensor is constantly changing between finger temperature (ridge) and air temperature (valley).

Pressure

Purely pressure-based scanners that utilize silicon chips to convert pressure to an electrical signal have been developed. Due to the natural pressure that is applied when a finger is placed on a scanner, this is probably the most intuitive method to use. However, as these scanners generally have low sensitivity, which is further lowered when a protective layer is added; the resulting images have quite low detail. As a result, no known systems utilize this method.

Ultrasound

Ultrasound scanners are the least common commercial option for fingerprint scanners. They operate by using acoustic energy that is partially reflected at each interface between different materials. The time between the reflections can be accurately measured to determine the depth at which the reflection occurred (see Fig 0 4). The advantage of using ultrasound is that the images returned are uncontaminated by any dirt or grease on the surface on which the finger rests, or on the finger itself (from [Ultra-Scan Corporation]). This, according to the manufacturer, results in higher quality images than is possible with optical systems. It also renders a physical replay attack (using a latent print) conceptually impossible as a latent print should not reflect sounds waves.

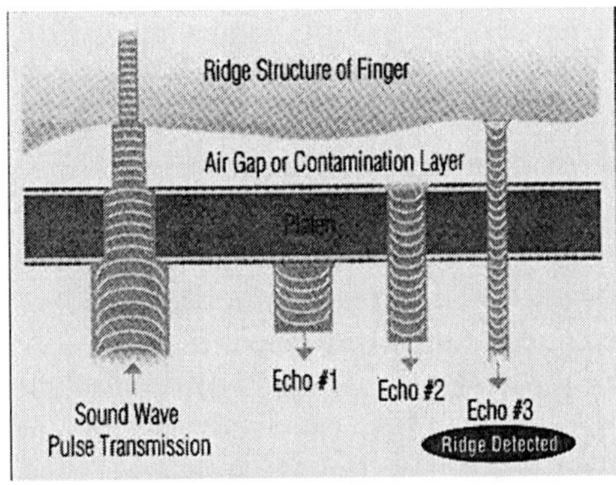

Fig. Operation of ultrasound scanner, sound waves return a partial echo at each change in material, from Ultra-Scan

However, despite these advantages, ultrasound scanners have a number of major drawbacks. Firstly, they are a large device (approximately 15 x 15 x 20 cm), preventing integration into small or portable components. In addition they are noisy (due to internal mechanical components), expensive and slow, with a single scan (at the highest quality setting) taking up to 4.60 seconds, according to [Ultra-Scan Corporation].

SCANNING A FINGERPRINT

Many of the algorithms require a linear scan of the fingerprint image. Scanning is achieved by moving a fixed size window across the picture in a grid-like pattern. This can be seen in the Fig 0 5. However, it is possible that areas of interest do not lie squarely in the one of the windows. To account for this the window is then shifted just vertically, just horizontally, and then vertically and horizontally by half the window size and the grid scan is completed again. Therefore, it takes four scans of the image to do the linear scan. This is not a problem because it is used for the preprocessing of a fingerprint image (which occurs only once) and is done in a linear manner.

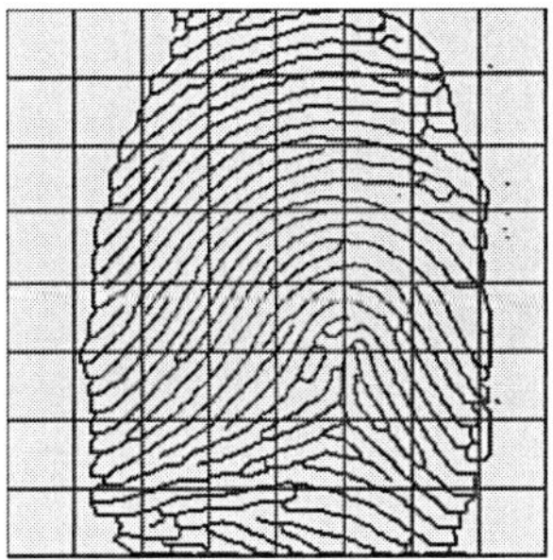

Fig. Scanned image

STEPS FOR FINGERPRINT ENHANCEMENT

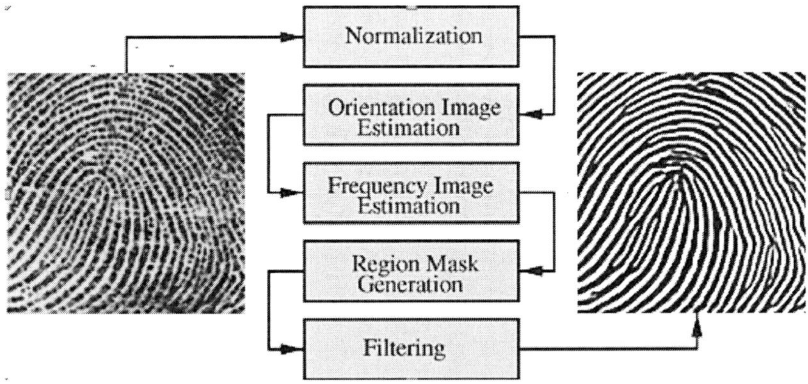

Fig. A flowchart of the proposed fingerprint enhancement algorithm

The flowchart of the fingerprint enhancement algorithm:

1. The image is first normalized to have desired mean and variance
2. It is then divided into non-overlapping blocks
3. Dominant ridge orientation is determined for each block
4. They are then smoothed and subsequently the block direction image is formed
5. Average ridge distance for the whole input image is determined
6. Directional filtering is used to enhance the image (Fig 0 6)

LRF (Local Ridge Frequency)

1. Project gray values of all the pixels located in each block along a direction orthogonal to the local orientation computed above. The projection forms 1D wave with the local extrema corresponding to the ridges and valleys.
2. L(i,j) is the average number of pixels between two consecutive peaks in 1D wave. The frequency f(i,j) is calculated as:

$$f(i,j)=1/L(i,j)$$

ALGORITHM FOR FINGERPRINT ENHANCEMENT:

1. The fast Fourier transform (FFT) of the fingerprint image is computed. Fast Fourier transform is a discrete Fourier transform algorithm which reduces the number of computations needed for N points from 2N2 to 2Nlog2N.

For directional filtering, Gabor filter is proposed:

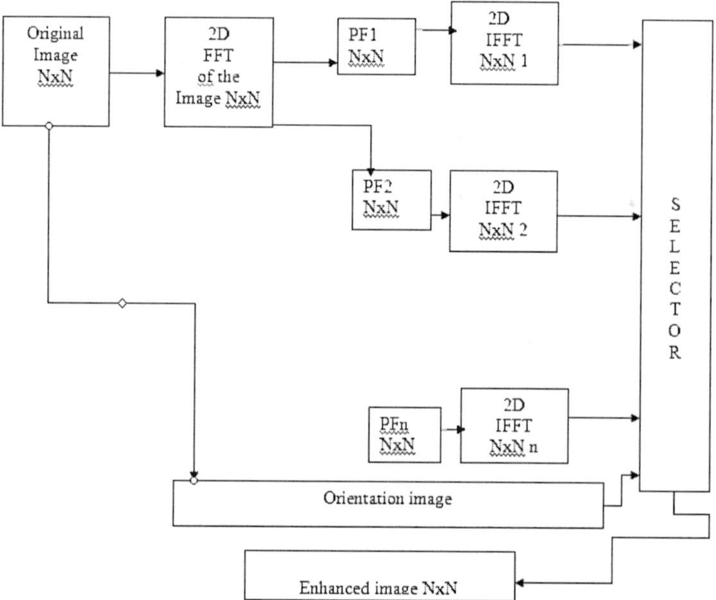

Fig. Algorithm for fingerprint enhancement

2. Fourier transform of the filter components Pi depends upon the orientation values. Each directional filter Pi is point by point multiplied by F, obtaining n filtered image transforms PFi, i=1, 2,……, n.
3. Inverse FFT is calculated for each PFi resulting in n filtered images Ii, i=1,2,…..,n (spatial domain).Inverse FFT is calculated which is exactly the reverse process of FFT
4. The filtered image is then formed by selecting, for each pixel position, the pixel value from the pre-filtered image whose direction of filtering corresponds most closely to the actual ridge orientation at that position.

The sinusoidal shaped waves of ridges and valleys vary slowly in a local constant orientation. Therefore a bandpass filter tuned to the corresponding frequency and orientation can effectively remove the noise and preserve the true ridge and valley structures.

7

Fingerprint Development Techniques

Fingerprints are made up of tiny droplets ranging in size from 1-20μm, which consist mainly of sweat and approximately one millionth of a gram of chemical material in total. A 'hidden' fingerprint, for example, one left at the scene of a crime, is called a 'latent' fingerprint ('latent' is the Latin word for 'hidden'). One way of making these fingerprints visible is called the 'powder and brush' technique. In this method, the surface is brushed with a very fine powder that sticks to these droplets. Some surfaces, however, absorb this powder and the fingerprints are not distinguishable. This problem can be overcome by the use of electrostatics. The article to be analysed is attached to a negatively charged electrode connected to a high voltage. A very fine powder is placed on the adjacent positive electrode, becomes charged and is attracted to the negatively charged specimen. Due to the high voltage, the particles travel quickly and stick firmly to the fingerprint. Particles hitting the electrode lose their positive charge and return to the positive electrode where they are recharged. This process continues until the fingerprint has been sufficiently built-up.

This technique is not appropriate for use on all surfaces, but there are several other methods that can be applied:

- Laser luminescence: Involves the illumination of fingerprints due to fluorescing particles picked up during everyday life from paints, inks and oil. It can be used on painted walls, metals, plastic and rubber, cloth and wood.
- Metal evaporation: The fingerprint is developed by first evaporating a thin layer of gold onto the specimen, followed by a layer of cadmium which fills in the print and provides a contrast.
- Silver nitrate: Used for fingerprints on paper, silver nitrate is sprayed onto the fingerprint where it reacts with the chlorides, to give the insoluble silver salt, silver chloride. Not suitable for fabrics or rough surfaces such as wood.
- Ninhydrin test: Indantrione hydrate reacts with the amino acids in the fingerprint, giving a visible deposit. Also not suitable for fabrics or rough surfaces.

- Iodine vapour: Can be used to develop fingerprints on fabrics and rough surfaces. Iodine vapour alone is useful only for prints up to 24 hours old, however a mixture of the vapour and steam allows this method to be effective for up to two months.
- Bacteria: Certain bacteria, for example acinetobacter calciacatieus, can be used to develop prints on valuable oil paintings, without harming the painting in the process. The bacteria in a nutrient gel are pasted onto the surface of the painting, making the print visible as they multiply. The gel can then simply be wiped off, leaving the painting unaffected.
- Autoradiography: Radioactive atoms are incorporated into the fingerprint by placing the piece of fabric into a container containing radioactive gases, such as iodine or sulphur dioxide, at a humidity of less than 50%. The fabric is then put into contact with photographic film, and the radioactive atoms cause a picture to become clear.

After the prints have been developed they can then be entered into a computer, which allows them to be quickly and easily recalled and compared to the fingerprint of a suspect. It should be noted that this method is as useful for proving the innocence of a suspect as it is for convicting criminals.

SENSITIVE FINGERPRINT TECHNIQUE DEVELOPED WITH THE HELP OF NEUTRONS

Fig. Bright hills and dark valleys

Researchers in the UK and France have developed a new and extremely sensitive method for visualizing fingerprints left on metal surfaces such as guns, knives and bullet casings. The technique utilizes colour-changing fluorescent films and the team says that it can be used to complement existing forensic processes. The chance that two people will have identical fingerprints is about 64 billion to 1, which is why law-enforcement agencies rely on fingerprint evidence. Despite advances in detection since the 19th century, only about 10% of crime-scene fingerprint images are of sufficient quality to lead to the unambiguous identification of an individual that is good enough to satisfy a court.

Fingerprints are essentially deposits of sweat and natural oils. Traditional visualization techniques involve applying a coloured powder, chemical or biological reagent that adheres to or interacts with the residue and creates a visual contrast to the underlying surface. A major limitation of the technique is that these deposits can degrade with time or exposure to water or other materials.

BARE METAL

Instead of focusing on the residue itself, Robert Hillman and colleagues at the University of Leicester, the Institut Laue-Langevin (ILL) and ISIS at the Rutherford Appleton Laboratory have decided to work with the bare surface between the ridges of a fingerprint. "Think of the deposits on the surface to be like little 'hills', we've decided to go for the bare metal at the bottom of the 'valleys'," explains Hillman.

Recently, the team has been experimenting with an electrochromic polymer that changes colour when an electrical voltage is applied. "We used electrochemistry to deposit a polymer from a monomer solution and subsequently we replaced the monomer deposition solution with a background electrolyte," explains Hillman. The invisible residue left by a human finger does not conduct electricity, so it acts like a stencil. When the polymer is deposited on a fingerprint and the voltage is applied, the sticky deposit blocks the current, directing the film to the "valleys" in between the "hills". The voltage changes the film's colour, optimizing the visual contrast and essentially creating a negative image of the print.

Now, the researchers have taken their technique one step further by adding fluorescent molecules (fluorophores), which cause the film to emit light of a certain colour when exposed to ultraviolet light. This approach broadens the colour palette of the polymer films, says Hillman, and also allows more control in terms of tuning the colouration to get the best possible contrast with the underlying metal surface.

NEUTRON REFLECTIVITY

For the technique to work, the fluorophores must completely permeate the film without reaching the underlying metal surface, where their fluorescence deteriorates. To ensure that this happens, the researchers at the ILL and ISIS used a technique called neutron reflectively. This involves firing a neutron beam at the film and measuring the reflected neutrons. Neutron scattering can be sensitive to the specific isotopes present in the sample. To take advantage of this, selected parts of the system were labelled using the hydrogen isotope deuterium and the measurements were used to determine the ideal conditions for the introduction of the fluorophores.

The researchers say that the fingerprinting method is extremely sensitive and only tiny amounts of the residue are required to make it work – much less than is typical for conventional approaches. It is also well suited for use in combination with existing approaches, which often involve using a succession of reagents to try to reveal a print. If conventional reagents fail to reveal the pattern, then the bare surface regions should still be free for polymer deposition, according to Hillman.

Despite the advantages, the new technique is limited because it only works on metal samples, says Paul Kelly of the UK's Loughborough University, who was not involved in the study. "It is certainly not a catch-all solution to fingerprint issues," he says. But he adds that "the further enhancements discussed in this latest phase of the work certainly bode well for maximizing the prowess of the technique with respect to items such as knives and cartridge casings". Kelly says that while scientists can develop new fingerprinting techniques, "in the end, though, it's down to forensic practitioners to assess their utility and applicability".

AMIDO BLACK

Amido black 10B is an amino acid staining diazo dye used in biochemical research to stain for total protein on transferred membrane blots. It is also used in criminal investigations to detect blood present with latent fingerprints. It stains the proteins in blood a blue-black color. Amido Black can be either methanol or water based as it readily dissolves in both. With picric acid, in a van Gieson procedure, it can be used to stain collagen and reticulin.

Application : Blood-contaminated latent fingerprints. Amido Black stains the protein in blood a bluc/ black color. A water-based formula is shown for substrates that are affected by methyl alcohol.

Formula : Methyl Alcohol Working solution

- 2.0 grams amido black 10B
- 100 mL glacial acetic acid
- 900 mL methyl alcohol
  - Place 2 grams of amido black into a clean, dry 2 L glass beaker.
  - Measure out 100 mL glacial acetic acid and add to amido black.
  - Measure out 900 mL of methyl alcohol and add to amido black solution.
  - Stir for 30 minutes to produce a blue-black working solution.
  - Store in a clean, dry, brown-glass bottle. Seal and label.

De-staining solution

- 100 mL glacial acetic acid
- 900 mL methyl alcohol
  - Measure out 100 mL of glacial acetic acid. Pour into a clean, dry 2 L glass beaker.
  - Measure out 900 mL of methyl alcohol and add to the glacial acetic acid.
  - Stir until solution is clear.

- Store the de-staining solution in a clean, dry brown-glass bottle. Seal and label.

Water-Based Working Solution

- 2.0 grams amido black 10B
- 20.0 grams citric acid
- 1 L distilled water
  - Place 2 grams of amido black into a clean, dry 2 L glass beaker.
  - Weigh out 20.0 grams citric acid and add to amido black.
  - Measure out 1 L of distilled water and add to amido black mixture.
  - Stir for 30 minutes to produce a blue-black working solution.
  - Store in a clean, dry, brown-glass bottle. Seal and label.

Fixing Solution

- 20.0 grams 5-sulphosalicylic acid
- 1 L distilled water
  - Place 20.0 grams of 5-sulphosalicylic acid in clean, dry 2 L glass beaker.
  - Measure out 1 L distilled water and add to sulphosalicylic acid.
  - Stir until solution is clear.
  - Store in a clean, dry glass bottle. Seal and Label.

Development :

Methyl Alcohol-Based Working Solution

- Immerse exhibit in methyl alcohol pre-wash for 5 minutes to fix blood.
- Immerse exhibit in working solution for 3 minutes.
- Immerse exhibit in the de-staining solution for a few minutes until excess dye has been removed.
- Repeat de-staining step in clean solution if necessary.
- Allow to air dry before examination.

Water-Based Working Solution

- Immerse exhibit in 5-sulphosalicylic fixing solution for 5 minutes.
- Immerse exhibit in working solution for 3 minutes.
- Immerse the exhibit in the distilled water for a few minutes until excess dye has been removed.
- Repeat de-staining step in clean water if necessary.
- Allow to air dry before examination.

Equipment :

- glass measuring cylinders; 100 mL, 1 L
- glass beaker 2 L
- brown-glass bottle 2 L (x2)
- Magnetic stirrer plate and stirrer bars

Chemicals :

- Acetic Acid, Glacial: A490-212 (Fischer Scientific); B27013-78 (VWR Canlab).
- Amido Black 10B: BP124-10 (Fischer Scientific); MO1167-30 (VWR Canlab).
- Methyl Alcohol: A412-4 (Fischer Scientific); ACS531-82 (VWR Canlab).
- 5-Sulphosalicylic Acid: S740-8 (Aldrich); B30314-34 (VWR Canlab); A297-500 (Fischer).

GENTIAN VIOLET

Crystal violet or Gentian violet (also known as Methyl Violet 10B, hexamethyl pararosaniline chloride, or pyoctanin(e)) is atriarylmethane dye. The dye is used as a histological stain and in Gram's method of classifying bacteria. Crystal violet hasantibacterial, antifungal, and anthelmintic properties and was formerly important as a topical antiseptic. The medical use of the dye has been largely superseded by more modern drugs, although it is still listed by the World Health Organization.

The name "gentian violet" was originally used for a mixture of methyl pararosaniline dyes (methyl violet) but is now often considered a synonym for crystal violet. The name refers to its colour, being like that of the petals of a gentian flower; it is not made from gentians or from violets.

Application : Grease contaminated surfaces. Gentian violet stains the lipids a purple color.

Formula :

- Stock solution 5.0 grams gentian violet (SUSPECTED CARCINOGEN)
- 10 grams phenol (TOXIC)
- 50 mL absolute ethyl alcohol
  - Place 5.0 grams of gentian violet into a 100 mL clean, dry glass beaker.
  - Add 10 .0 grams of phenol to gentian violet.
  - Measure out 50 mL of ethyl alcohol and add to gentian violet and phenol.
  - Stir until dissolved and store in a clean, dry, brown-glass bottle. Seal and label.

Working solution 1 mL of stock solution

- Distilled water until gold film disappears (approx. 30 mL)
  - Syringe 1 mL of stock solution into a clean, dry 100 mL glass beaker.
  - Add distilled water slowly to beaker constantly swirling the mixture until the gold film on the surface disappears.

Development :

- Pour sufficient solution into a clean, dry glass tray.
- For adhesive tapes; pull exhibit over the surface of the solution.
- For other exhibits immerse into solution.
- Rinse under cold running water if necessary.
- Allow to dry before examination

Equipment :

- glass measuring cylinders; 100 mL
- glass beaker 100 mL
- brown-glass bottle 100 mL
- glass syringe 1 mL
- glass tray
- magnetic stirrer plate and stirrer bars

Chemicals :

- Absolute Ethyl Alcohol: PO16EAAN (Commercial Alcohol)
- Crystal Violet: 86-099-9 (Aldrich)
- Phenol: 24,232-2 (Aldrich)

HUNGARIAN RED

Hungarian Red is a dye solution in water/acetic acid mixture that is used for staining fingerprints and footprints made in blood. Prints in blood are colored red after treatment with Hungarian red. Hungarian Red should not be used on absorbent surfaces like paper, carton materials, bed sheets, or carpet. It works very well on non-absorbent backgrounds like linoleum, glass, tiles, painted surfaces, or PVC floor covering.

Application: Blood-contaminated latent fingerprints. Hungarian Red stains the protein in blood a pink/ red color.

Formula : Fixing solution

- 2.0 grams 5-sulphosalicyclic acid
- 100 mL of distilled water

Working solution

- Commercial product use as supplied.

De-staining solution

- Add 10 mL of glacial acetic acid to 190 mL of distilled water.

Development :

- Place 2.0 grams of sulfosalicyclic acid in a 200 mL beaker.
- Measure out 100 mL distilled water and add to sulfosalicyclic acid.
- Stir until dissolved.
- Immerse exhibit in 2% sulfosalicyclic acid solution for 5 minutes.
- Immerse exhibit in Hungarian Red solution for 1 minute.
- Immerse exhibit in the de-staining solution until good contrast between blood and background is observed.
- Allow to air dry before examination

Equipment :

- glass measuring cylinders; 100 mL, 250 mL
- glass beaker 200 mL
- plastic wash bottle 250 mL (x2)
- magnetic stirrer plate and stirrer bars

Chemicals :

- Acetic Acid, Glacial: A490-212 (Fischer Scientific); B27013-78 (VWR Canlab).
- Hungarian Red: Forensi-Tech
- 5-Sulphosalicylic Acid: S740-8 (Aldrich); B30314-34 (VWR Canlab); A297-500 (Fischer).

LUMINOL

Luminol (C8H7N3O2) is a versatile chemical that exhibits chemiluminescence, with a striking blue glow, when mixed with an appropriate oxidizing agent. Luminol is a white-to-pale-yellow crystalline solid that is soluble in most polar organic solvents but is insoluble in water.

Luminol is used by forensic investigators to detect trace amounts of blood left at crime scenes, as it reacts with iron found in hemoglobin. It is used by biologists in cellular assays for the detection of copper, iron, and cyanides, as well as of specific proteins by western blot.

Luminol can be sprayed evenly across the area, and trace amounts of an activating oxidant will cause the luminol to emit a blue glow that can be seen in a darkened room. The glow lasts for about 30 seconds, but the effect can be documented by a long-exposure photograph. It is important that the spraying be evenly applied to avoid a biased impression, such as blood traces appearing to be more concentrated in areas that received more spray. The intensity of the glow does not indicate the original amount present but indicates only the distribution of trace amounts of substances left in the area.

Application : Crime scene examination of latent bloodstains and blood-contaminated imprints. An alkaline solution of Luminol will oxidize in the presence of an oxidizing agent such as hydrogen peroxide or sodium perborate and a Hematin-catalyzed peroxidase system, such as that found in blood. This oxidation reaction produces a blue chemiluminescence. False positive reactions are common.

Formula : Working solution

- 0.5 grams Luminol
- 25.0 grams sodium carbonate
- 500 mL distilled water
  - Place 0.5 grams of luminol into a clean, dry 1 L glass beaker.
  - Weigh out 25.0 grams sodium carbonate and add to luminol.
  - Measure out 500 mL of distilled water and add to luminol mixture.

- Stir until completely dissolved.
- Store in a clean, dry, brown-glass bottle. Seal and label.
- Weigh out 3.5 grams sodium perborate and immediately prior to use add to luminol solution. Stir until dissolved.
- Decant into a clean, dry pump-action plastic spray bottle.

Development :

- Darken crime scene.
- Set-up camera for photography.
- Wearing face shield and officer protection suits spray area with luminol solution.
- Immediately photograph chemiluminescence.
- Repeat spraying step if chemiluminescence fades.

Equipment :

- glass measuring cylinders; 500 mL
- glass beaker 1 L
- brown-glass bottle 1 L
- plastic pump-action spray bottle 1 L
- magnetic stirrer plate and stirrer bars

Chemicals :

- Luminol: 12,307-2 (Aldrich)
- Sodium Carbonate: 22,353-0 (Aldrich); S263-500 (Fischer Scientific) B30121-34 (VWR Canlab)
- Sodium Perborate: 24,412-0 (Aldrich); B30196-34 (VWR Canlab).

MOLYBDENUM DISULFIDE

Molybdenum disulfide is the inorganic compound with the formula MoS2. The compound is classified as a di-chalcogenide. It is a silvery black solid that occurs as the mineral molybdenite, the principal ore for molybdenum. MoS2 is relatively unreactive. It is unaffected by dilute acids and oxygen. In appearance and feel, molybdenum disulfide is similar to graphite. It is widely used as a solid lubricant because of its low friction properties and robustness. Application : Detection of latent fingerprints on non-porous exhibits that have been wet. Molybdenum disulfide adheres to fats to form a grey colored deposit.

Formula :

- Stock solution 50.0 grams molybdenum disulfide
- 500 mL water
- 7.5 mL 10% Aerosol OT solution*
  - Measure out 500 mL of water. Pour into a clean, dry 1 L glass beaker.
  - Measure out 7.5 mL of 10% Aerosol OT solution using a clean, disposable syringe or 10 mL clean glass measuring cylinder. Add to water.

- Stir with a spatula to mix detergent solution.
- Place 50 grams of molybdenum disulfide into a clean, dry 1 L glass beaker.
- Add a small amount of the detergent solution to molybdenum disulfide to form a paste.
- Add half the detergent solution to the paste whilst stirring with a spatula to create a slurry.
- Pour the slurry into a clean 1 L glass bottle.
- Pour remaining detergent solution into the glass beaker and pour into the bottle transferring remaining molybdenum disulfide in the process. Seal and label.

*NOTE: 10% Aerosol OT solution can be prepared by dissolving 10 grams of dioctyl sulfosuccinate, sodium salt in 100 mL of water.

Working solution Stock Solution

- 3.0 L water
  - Shake stock solution thoroughly and pour into a plastic 5.0 L bottle.
  - Measure out 3.0 L of water and add to stock solution.
  - Shake solution thoroughly to mix.
  - Pour into a suitable spray system such as those used for gardening (remove any internal filters prior to use).

Development :

1. Adjust spray system to give a cone-shaped jet.
2. Direct spray towards the top of the item and move downwards.
3. When an area shows fingerprint development continue to direct jet above the area until maximum contrast is obtained.
4. Excess powder can be removed by gently spraying water above the area of interest.
5. Allow to air dry before examination.
6. Developed fingerprints should be photographed and then lifted using transparent, fingerprint lifters.

Equipment :

- glass measuring cylinders; 10 mL, 500 mL, 1 L
- glass beaker 1 L (x2)
- disposable syringe 10 mL
- spatula
- glass bottle 1 L
- plastic bottle 5 L
- plastic spraying system 5 L

Chemicals :

- Molybdenum Disulfide: 23,484-2 (Aldrich)
- 10% Aerosol OT solution: SA292-4 (Fischer Scientific)
- Dioctyl Sulfosuccinate, Sodium Salt: 32,358-6 (Aldrich)

LEUCOMALACHITE GREEN

Malachite green is an organic compound that is used as a dyestuff and has emerged as a controversial agent in aquaculture. Malachite green is traditionally used as a dye for materials such as silk, leather, and paper. Although called malachite green, the compound is not related to the mineral malachite — the name just comes from the similarity of color.

Application : Blood-contaminated latent fingerprints. Leucomalachite green reacts with the protein in blood to produce a green color.

Formula : Working solution

- 0.2 grams leucomalachite green
- 0.67 gram sodium perborate
- 33 mL glacial acetic acid
- 67 mL methyl alcohol
- 300 mL Vertrel XF or HFE7100
  - Place 0.2 grams of leucomalachite green into a clean, dry 200 mL glass beaker.
  - Measure out 67 mL of methyl alcohol and add to leucomalachite green.
  - Measure out 33 mL glacial acetic acid and add to leucomalachite green.
  - Weigh out 0.67 gram of sodium perborate and add to solution.
  - Stir until dissolved. Pour solution into a 1 L beaker.
  - Measure out 300 mL of Vertrel XF or HFE7100 and add to solution.
  - Store in a clean, dry, brown-glass bottle. Seal and label.

Development :

1. Immerse the exhibit in methyl alcohol for 5 minutes.
2. Apply several drops of leucomalachite green solution to cover the blood stain.
3. Allow to air dry before examination

Equipment :

- glass measuring cylinders; 100 mL, 1 L
- glass beaker 200 mL, 1 L
- brown-glass bottle 1 L
- magnetic stirrer plate and stirrer bars

Chemicals :

- Acetic Acid, Glacial: A490-212 (Fischer Scientific); B27013-78 (VWR Canlab).

- Leucomalachite Green: 12,566-0 (Aldrich).
- Methyl Alcohol: A412-4 (Fischer Scientific); ACS531-82 (VWR Canlab).
- HFE7100: CA34210-002 (VWR Canlab).
- Sodium Perborate: 24,412-0 (Aldrich); B30196-34 (VWR Canlab).
- Identi-sol: 905-9220001 (Paragon).

NINHYDRIN

Ninhydrin (2,2-Dihydroxyindane-1,3-dione) is a chemical used to detect ammonia or primary and secondary amines. When reacting with these free amines, a deep blue or purple color known as Ruhemann's purple is produced. Ninhydrin is most commonly used to detect fingerprints, as the terminal amines of lysine residues in peptides and proteins sloughed off in fingerprints react with ninhydrin. It is a white solid which is soluble in ethanol and acetone at room temperature. Ninhydrin can be considered as the hydrate of indane-1,2,3-trione.

Application : Detection of latent fingerprints on porous exhibits. Ninhydrin reacts with amino acids to produce a purple colored product.

Formula :

Stock solution

- 25.0 grams ninhydrin
- 225 mL ethyl alcohol
- 10 mL ethyl acetate
- 25 mL glacial acetic acid

1. Place 25 grams of ninhydrin into a clean, dry 500 mL glass beaker.
2. Measure out 225 mL of ethanol and add to ninhydrin.
3. Measure out 10 mL ethyl acetate and add to ninhydrin solution.
4. Measure out 25 mL glacial acetic acid and add to solution. Stir until dissolved.

Working solution

- 52 mL stock solution
- 1 L HFE7100 or Vertrel XF
  - Measure 52 mL of stock solution and pour into a 2 L beaker.
  - Measure out 1 L of HFE7100 and add to solution.
  - Store in a clean, dry, brown-glass bottle. Seal and label.

Development :

1. Pour solution into clean, dry glass, metal or plastic tray.
2. Immerse exhibit into solution until wetted.
3. Allow exhibit to air dry before placing in an oven set at 100 Celsius with relative humidity set to 80%.
4. After 20 minutes remove exhibits from the oven.

Equipment :

- glass measuring cylinders; 50 mL, 250 mL, 1 L

- Glass beaker 200 mL, 1 L
- Brown-glass bottle 1 L
- Magnetic stirrer plate and stirrer bar

PHYSICAL DEVELOPER

Physical developer is a silver-based aqueous reagent that reacts with the components of sebaceous sweat in latent fingerprints to form a silver-gray deposit. Physical developer is superior in sensitivity to many silver nitrate products presently in use. It can be used to develop latent prints on paper and other porous materials. Normally, physical developer is used on latent prints after they've been treated with ninhydrin. Just mix the solutions and immerse documents in the mixture. Rinse and wait for prints to develop. A technical information sheet and procedure recommendations are included with each set, which includes A and B solutions.

Application : Detection of latent fingerprints on porous exhibits that have been wet. Physical Developer reacts with fats to produce a silver-grey colored deposit.

Formula :

- Acid Pre-wash 25 grams maleic acid
- 1 L distilled water
  - Measure out 1 L distilled water and pour into a clean 2 L glass beaker.
  - Weigh out 25 grams of maleic acid and add to water. Stir until dissolved.
  - Store in a clean glass bottle. Seal and label.

Detergent solution

- 2.8 grams n-dodecylamine acetate
- 2.8 grams Synperonic N
- 1 L distilled water
  - easure out 1 L of distilled water and pour into a clean 2 L glass beaker.
  - Weigh out 2.8 grams of n-dodecylamine acetate and add to water (weigh boat included).
  - Weigh out 2.8 grams Synperonic N and add to solution. Stir until dissolved.
  - Store in a clean 1 L bottle. Seal and label.

Working solution

- 10 grams silver nitrate
- 30 grams iron nitrate
- 80 grams ammonium iron (II) sulfate
- 20 grams citric acid
- Distilled water
- 40 mL detergent solution

1. Measure 50 mL of distilled water and pour into a 100 mL glass beaker.
2. Weigh out all dry chemicals.
3. Add 10 grams of silver nitrate to distilled water. Stir until dissolved. Store in the dark
4. Measure out 900 mL of distilled water and pour into a clean 2 L glass beaker.
5. Add 30 g iron (III) nitrate, 80 g ammonium iron (II) sulfate and 20 g citric acid to distilled water quickly in sequence. Stir until dissolved.
6. Measure out 40 mL of detergent solution and add to iron solution. Stir for a few minutes
7. Add silver nitrate solution to iron solution and stir for a few minutes. Use immediately.

Development :

1. Pour acid pre-wash solution into a clean, dry glass tray (or cover a glass tray with a large clear clean plastic bag and pour solution into the plastic-covered tray).
2. Immerse exhibit into acid pre-wash solution until no more bubbles are apparent or for a minimum of ten minutes.
3. Pour working solution into a clean, dry glass tray (or cover a glass tray with a large clear clean plastic bag and pour solution into the plastic-covered tray).
4. Immerse exhibit into working solution and gently rock the dish until dark grey fingerprints appear or for a maximum of 20 minutes depending on background development.
5. Pour distilled water into three glass trays.
6. Immerse exhibit into first tray of distilled water and wash for five minutes, rocking dish occasionally. Repeat with remaining two trays of distilled water.

Equipment :

- Glass measuring cylinders; 50 mL, 1 L
- Glass beaker 100 mL, 2 L (x2)
- Glass bottle 1 L
- Glass tray (x5)
- Magnetic stirrer plate and stirrer bars

Chemicals :

- Ammonium Iron (II) Sulfate: 21,540-6 (Aldrich)
- Citric Acid: 25,127-5 (Aldrich)
- n-Dodecylamine Acetate: D56270 (Pfaltz and Bauer)
- Iron (II) Nitrate Nonahydrate: 25,422-3 (Aldrich)
- Maleic Acid: M15-3 (Aldrich)

- Synperonic-N: contact FIOSS
- Silver Nitrate: 20,913-9

Supplier :

- Aldrich Chemical Co., tel: 1-888-565-1400
- FIOSS, tel: 1-613-998-6188

BRILLIANT YELLOW 40

Application : Detection of latent fingerprints on non-porous exhibits that have been pre-treated with cyanoacrylate. The cyanoacrylate reacts with moisture and other bases within the fingerprint residue to produce a polymer that may or may not be apparent to the unaided eye. Brilliant Yellow 40 will dye the poly-cyanoacrylate yellow and cause a yellow fluorescence when viewed under blue light using an orange KV 550 NM filter.

Formula :

Working solution

- 2.0 g brilliant yellow 40
- 1L ethyl alcohol
  - Weigh out 2 g of Brilliant Yellow 40 and place in a 2 L glass beaker.
  - Measure out 1 L of ethyl alcohol and add to Brilliant Yellow 40. Stir for 5 minutes.
  - Store in a clean, dry glass bottle. Seal and label.

Development :

1. Pour solution into a clean, dry glass, metal or plastic tray.
2. Immerse exhibit into solution and rock gently for several seconds.
3. Remove exhibit and rinse under gently running tap water
4. Air dry before examining under blue light using KV 550 NM viewing filters.
5. Areas of interest that are covered with water marks can be further washed with methyl alcohol.

Equipment :

- glass measuring cylinders; 1L
- glass beaker 2L
- glass bottle 1L
- magnetic stirrer plate and stirrer bars

Chemicals :

- Brilliant Yellow 40: Forensi-Tech
- Ethyl Alcohol: 36,280-8 (Aldrich); A962-4 (Fischer).

RHODAMINE 6G

Rhodamine 6G /'ro?d?mi?n/ is a highly fluorescent Rhodamine family dye. It is often used as a tracer dye within water to determine the rate and direction of flow and transport. Rhodamine dyes fluoresce and can thus be detected easily and inexpensively with instruments called fluorometers. Rhodamine dyes are used extensively in biotechnology applications such asfluorescence microscopy, flow cytometry, fluorescence correlation spectroscopy and ELISA.

Fig. Rhodamine 6G Chloride powder mixed with methanol, emitting yellow light under green laser illumination

Fig. Rhodamine 6G-based dye laser. The dye solution is the orange fluid in the tubes

Rhodamine 6G is also used as a laser dye, or gain medium, in dye lasers, and is pumped by the 2nd (532 nm) harmonic from an Nd:YAG laser or nitrogen laser. The dye has a remarkably high photostability, high fluorescence quantum yield (0.95), low cost, and its lasing range has close proximity to its absorption maximum (approximately 530 nm). The lasing range of the dye is 555 to 585 nm with a maximum at 566 nm.

Rhodamine 6G usually comes in three different forms. Rhodamine 6G chloride is a bronze/red powder with the chemical formula $C_{27}H_{29}ClN_2O_3$. Although highly soluble, this formulation is very corrosive to all metals except stainless steel. Other formulation are less soluble, but also less corrosive. Rhodamine 6G Perchlorate, ($C_{27}H_{29}ClN_2O_7$), comes in the form of red crystals, while rhodamine 6G tetrafluoroborate, ($C_{27}H_{29}BF_4N_2O_3$), appears as maroon crystals.

Application : Detection of latent fingerprints on non-porous exhibits that have been pre-treated with cyanoacrylate. The cyanoacrylate reacts with moisture and other bases within the fingerprint residue to produce a polymer that may or may not be apparent to the unaided eye. Rhodamine 6G will dye the poly-cyanoacrylate pink and cause a yellow fluorescence when viewed under blue light using an orange KV 550 nM filter.

Formula :

Stock solution

- 1.0 g Rhodamine 6G
- 1 L methyl alcohol
  - Weigh out 1.0 g of Rhodamine 6G and place in a 2 L glass beaker.
  - Measure out 1 L of methyl alcohol and add to Rhodamine 6G. Stir until dissolved.
  - Store in a clean, dry glass bottle. Seal and label.

Working solution

- 30 mL stock solution
- 1 L methyl alcohol
  - Measure out 30 mL of stock solution and pour into a 2 L glass beaker.
  - Measure out 1 L of methyl alcohol and add to Rhodmine 6G. Stir until dissolved.
  - Store in a clean, dry glass bottle. Seal and label.

Development :

1. Pour working solution into a clean, dry glass, metal or plastic tray.
2. Immerse exhibit into solution and rock gently for several seconds.
3. Remove exhibit and rinse under gently running tap water.
4. Air dry before examining under blue light using KV 550 NM viewing filters.
5. Areas of interest that are covered with water marks can be further washed with methyl alcohol.

Equipment :

- glass measuring cylinders; 1 L
- glass beaker 2 L (x2)
- glass bottle 1 L (x2)
- magnetic stirrer plate and stirrer bars

Chemicals :

- Methyl Alcohol: A412-4 (Fischer Scientific); ACS531-82 (VWR Canlab)
- Rhodamine 6G: 20,132-4 (Aldrich); Forensi-Tech

ARDROX

Application : Detection of latent fingerprints on non-porous exhibits that have been pre-treated with cyanoacrylate. The cyanoacrylate reacts with moisture and other bases within the fingerprint residue to produce a polymer that may or may not be apparent to the unaided eye. Ardrox will dye the poly-cyanoacrylate yellow and cause a yellow fluorescence when viewed under long-wave ultra-violet light using an orange KV 550 nM filter.

Formula :

Working solution 10 mL Ardrox

1 L methyl alcohol

1. Measure out 10 mL of Ardrox and place in a 2 L glass beaker.
2. Measure out 1 L of methyl alcohol and add to Ardrox. Stir for 2 minutes.
3. Store in a clean, dry glass bottle. Seal and label.

Development :

1. Pour solution into a clean, dry glass, metal or plastic tray.
2. Immerse exhibit into solution and rock gently for several seconds.
3. Remove exhibit and rinse under gently running tap water.
4. Air dry before examining under long-wave ultra-violet light using KV 550 nM viewing filters.
5. Areas of interest that are covered with water marks can be further washed with methyl alcohol.

Equipment :

- glass measuring cylinders; 10 mL, 1 L
- glass beaker 2 L glass bottle 1 L
- magnetic stirrer plate and stirrer bars

Chemicals :

- Methyl Alcohol: A412-4 (Fischer Scientific); ACS531-82 (VWR Canlab).
- Ardrox: Forensi-Tech.

TEC

Application: Detection of latent fingerprints on non-porous exhibits that have been pre-treated with cyanoacrylate. The cyanoacrylate reacts with moisture and other bases within the fingerprint residue to produce a polymer that may or may not be apparent to the unaided eye. TEC will dye the poly-cyanoacrylate and cause a red fluorescence when viewed under long-wave ultraviolet light using an orange KV 550 NM filter.

Formula :

Stock solution

- 1 g thenoyltrifluoroacetone
- 200 mL methyl ethyl ketone
- 0.5 g europium trichloride hexahydrate
- 800 mL distilled water
  - Weigh out 1 gram thenoyltrifluoroacetone and place in 500 mL glass beaker.

- Measure out 200 mL of methyl ethyl ketone and add to beaker. Cover with plastic wrap and stir until dissolved.
- Weigh out 0.5 g of europium trichloride hexahydrate and place in 1 L glass beaker.
- Measure out 800 mL distilled water and add to europium. Stir until dissolved.
- Pour methyl ethyl ketone solution into water solution. Seal and stir for several minutes.
- Store in a brown, clean glass bottle. Seal and label.

Working solution

- 200 mL stock solution
- 180 mL methyl ethyl ketone
- 720 mL distilled water
  - easure out 200 mL of Stock solution and pour into a 2 L glass beaker.
  - Measure out 180 mL of methyl ethyl ketone and add to stock.
  - Measure out 720 mL of distilled water and add to stock solution. Seal and stir for several minutes.
  - Store in a clean, dry glass bottle. Seal and label.

Wash solution

- 700 mL methyl alcohol
- 300 mL distilled water
  - Measure out 700 mL methyl alcohol and add to plastic wash bottle.
  - Measure out 300 mL distilled water and add to plastic wash bottle.
  - Seal and label.

Development :

1. Pour working solution into a clean, dry sealable plastic tray.
2. Immerse exhibit into solution and rock gently for several seconds.
3. Remove exhibit and rinse gently with methyl alcohol wash solution.
4. Air dry before examining under long-wave ultraviolet light using KV 550 nM viewing filters.

Equipment :

- glass measuring cylinders; 1 L
- glass beaker 2 L
- glass bottle 1 L
- magnetic stirrer plate and stirrer bars

Chemicals :

Europium (III) Trichloride Hexahydrate: 21,288-1 (Aldrich).

Methyl Ethyl Ketone: 36,047-3 (Aldrich).

Thenoyltrifluoroacetone: T2,700-6 (Aldrich).

PETROLEUM HYDROCARBON FINGERPRINTING - NUMERICAL INTERPRETATION DEVELOPMENTS

Hydrocarbon fuels and derivative products discovered in soils and groundwater at environmental release sites are often characterized by use of a laboratory technique known as capillary column gas chromatography (also referred as hydrocarbon fingerprinting, gas chromatography or GC fingerprinting). GC fingerprinting is an extremely useful tool in the investigation of subsurface contamination of soil and groundwater. (Bruce and Schmidt) GC fingerprints are used to obtain information from liquid hydrocarbon samples (free product) by determining the composition of the hydrocarbons present. The identification and interpretation of GC fingerprints, however, is largely a qualitative practice and dependent upon the skill and experience of the individuals(s) involved.

PETROLEUM HYDROCARBON CHEMISTRY

Petroleum hydrocarbons consist of a very large number of compounds that, by definition, are found in crude oil, as well as other sources of petroleum such as natural gas, coal, and peat. Petroleum hydrocarbons consist of three major groups of compounds. These are alkanes (paraffins), alkenes (olefins), and aromatics.

Paraffins, are one of the major constituents of crude oil and are found in refined petroleum products such as gasoline, kerosene, diesel fuel, heating oil, etc. There are three major classes of paraffins; these are linear alkanes, branched alkanes, and naphthenes. The linear alkanes have carbon atoms arranged in a line and there are only two ends to these molecules. Linear alkanes are also referred to in the literature as n-alkanes. Branched alkanes have the carbon atoms arranged similar to the n-alkanes, however, some of the carbon atoms are branched, thus creating many differing configurations. Naphthenes are molecules in which the carbon atoms are arranged in one or more rings.

Olefins are formed during the refining process of creating petroleum products from crude oil. These molecules have a double bond and two less hydrogen atoms than their corresponding alkane.

Aromatics contain one or more 6 carbon rings with 3 of the carbons containing double bonds. Examples of 1 ring (or mononuclear) aromatics are Benzene, Toluene, Ethylbenzene, and Xylene (BTEX). Multiple ring aromatics (polynuclear) are aromatic compounds with multiple 6 carbon ring molecules. Examples of these are naphthalene, anthracene, pyrene, and many more.

Hydrocarbon products such as gasoline, diesel fuel, and asphalts are all created from crude oil by a variety of refining and distillation processes. Each product is produced by the combination of multiple individual hydrocarbon compounds all of which have slightly different vaporization and boiling temperatures. For example, gasoline is the combination of many lower boiling range compounds including C4 to C12 alkanes, C4 to C7 alkenes, and aromatics BTEX. The middle boiling range compounds are used in differing proportions to create products such as kerosene, diesel, and heating oil. These products predominantly contain C10 to C24 alkanes, and polynuclear aromatics with little to no olefins. (Zemo, Graff, and Bruya)

HYDROCARBON FINGERPRINTING

GC fingerprints are created by injecting a small portion of the sample into a gas chromatograph. Once injected, the product is heated and vaporized and carried into a capillary column by a flow of inert gas. After injection the temperature of the column is slowly raised. As the temperature increases the compounds begin to move through the column, in general the more volatile and lower boiling compounds start moving first. A flame ionization detector connected to the end of the column detects the components of the product as they elute from the column. The time that it takes for individual components to go through the column depends on the temperature, length of column, column characteristics, and the character of the compound itself. Five GC fingerprints of various hydrocarbon products (gasoline, kerosene, diesel, JP-8 jet fuel and crude oil) are presented. A few of the peaks have been labeled identifying some of the compounds in each of the products.

MATERIALS & METHODS

Gas Chromatography

Hydrocarbon samples were analyzed on a Hewlett Packard 5890 instrument equipped with a split/splitless injector, J&W 30 meter DB-1 column and an FID detector: All gas flow rates were set to manufacturer specifications. Injections were made in split mode with a split ratio of 1:100. The column oven was programmed from -10 o to 350 o C at 10 o C/minute with 4 minute hold at 350o C. The injector temperature is set at 350o C and the detector temperature is 360o C. Data was acquired and processed with an EZChrom Chromatography data system.

Gasoline Weathering Simulation

One of the weathering processes that can affect released gasoline is evaporation. To simulate evaporation under controlled conditions, three different grades of gasoline were obtained from a local retailer and allowed to evaporate to controlled volumes.

The gasoline components with the lowest boiling points tend to volatilize more rapidly than the components with higher boiling points. On the GC fingerprint the components that have the shortest retention times (left side of the GC fingerprint) are the most volatile and will tend to decrease in peak intensity preferentially as more volatilization takes place. GC fingerprints from the same gasoline are shown under differing levels of volatilization.

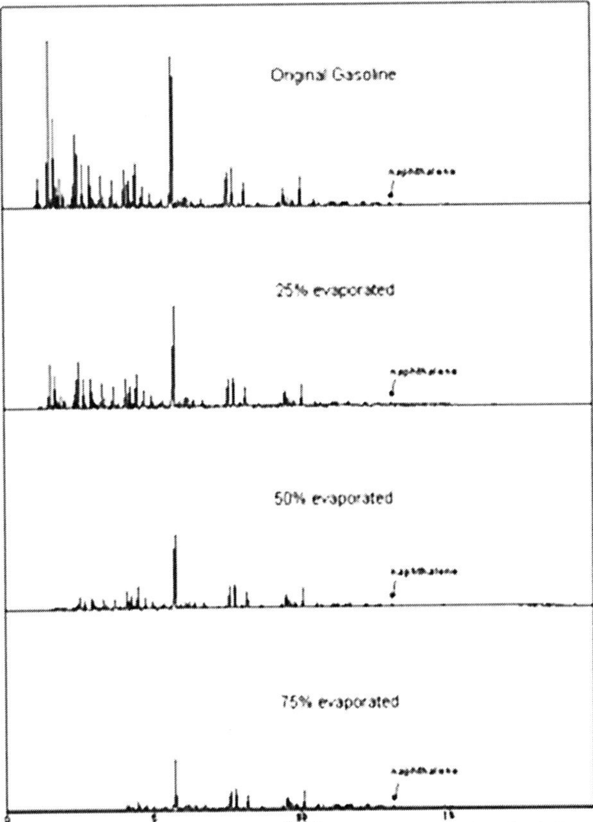

Fig. Controlled Evaporation Of Regular Grade Of Unleaded Gasoline (Note: Chromatograms Have Been Normalized To Make The Heights of Naphthalene Peaks Equal)

Evaporation Procedures

The evaporation procedure that was used is described below:

- Samples of three grades of gasoline (87, 89, and 93 octane) were acquired at a local service station.
- Equal volumes of each grade of gasoline were placed in four (4) 40 milliliter vials, making a total of 12 vials.
- Three (3) vials of each sample were uncovered (total of 9) and allowed to volatilize at room temperature.
- The uncovered vials were closely monitored and capped when the gasoline was reduced to the desired volume resulting in one vial each at 75%, 50%, and 25% of original volume for each of the three grades of gasoline.

ALGORITHMS DESCRIBED

Correlation Coefficient

The correlation coefficient, denoted by , measures the relationship between two data sets that are scaled to be independent of the unit of measure and is given by the formula:

$$r = \frac{\sum xy - \frac{(\sum x)(\sum y)}{n}}{\sqrt{\left[\sum x^2 - \frac{(\sum x)^2}{n}\right] \cdot \left[\sum y^2 - \frac{(\sum y)^2}{n}\right]}}$$

Where x and y are values in each corresponding data set.

The value of the correlation coefficient is always between -1 and +1. A value of equal to -1 indicates a perfect linear relationship between the sample values of x and y, with the value of y decreasing as the value of x increases. A value of equal to +1 also indicates a perfect linear relationship between the sample values, but one in which the value of y increases as x increases. Larger values of y are associated with larger values of x; and smaller values of y are associated with smaller values of x. If there is no linear relationship between the sample values of x and y, then will have a value near or equal to zero (Hayslett).

The correlation coefficient determines whether two data sets move together; that is, whether large values of one set are associated with large values of the other (positive correlation), whether small values of one set are associated with large values of the other (negative correlation), or whether the values in the two sets are unrelated.

In this study, 71 hydrocarbon chromatogram peaks, each representing a different hydrocarbon compound, were used in the analysis. presents a list of the compounds. Integrated peak areas were measured and then tabulated for each of the five hydrocarbon samples in figures 1 through 5. The integrated peak is a measure of the intensity of the response of the flame ionization detector to each of the individual compounds measured in millivolt seconds. The library samples includes gasoline, kerosene, diesel, JP-8 jet fuel, and crude oil. Once the peak area data were collected and tabulated for these hydrocarbon samples, three additional hydrocarbon samples were also collected. The first sample collected had been identified as a kerosene from the provider, however, the GC fingerprint clearly illustrated a much broader range of hydrocarbons than the kerosene run earlier. The second sample was a laboratory standard mixture of gasoline and diesel fuel. The third sample was a regular grade of unleaded gasoline from a different service station.

Table. Hydrocarbon Compounds Used For Analysis

1	iC4 = Isobutane	37	IP14 = C14 Isoprenoid
2	nC4 = Butane	38	1 M naph = 1 Methylnaphthalene
3	iC5 = Isopentane	39	nC13 = Tridecane
4	nC5 = Pentane	40	IP15 = Farnesane
5	2 M Pent = 2 Methylpentane	41	nC14 = Tetradecane
6	3 M Pent = 3 Methylpentane	42	IP16 = C16 Isoprenoid
7	nC6 = Hexane	43	nC15 Pentadecane
8	C6 Olefin = C6 Olefin	44	nC16 = Hexadecane
9	M Cycl Pent = Methyl cyclopentane	45	IP18 = Norpristane
10	2,4 DMP = 2,4 Dimethylpentane	46	nC17 = Heptadecane
11	Bnz = Benzene	47	Pristane = Pristane
12	Cyclo Hexane = Cyclohexane	48	nC18 = Octadecane
13	2 M Hex = 2 Methylhexane	49	Phytane = Phytane
14	3 M Hex = 3 Methylhexane	50	nC19 = Nonadecane
15	Isooctane = Isooctane or 2,2,4 Trimethylpentane	51	nC20 = Eicosane
16	nC7 = Heptane	52	nC21 = Heneicosane
17	MCHX = Methylcyclohexane	53	nC22 = Docosane
18	Tol = Toluene	54	nC23 = Tricosane
19	nC8 = Octane	55	nC24 = Tetracosane
20	EB = Ethylbenzene	56	nC25 = Pentacosane
21	m/p-xyl = m/p Xylene	57	nC26 = Hexacosane
22	o-xyl = o Xylene	58	nC27 = Heptacosane
23	nC9 = Nonane	59	nC28 = Octacosane
24	propylbenzene = n Propylbenzene	60	nC29 = Nonacosane
25	1M 3E benz = 1 Methyl 3 ethylbenzene	61	nC30 = Triacontane
26	1M 4E benz = 1 Methyl 3 ethylbenzene	62	nC31 = Hentriacontane
27	1,3,5 T M Benz = 1,3,5 Trimethylbenzene	63	nC32 = Dotriacontane
28	3,3,4 T M Hept = 3,3,4 Trimethylheptane	64	nC33 = Tritriacontane
29	1,2,4 T M Benz = 1,2,4 Trimethylbenzene	65	nC34 = Tetratriacontane
30	nC10 = Decane	66	nC35 = Pentatriacontane
31	1,2,3 T M Benz = 1,2,3 Trimethylbenzene	67	nC36 = Hexatriacontane
32	nC11 / 1,2,4,5 TeMB = Undecane and 1,2,4,5 Tetramethlybenzene	68	nC37 = Heptatriacontane
33	Naph = Naphthalene	69	nC38 = Octatriacontane
34	nC12 = Dodecane	70	nC39 = Nonatriacontane
35	IP13 = C13 Isoprenoid	71	nC40 = Tetracontane
36	2 M naph = 2 Methylnaphthalene		

Once the peak areas for all of the 71 individual components were established for each of the samples, the correlation coefficient was calculated between the samples presented in Figures.

Table. Results of Correlation Coefficient Determinations

	Gasoline I Figure 1	KeroseneI Figure 2	Diesel Figure 3	JP-8 Jet Fuel Figure 4	Crude Oil Figure 5
Kerosene II Figure 7	-0.156	0.732	0.882	0.932	0.379
Gas/Diesel Mixture Figure 8	0.638	0.333	0.528	0.505	0.440
Gasoline II Figure 9	0.894	-0.112	-0.213	-0.065	0.134

To evaluate the reproducibility of this process, the regular unleaded gasoline presented in Figure 9, was run on the GC five separate times, thus creating five GC fingerprints and five slightly differing numerical data sets. The data collected for all 71 compounds were then compared to each other, thus creating a total of twenty (20) correlation coefficient comparisons.

Gasoline Weathering

An algorithm was developed to model the volatilization process of gasoline released into the environment. This was accomplished by using experimental data obtained from the controlled evaporation of the three different grades of gasoline described earlier. GC fingerprint data were used to create a numerical function that describes the volatilization process. This numerical function can then be applied to fresh gasoline samples to predict what the product GC fingerprint would look like if weathered in an environmental release.

As discussed earlier, the gasoline components with the lowest boiling points tend to volatilize more rapidly than the rest of the components. The components with the highest boiling points (components at the right hand side of the GC fingerprints with retention times > 10 minutes) experience little volatilization under the weathering conditions described above.

With the above in mind, it was assumed that the actual volume of the naphthalene stayed relatively constant during the weathering simulation and can be used similar to an internal standard. By utilizing this, the GC data from each stage of the weathering process were normalized to the naphthalene peak. Once each GC fingerprint was normalized to naphthalene, each component was then evaluated as the total volume of product decreased. Once this process was completed for all components, then a matrix of volatilization multipliers was created. This matrix consists of a table of factors ranging from 0.0 to 1.0 describing the amount of volatilization of each of the 71 components at differing stages of evaporation of the total product. To demonstrate how the matrix was created, the integrated peak areas for the first 8 of the 71 components from the premium grade unleaded gasoline used in the experiment. The same component integrated peak areas after they have been normalized to naphthalene. Each component integrated peak area from table 4 normalized from 0 to 1. Table 5 represents the matrix of multipliers. Two other similar tables were also produced for the mid-grade and regular grades of gasoline. The entire matrix of multipliers for each of the grades of gasoline was not presented because of the size of the tables.

Table. Peak Areas for Components at Differing % Volatilization (Premium Grade Gasoline)

Sample Id	iC4	nC4	iC5	nC5	2 M Pent	3 M Pent	nC6	C6 Olefin
Gasoline % Vol.	Peak Area	Peak Area	Peak Area	Peak Area	Peak Area	Peak Area	Peak Area	Peak Area
Prem-0	5139	53846	129407	21998	34183	18112	14922	2865
Prem-25	0	737	23654	6196	20421	11421	10368	1940
Prem-50	0	0	0	0	1666	2305	2894	547
Prem-75	0	0	0	0	0	0	0	0

Table. Peak Areas for Components At Differing % Volatilization After Normalizing with Naphthalene (Premium Grade Gasoline)

Sample Id	iC4	nC4	iC5	nC5	2 M Pent	3 M Pent	nC6	C6 Olefin
Gasoline % Vol.	Peak Area	Peak Area	Peak Area	Peak Area	Peak Area	Peak Area	Peak Area	Peak Area
Prem-0	5139	53846	129407	21998	34183	18112	14922	2865
Prem-25	0	531	17047	4465	14717	8231	7472	1398
Prem-50	0	0	0	0	1165	1612	2024	383
Prem-75	0	0	0	0	0	0	0	0

Table. Matrix of Multipliers For Individual Components of Gasoline Under Differing Percentages of Overall Product Volatilization (Premium Grade Gasoline)

Sample Id	iC4	nC4	iC5	nC5	2 M Pent	3 M Pent	nC6	C6 Olefin
Gasoline % Vol.	Peak Area	Peak Area	Peak Area	Peak Area	Peak Area	Peak Area	Peak Area	Peak Area
Prem-0	1	1	1	1	1	1	1	1
Prem-25	0	0.01	0.132	0.203	0.4305	0.4544	0.501	0.488
Prem-50	0	0	0	0	0.0341	0.089	0.136	0.1336
Prem-75	0	0	0	0	0	0	0	0

RESULTS & DISCUSSION

Reproducibility

The reproducibility of the gas chromatography analysis technique was evaluated by analyzing the gasoline sample presented in Figure 9 a total of 5 times. The 20 correlation coefficients calculated between each of the 5 analyses and the other four had a minimum of 0.99545, a maximum of 0.999989, an average of 0.997921 and a standard deviation of 0.001704. From this evaluation, truly alike GC chromatograms will probably have correlation coefficients of 0.99 or better. It is possible that other product types may have different reproducibilities since their data may include different peaks that come from a different part of the GC fingerprint.

Correlation

The results of the correlation coefficients calculated when comparing the samples. Prior to calculating the correlation coefficients it was expected that similar products, for example gasolines, would show higher correlations among themselves and less correlation when compared to other product types. The exact numbers, however, could not be anticipated nor how the correlation coefficients would vary between similar and different product types. Significant, logical, and reproducible differences and similarities in the correlation coefficient numbers are crucial for this process to be a useful tool. The correlation coefficient results must also make sense and compare favorably with visual inspection of the GC fingerprints. From this feasibility study, it appears that there are meaningful similarities and differences in correlation coefficient numbers calculated using GC fingerprint data. This study suggest that similar product types such as gasolines could be expected to have correlation coefficients of about 0.9 or better. Dissimilar product types have a much lower correlation coefficient of perhaps 0.6 or 0.5 or even less. The correlation coefficients shown here also make sense when compared to the visual evaluation of the GC fingerprints.

An unexpected and interesting result of the correlations was that the JP-8 jet fuel and the kerosene II sample had a high correlation coefficient. At first this seemed unusual, but it must be remembered that JP-8 and kerosene are often times from the same distillation range of the crude oil. Visual comparison of the two GC chromatograms confirms the rather high degree of similarity of the two products.

Gasoline Weathering

The matrix of multipliers created for the evaporation sequences for the three different grades of gasoline numerically models how the volatile components tend to evaporate from the sample. The matrix of multipliers can be used to numerically alter ("evaporate") the data from a fresh sample in an attempt to estimate the composition of the sample after a weathering process.

Once the sample has been artificially altered, it can then be numerically compared to other controlled weathered samples.

This weathering algorithm can also be used in the inverse. For example, if one had a hydrocarbon sample from a site but did not know the extent of weathering that has already taken place, the sample could be correlated with the library of samples of known weathered gasolines. Once a library sample with the highest correlation has been determined, a matrix of multipliers of the sample with the highest correlation could be used to reconstruct the composition of the original gasoline sample when fresh. This matrix of multipliers needed to estimate the original gasoline composition would be constructed by simply using the inverse of the individual compounds within the matrix. that most closely correlated with the weathered sample. (For example, if Benzene's multiplier is 0.25, then to reconstruct the original amount of Benzene, one would multiply the peak area by 1/0.25 = 4.0.)

Future Work

The information and techniques presented in this paper represent some beginning examples of the types of analysis that are possible with GC fingerprint numerical data. Listed below are a number of additional numerical techniques and experiments that are under consideration for future work:

1. Investigate algorithms that numerically weight key compounds,
2. Develop algorithms that use the presence of unique compounds to indicate specific characteristics. For example olefins indicate the presence of catalytic cracked hydrocarbons.
3. Develop algorithms that further refine gasoline compound recognition,
4. Investigate additional controlled weathering experiments on other types of hydrocarbon products,
5. Investigate other weathering processes such as water washing, biodegradation, volatilization, and chemical speciation,
6. Expand the GC fingerprint library to include other petroleum products and hydrocarbon samples from the extraction of both soils and groundwater,
7. Investigate correlation experiments using isolated ranges of hydrocarbons; therefore, allowing mixtures of products (say gasoline and diesel fuel) to be evaluated separately and then numerically added together.

8

Fingerprint Structure Imaging Based on an Ultrasound Camera

INTRODUCTION

Over the last few years a new area of engineering science has been established whose products are likely to create a large market in the near future. It has been called "biometrics". The pioneers of this new domain intend to construct devices which would allow identification of a person on the basis of his/her "biological" characteristics: voice, dynamics of movements, features of face and other parts of the body, retina or iris pattern. However, the greatest hope seems to be lying in the possibility of the fingertip structure recognition (this structure is reflected in the fingerprint pattern). It is well known that the finger ridge pattern is different for each individual and that it does not change over the life time. Touching of a sensor surface is a simple act. Many inventors of biometric devices hope to develop a button which would "know" by whom it has been pressed and which finger has been used. A button used for the door unlocking would of course let in only authorized people and this is what the whole new area wants to live.

Systems for the ridge pattern imaging with the optical acquisition of data have been investigated for a number of years. They show "live" fingerprint images directly from a finger without the need for ink and paper which have been traditionally used by policemen since Galton times. The systems with optical data acquisition, however, have a number of drawbacks: the direct image of the fingertip has a very low contrast and it is easier to see the dirt than the ridge pattern. In turn, methods employing the reflection from the surface are very sensitive to grease, dirt, and water. Three dimensional image is difficult to create and does not provide satisfactory results for damaged fingers. Furthermore, no method allows to decide in an easy way whether the object under observation is a real finger, an imitation, or perhaps a greasy residue of a finger on the sensor. The description of a typical optical fingerprint imaging system is given in.

Hence, it should not be surprising that there has been interest in alternative methods of the ridge pattern imaging. For instance, Constantine Tsikosa proposed a capacitive method [7], further developed recently by SGS-Thomson [8] and Siemens [3], [9]. So far only prototypes of such devices have been presented and there is little known about their practical usefulness.

PERSPECTIVES OF ULTRASOUND DEVICES DEVELOPMENT

In 1986, the author of this paper proposed a method based on ultrasound data acquisition. This approach allows to distinguish between real fingers and any imitations. Furthermore, it is not sensitive to any dirt, grease etc. There is also a completely new perspective, unthinkable in the case of other methods. It is possible to create a device with a surface reacting to a finger touch (or a number of fingers) which would be able to decide where the finger has been placed, identify it and register its movements. Such a device would not have any moving parts and could replace today's keyboards, mice, graphic pads, and fingerprint identification systems, though this is not the end of its potential applications. To complete the picture, it is worth knowing that it is feasible to create a device which would be small, inexpensive (a kind of a chip), and could really fit in a button. Such a device would have another interesting property. It would enable us to devise a system for remote people identification (through a network) which cannot be cheated, even if a person sitting at a remote terminal has unlimited possibilities of carrying out a fraud.

A number of papers have been published describing our method, a few patents have been granted and a few other patents are pending (the owner of the patents and commercial rights to the device is Sonident, Vaduz). This work is aimed to be a brief presentation of the key aspects of the method employed by us which have not been described in detail in the previous papers. The paper is also intended to present the subject to the readership of "Archives of Acoustics".

The idea of the ultrasonic camera operation

The operation of our devices is possible thanks to the phenomenon, which apparently has not so far been employed by anyone and perhaps not even noticed (to the best of our knowledge). It can be summarized in the form of the following rule: Consider a surface of a solid object against which another object has been placed, so that the contact between the two objects is not ideal, i.e. there are some inhomogeneities. The sound wave which reaches such a place does not only pass from one environment to the other, get reflected and diffracted in the contact area as described by classical theory but it also is subject to some additional scattering and transformation to a different kind of waves. This phenomenon is the effect of disturbance in the sound propagation conditions in the contact area between two objects, hence it will be referred to as the contact scattering. It is sure that this kind of scattering is the result of not only the contact area of the two objects but also the area near the objects' surface (henceforth it will be referred to as the near surface structure). It is likely that for this reason, the contact scattering is strongly dependent upon the substance of the placed object.

Experiments show that the transition of the wave from one environment to the other may practically not occur at all and observed are only the contact scattering and generation of other types of waves (it is particularly conspicuous for transversal waves). It is likely that the disturbances of the wave occurring in the contact areas are mainly in the phase (the phase front is spatially distorted) and they are responsible for the observed contact scattering. At the moment, the research is being carried out to develop a theory adequately describing this phenomenon. We shall devote further publications to this subject.

The design of the ultrasonic camera

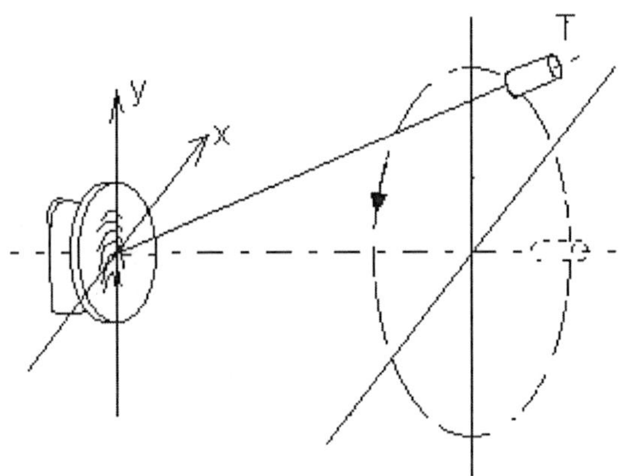

Fig. Schematic diagram of the device

Employing the phenomenon described in the previous section we have designed a device for measuring and analysis of the signals being the result of the contact scattering of objects placed against a plastic window. The device is designed mainly for the near surface observation of the finger ridge patterns. A detailed description of our device has been presented in the aforementioned papers. For all those readers who are not familiar with the subject, we offer a brief description:

An acoustic wave is sent in the direction of the surface against which an object has been placed (see Fig. 1). The signals which are scattered by the object are received by the transducer (T), which is moving along a circular trajectory whose axis is perpendicular to the contact surface (x-y). The same element can be used both as an emitter and a receiver. Alternatively, instead of one moving transducer it is possible to employ a number of fixed transducers. For the object analysis with the resolution of around 0.1 mm, it is necessary to collect scattered signal data from about 256 different angles. At the moment, our device sends in each of the 56 directions a short pulse and receives the impulse response (in the case of a finger, the signal spectrum is in the range from 4 to 16 MHz and it is dependent on the device design).

Impulse response of a ball

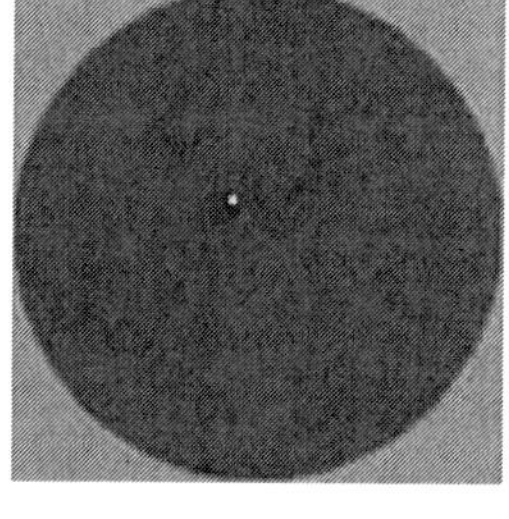

Reconstruction of a ball

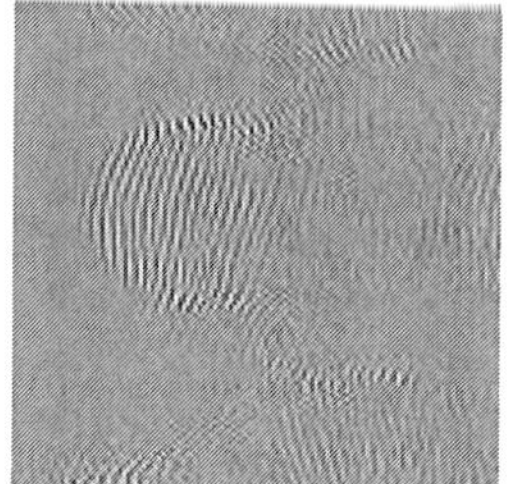

Impulse response of a finger

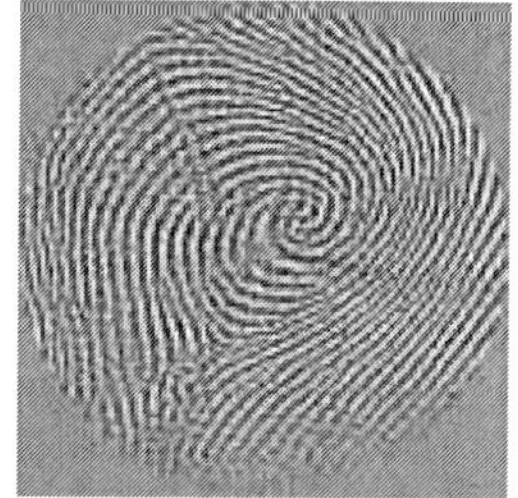

Reconstruction of a finger.

Reconstruction of a stamp

The set of impulse responses for a small ball, whereas Fig. for a finger (vertical axis corresponds to time, horizontal axis corresponds to angle, the value of the signal is represented through the grey level). In order to obtain the observed structure from the collected data, a reconstruction procedure is used which is similar to methods used in ultrasound reflection tomography. A set of programs have been written, aimed at achieving high quality and high speed reconstruction. The algorithms developed at Optel enable image reconstruction based on a set of 256 impulse responses each containing 256 samples in about 50 ms (using a standard PC based on the Cyrix 6x86 P200+ processor). We expect to develop in the near future an improved algorithm which would allow cutting the computation time to about 20 ms.

TECHNICAL SOLUTIONS EMPLOYED IN THE CAMERA'S DESIGN

The Picture of the actual existing camera

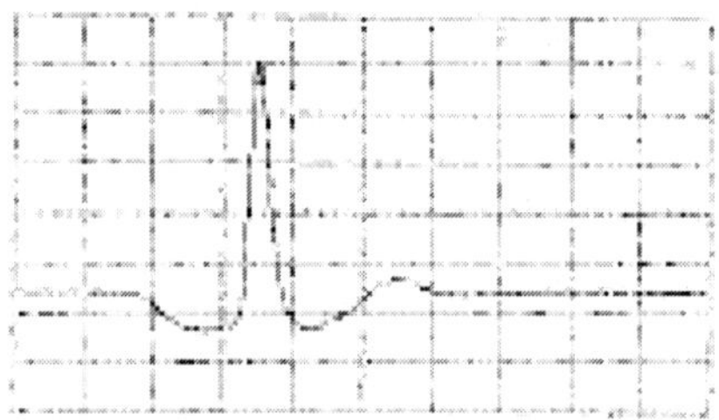

50ns/div

The shape of the pulse generated by of our transducer. The use of the contact scattering phenomenon discovered by us and computer tomography methods was not enough to construct an ultrasound camera. We had to solve a few other problems: In order to obtain the required resolution it was necessary to develop a circuit which having a relatively small diameter would emit a gaussian ultrasound beam of high amplitude and have a high sensitivity as a receiver. Such a circuit has been developed and patented [16] and we intend to present its construction in a separate paper.

It was also necessary to develop a transducer which would be able to emit a short pulse and as a receiver would have the required bandwidth (4-16 MHz). Moreover, its phase function was required to have the smallest possible variance. It was also important that such a transducer would have to be cheap and have repeatable parameters. The final effect of our research is to be a device suitable for mass production whose price has to be reasonable. The researchers at Optel managed to develop a transducer which has a completely new design (a patent application has been submitted). It is able to emit very short pulses and has very wide bandwidth as receivers (ca 4-25 MHz). The amplitude of the signal emitted by the new transducers is about two times higher than for classical pulse transducers. Their sensitivity as receivers is however slightly lower which in the measurement cycle gives a comparable result. Nevertheless, the idea behind the new transducers opens a new path in the design of the ultrasound transducers and it is fair to expect significant improvement in their parameters. Again, we wish to devote a separate paper to this subject.

The design of our ultrasound camera would not have been possible, had we not developed our own electronic circuitry which includes the transceiver circuit and an oscilloscope card. These elements are also based on our own original ideas: The pulse generator is capable of generating pulses as short as 20 ns which have the amplitude of ca 600 V; the receiver has the sensitivity of 5m V for the frequencies in the range 4-16 MHz, and the dynamic range of 60 dB. The oscilloscope card enables sampling at up to 200 MS/s and is specifically dedicated for processing sets of ultrasound signals (it satisfies some strict timing parameters).

Observations with the use of the camera

- Objects of similar structure but made of different substances give significantly different signals (both in amplitude and in character). The structure of the objects is nevertheless visible. Hence, it is possible to distinguish between "real" and "artificial" fingers.
- Spreading gel on the surface of an object, soaking it in water or covering with dirt does not result in significant changes of the signal.
- A fingerprint is hardly noticeable because the signal level it gives is at least 30 dB lower than the the signal given by a real finger (in contrast to this, for optical devices this level does not change significantly). The above observation is also true when soot or metal powder is used in order to enhance the fingerprint.
- A fingerprint left on a thick (ca 0.5 mm) layer of gel or grease is noticeable but it is very different when observed directly.
- Fingers which have damaged surface still give relatively clear image. Their internal structure seems to be visible, since the phenomenon on which our observations are based applies to the near surface layer.

FUTURE WORK

In the near future, we plan to develop a new model of the camera, which will be based on fixed transducers and will be capable of showing "live" pictures of objects at 25 frames per second. It will be a kind of a "real-time" ultrasound camera which can see the near surface structures of objects placed against its sensitive surface. The camera will contain its own electronic circuit for reconstruction and it will have an output for a standard monitor. The camera used at present is based on a moving transducer and can produce a few frames per second. It also needs a computer which does the signal processing and displays the image on its screen. In 1998, we plan to develop an integrated version of the device. Eventually, we hope to implement it in a kind of a chip.

DIGITIZATION AND PROCESSING OF FINGERPRINTS

ALGORITHMS

Demands imposed by the painstaking attention needed to visually match the fingerprints of varied qualities, the tedium of the monotonous nature of the manual work, and increasing workloads due to a higher demand on fingerprint recognition services prompted law enforcement agencies to initiate research into acquiring fingerprints through electronic media and to automate fingerprint indi¬vidualization based on digital representation of fingerprints. As a result of this research, a large number of computer algorithms have been developed during the past three decades to automatically process digital fingerprint images. An algorithm is a finite set of well-defined instructions for accomplishing some task which, given an initial state and input, will terminate in a corresponding recognizable end-state and output. A computer algorithm is an algorithm coded in a programming language to run on a computer. Depending upon the application, these computer algo¬rithms could either assist human experts or perform in lights-out mode. These algorithms have greatly improved the operational productivity of law enforcement agencies and reduced the number of fingerprint technicians needed. Still, algorithm designers identified and investigated the following five major problems in designing automated fin¬gerprint processing systems: digital fingerprint acquisition, image enhancement, feature (e.g., minutiae) extraction, matching, and indexing/retrieval.

Image Acquisition

Known fingerprint data can be collected by applying a thin coating of ink over a finger and rolling the finger from one end of the nail to the other end of the nail while press¬ing the finger against a paper card. This would result in an inked "rolled" fingerprint impression on the fingerprint card. If the finger was simply pressed straight down against the paper card instead of rolling, the resulting fingerprint impression would only contain a smaller central area of the finger rather than the full fingerprint, resulting in an inked "flat" or "plain" fingerprint impression. The perspiration and contaminants on the skin result in the impression of a finger being deposited on a surface that is touched by that finger. These "latent" prints can be chemi¬cally or physically developed and electronically captured or manually "lifted" from the surface by employing certain chemical, physical, and lighting techniques. The developed fingerprint may be lifted with tape or photographed. Often these latent fingerprints contain only a portion of the friction ridge detail that is present on the finger, that is, a "partial" fingerprint. Fingerprint impressions developed and preserved using any of the above methods can be digitized by scanning the inked card, lift, item, or photograph. Digital images acquired by this method are known as "off-line" images. (Typically, the scanners are not designed specifically for fingerprint applications.)

Since the early 1970s, fingerprint sensors have been built that can acquire a "livescan" digital fingerprint image directly from a finger without the intermediate use of ink and a paper card. Although off line images are still in use in certain forensic and government applications, on line fingerprint images are increasingly being used. The main parameters characterizing a digital fingerprint image are resolution area, number of pixels, geometric accuracy, contrast, and geometric distortion. CJIS released specifica¬tions, known as Appendix F and Appendix G, that regu¬late the quality and the format of fingerprint images and FBI-compliant scanners. All livescan devices manufactured for use in forensic and government law enforcement ap¬plications are FBI compliant. Most of the livescan devices manufactured to be used in commercial applications, such as computer log-on, do not meet FBI specifications but, on the other hand, are usually more user-friendly, compact, and significantly less expensive. There are a number of livescan sensing mechanisms (e.g., optical, capacitive, thermal, pressure-based, ultrasound, and so forth) that can be used to detect the ridges and valleys present in the fingertip. However, many of these methods do not provide images that contain the same representation of detail necessary for some latent fingerprint comparisons. For example, a capacitive or thermal image may represent the edges and pores in a much different way than a rolled ink impression. Figure 6–6 shows an off line fingerprint image acquired with the ink technique, a latent fingerprint image, and some livescan images acquired with different types of commercial livescan devices.

Image Enhancement

Fingerprint images originating from different sources may have different noise characteristics and thus may require some enhancement algorithms based on the type of noise. For example, latent fingerprint images can contain a variety of artifacts and noise. Inked fingerprints can contain blobs or broken ridges that are due to an excessive or inadequate amount of ink. Filed paper cards may contain inscriptions overlapping the fingerprints and so forth. The goal of finger¬print enhancement algorithms is to produce an image that does not contain artificially generated ridge structure that might later result in the detection of false minutiae features while capturing the maximum available ridge structure to allow detection of true minutiae. Adapting the enhance¬ment process to the fingerprint capture method can yield the optimal matching performance over a large collection of fingerprints. A fingerprint may contain such poor-quality areas that the local ridge orientation and frequency estimation algorithms are completely wrong. An enhancement algorithm that can reliably locate (and mask) these extremely poor-quality areas is very useful for the later feature detection and individualization stages by preventing false or unreliable features from being created.

Fingerprint images can sometimes be of poor quality be¬cause of noise introduced during the acquisition process. For example: a finger may be dirty, a latent print may be lifted from a difficult surface, the acquisition medium (pa¬per card or livescan) may be dirty, or noise may be intro¬duced during the interaction of the finger with the sensing surface (such as slippage or other inconsistent contact). When presented with a poor-quality image, a forensic ex¬pert would use a magnifying glass and try to decipher the fingerprint features in the presence of the noise. Automatic fingerprint image-enhancement algorithms can significantly improve the quality of fingerprint ridges in the fingerprint image and make the image more suitable for further manual or automatic processing. The image enhancement algorithms do not add any external information to the fingerprint image. The enhancement algorithms use only the information that is already present in the fingerprint image. The enhancement algorithms can suppress various types of noise (e.g., another latent print, background color) in the fingerprint image and highlight the existing useful features. These image enhancement algorithms can be of two types.

Enhancement of Latent Prints for AFIS Searching

In the case of latent searches into the forensic AFISs, the enhancement algorithm is interactive, that is, live feedback about the enhancement is provided to the forensic expert through a graphical user interface. Through this interface, the forensic expert is able to use various algorithms to choose the region of interest in the fingerprint image, crop the image, invert color, adjust intensity, flip the image, magnify the image, resize the image window, and apply compression and decompression algorithms. The forensic expert can selectively apply many of the available enhance¬ment algorithms (or select the parameters of the algorithm) based on the visual feedback. Such algorithms may include histogram equalization, image intensity rescaling, image intensity adjustments with high and low thresholds, local or global contrast enhancement, local or global background subtraction, sharpness adjustments (applying high-pass filter), background suppression (low-pass filter), gamma adjustments, brightness and contrast adjustments, and so forth. An example of local area contrast enhancement is shown in Figure 6–9. In this example, the fingerprint image enhancement algorithm enhances only a small, square,local area of the image at a time but traverses over the entire image in a raster scan fashion such that the entire image is enhanced. Subsequent fingerprint feature extrac¬tion can then be either performed manually or through automatic fingerprint feature extraction algorithms.

Automated Enhancement of Fingerprint Images

In the case of lights-out applications (frequently used in automated background checks and commercial applications for control of physical access), human assistance does not occur in the fingerprint individualization process. Enhance¬ment algorithms are used in the fully automated mode to improve the fingerprint ridge structures in poor-quality fingerprint images. An example of a fully automated fingerprint image- enhancement algorithm is shown in Figure 6–10. In this example, contextual filtering is used that has a low-pass (smoothing) effect along the fingerprint ridges and a band-pass (differentiating) effect in the direction orthogonal to the ridges to increase the contrast between ridges and valleys. Often, oriented band-pass filters are used for such filtering. One such type of commonly used filters is known as Gabor filters. The local context is provided to such contextual filters in terms of local orientation and local ridge frequency.

FEATURE EXTRACTION

Local fingerprint ridge singularities, commonly known as minutiae points, have been traditionally used by forensic experts as discriminating features in fingerprint images. The most common local singularities are ridge endings and ridge bifurcations. Other types of minutiae mentioned in the literature, such as the lake, island, spur, crossover, and so forth (with the exception of dots), are simply composites of ridge endings and bifurcations. Composite minutiae, made up of two to four minutiae occurring very close to each other, have also been used. In manual latent print processing, a forensic expert would visually locate the minutiae in a fingerprint image and note its location, the orientation of the ridge on which it resides, and the minutiae type. Automatic fingerprint feature-extraction algorithms were developed to imitate minutiae location performed by forensic experts. However, most automatic fingerprint minutiae-extraction algorithms only consider ridge end¬ings and bifurcations because other types of ridge detail are very difficult to automatically extract. Further, most algorithms do not differentiate between ridge endings and bifurcations because they can be indistinguishable as a result of finger pressure differences during acquisition or artifacts introduced during the application of the enhance¬ment algorithm.

One common approach followed by the fingerprint feature extraction algorithms is to first use a binarization algorithm to convert the gray-scale-enhanced fingerprint image into binary (black and white) form, where all black pixels correspond to ridges and all white pixels correspond to valleys. The binarization algorithm ranges from simple thresholding of the enhanced image to very sophisticated ridge location algorithms. Thereafter, a thinning algorithm is used to convert the binary fingerprint image into a single pixel width about the ridge centerline. The central idea of the thinning process is to perform successive (iterative) erosions of the outermost layers of a shape until a con¬nected unit-width set of lines (or skeletons) is obtained. Several algorithms exist for thinning. Additional steps in the thinning algorithm are used to fill pores and eliminate noise that may result in the detection of false minutiae points. The resulting image from the thinning algorithm is called a thinned image or skeletal image. A minutiae detection algorithm is applied to this skeletal image to locate the x and y coordinates as well as the orientation (theta) of the minutiae points. In the skeletal image, by definition, all pixels on a ridge have two neighboring pixels in the im-mediate neighborhood. If a pixel has only one neighboring pixel, it is determined to be a ridge ending and if a pixel has three neighboring pixels, it is determined to be a ridge bifurcation.

Each of the algorithms used in fingerprint image enhance¬ment and minutiae extraction has its own limitation and results in imperfect processing, especially when the input fingerprint image includes non-friction-ridge noise. As a result, many false minutiae may be detected by the minu¬tiae detection algorithm. To alleviate this problem, often a minutiae postprocessing algorithm is used to confirm or validate the detected minutiae. Only those minutiae that pass this postprocessing algorithm are kept and the rest are removed. For example, if a ridge length running away from the minutia point is sufficient or if the ridge direction at the point is within acceptable limits, the minutia is kept.

Matching

Fingerprint matching can be defined as the exercise of finding the similarity or dissimilarity in any two given fin¬gerprint images. Fingerprint matching can be best visual¬ized by taking a paper copy of a file fingerprint image with its minutiae marked or overlaid and a transparency of a search fingerprint with its minutiae marked or overlaid. By placing the transparency of the search print over the paper copy of the file fingerprint and translating and rotating the transparency, one can locate the minutiae points that are common in both prints. From the number of common minutiae found, their closeness of fit, the quality of the fingerprint images, and any contradictory minutiae match¬ing information, it is possible to assess the similarity of the two prints. Manual fingerprint matching is a very tedious task. Automatic fingerprint-matching algorithms work on the result of fingerprint feature-extraction algorithms and find the similarity or dissimilarity in any two given sets of minutiae. Automatic fingerprint matching can perform fingerprint comparisons at the rate of tens of thousands of times each second, and the results can be sorted accord¬ing to the degree of similarity and combined with any other criteria that may be available to further filter the candidates, all without human intervention.

It is important to note, however, that automatic fingerprint-matching algorithms are significantly less accurate than a well-trained forensic expert.

Even so, depending on the application and the fingerprint image quality, the automatic-fingerprint-matching algorithms can significantly reduce the work for forensic experts. For example, in the case of latent print matching where only a single, very poor quality partial fingerprint image is available for matching, the matching algorithm may not be very accurate. Still, the matching algorithm can return a list of candidate matches that is much smaller than the size of the database; the forensic expert then needs only to manually match a much smaller number of fingerprints. In the case of latent print matching when the latent print is of good quality, or in the case of tenprint-to-tenprint matching in a background check application, the matching is highly accurate and requires minimal human expert involvement.

Automatic fingerprint-matching algorithms yield imperfect results because of the difficult problem posed by large intraclass variations (variability in different impressions of the same finger) present in the fingerprints. These intra¬class variations arise from the following factors that vary during different acquisition of the same finger:

1. Ddisplace¬ment,
2. Rotation,
3. Partial overlap,
4. Nonlinear distortion because of pressing of the elastic three-dimensional finger onto a rigid two-dimensional imaging surface,
5. Pressure,
6. Skin conditions,
7. Noise introduced by the imaging environment, and (8) errors introduced by the automatic feature-extraction algorithms. A robust fingerprint-matching algorithm must be able to deal with all these intraclass variations in the various impressions of the same finger. The variations in displace¬ment, rotation, and partial overlap are typically dealt with by using an alignment algorithm.

The alignment algorithm should be able to correctly align the two fingerprint minutiae sets such that the corresponding or matching minutiae correspond well with each other after the alignment. Certain alignment algorithms also take into account the variability caused by nonlinear distortion. The alignment algorithm must also be able to take into consid¬eration the fact that the feature extraction algorithm is imperfect and may have introduced false minutiae points and, at the same time, may have missed detecting some of the genuine minutiae points. Many fingerprint alignment algorithms exist. Some may use the core and delta points, if extracted, to align the fingerprints. Others use point pattern-matching algorithms such as Hough transform (a standard tool in pattern recognition that allows recogni¬tion of global patterns in the feature space by recognition of local patterns in a transformed parameter space), relaxation, algebraic and operational research solutions, "tree pruning," energy minimization, and so forth, to align minutiae points directly. Others use thinned ridge matching or orientation field matching to arrive at an alignment.

Once an alignment has been established, the minutiae from the two fingerprints often do not exactly overlay each other because of the small residual errors in the alignment algorithm and the nonlinear distortions. The next stage in a fingerprint minutiae-matching algorithm, which estab¬lishes the minutiae in the two sets that are corresponding and those that are noncorresponding, is based on using some tolerances in the minutiae locations and orienta¬tion to declare a correspondence. Because of noise that is introduced by skin condition, recording environment, imaging environment, and the imperfection of automatic fingerprint feature-extraction algorithms, the number of corresponding minutiae is usually found to be less than the total number of minutiae in either of the minutiae sets in the overlapping area. So, finally, a score computation algo¬rithm is used to compute a matching score.

The matching score essentially conveys the confidence of the fingerprint matching algorithm and can be viewed as an indication of the probability that the two fingerprints come from the same finger. The higher the matching score, the more likely it is that the fingerprints are mated (and, conversely, the lower the score, the less likely there is a match). There are many score computation algorithms that are used. They range from simple ones that count the number of matching minutiae normalized by the total number of minutiae in the two fingerprints in the overlapping area to very complex probability-theory-based, or statistical-pattern-recognition-classifier-based algorithms that take into account a number of features such as the area of overlap, the quality of the fingerprints, residual distances between the matching minutiae, the quality of individual minutiae, and so forth. Figure 6–12 depicts the steps in a typical fingerprint match¬ing algorithm. Note that the stages and algorithms described in this sec-tion represent only a typical fingerprint minutiae-matching algorithm.

Many fingerprint minutiae-matching algorithms exist and they all differ from one another. As with the various extraction algorithms, matching algorithms use different implementations, different stages, and different orders of stages. For example, some minutiae-matching al-gorithms do not use an alignment stage. These algorithms instead attempt to prealign the fingerprint minutiae so that alignment is not required during the matching stage. Other algorithms attempt to avoid both the prealignment and alignment during matching by defining an intrinsic coordinate system for fingerprint minutiae. Some minutiae-matching algorithms use local alignment, some use global alignment, and some use both local and global alignment. Finally, many new matching algorithms are totally differ¬ent and are based on the non-minutiae-based features automatically extracted by the fingerprint feature-extraction algorithm, such as pores and texture features.

Indexing and Retrieval

In the previous section, the fingerprint matching problem was defined as finding the similarity in any two given fingerprints. There are many situations, such as controlling physical access within a location or affirming ownership of a legal document (such as a driver's license), where a sin¬gle match between two fingerprints may suffice. However, in a large majority of forensic and government applications, such as latent fingerprint individualization and background checks, it is required that multiple fingerprints (in fact, up to 10 fingerprints from the 10 fingers of the same person) be matched against a large number of fingerprints present in a database. In these applications, a very large amount of fin-gerprint searching and matching is needed to be performed for a single individualization. This is very time-consuming, even for automatic fingerprint-matching algorithms. So it becomes desirable (although not necessary) to use auto¬matic fingerprint indexing and retrieval algorithms to make the search faster.

Traditionally, such indexing and retrieval has been per¬formed manually by forensic experts through indexing of fingerprint paper cards into file cabinets based on finger¬print pattern classification information as defined by a particular fingerprint classification system. Similar to the development of the first automatic finger¬print feature extraction and matching algorithms, the initial automatic fingerprint indexing algorithms were developed to imitate forensic experts. These algorithms were built to classify fingerprint images into typically five classes (e.g., left loop, right loop, whorl, arch, and tented arch) based on the many fingerprint features automatically extracted from fingerprint images. (Many algorithms used only four classes because arch and tented arch types are often dif¬ficult to distinguish.)

Fingerprint pattern classification can be determined by explicitly characterizing regions of a fingerprint as belong¬ing to a particular shape or through implementation of one of many possible generalized classifiers (e.g., neural networks) trained to recognize the specified patterns. The singular shapes (e.g., cores and deltas) in a fingerprint image are typically detected using algorithms based on the fingerprint orientation image. The explicit (rule-based) fingerprint classification systems first detect the fingerprint singularities (cores and deltas) and then apply a set of rules (e.g., arches and tented arches often have no cores; loops have one core and one delta; whorls have two cores and two deltas) to determine the pattern type of the fingerprint image.

The most successful generalized (e.g., neural network-based) fingerprint classification systems use a combination of several different classifiers. Such automatic fingerprint classification algorithms may be used to index all the fingerprints in the database into distinct bins (most implementations include overlapping or pattern referencing), and the submitted samples are then compared to only the database records with the same classification (i.e., in the same bin). The use of fingerprint pattern information can be an effective means to limit the volume of data sent to the matching engine, resulting in benefits in the system response time. However, the auto¬matic fingerprint classification algorithms are not perfect and result in errors in classification. These classification errors increase the errors in fingerprint individualization because the matching effort will be conducted only in a wrong bin. Depending on the application, it may be feasible to manually confirm the automatically determined finger¬print class for some of the fingerprints where the auto¬matic algorithm has low confidence. Even so, the explicit classification of fingerprints into just a few classes has its limitations because only a few classes are used (e.g., five), and the fingerprints occurring in nature are not equally distributed in these classes (e.g., arches and tented arches are much more rare than loops and whorls).

Many of the newer automatic fingerprint classification algo¬rithms do not use explicit classes of fingerprints in distinct classifications but rather use a continuous classification of fingerprints that is not intuitive for manual processing but is amenable to automatic search algorithms. In continuous classification, fingerprints are associated with numerical vectors summarizing their main features. These feature vectors are created through a similarity-preserving transfor¬mation, so that similar fingerprints are mapped into close points (vectors) in the multidimensional space. The retrieval is performed by matching the input fingerprint with those in the database whose corresponding vectors are close to the searched one. Spatial data structures can be used for indexing very large databases.

A continuous classification approach allows the problem of exclusive membership of ambiguous fingerprints to be avoided and the system's efficiency and accuracy to be balanced by adjusting the size of the neighborhood considered. Most of the continu¬ous classification techniques proposed in the literature use the orientation image as an initial feature but differ in the transformation adopted to create the final vectors, and in the distance measure. Some other continuous indexing methods are based on fingerprint minutiae features using techniques such as geo¬metric hashing. Continuous indexing algorithms can also be built using other non-minutiae-based fingerprint features such as texture features.

9

Scientific Validation of Fingerprint Evidence

INTRODUCTION

A fingerprint comparison was first used as evidence against a defendant in 1892 in Argentina. The first case in the United States was 1904 (Cole, 2001). There are no records of how many criminal defendants have been charged and how many convicted in federal and state courts in the United States since 1904 based totally or partially on fingerprint evidence. Speculative estimates point to over 100,000 indictments in the United States in 100 years (Cole, 2005, suggests even more). Most fingerprint cases escape the scrutiny of a trial. The defendant pleads guilty (sometimes to a lesser charge that may not involve the fingerprint evidence) and waives trial. Altschuler (2005) estimated that 95% of such felony cases are adjudicated in this fashion. Fingerprint evidence has been accepted virtually without challenge or question over this 100 year history.

The United States Supreme Court's Daubert ruling (Daubert v. Merrill Dow Pharmaceuticals, 1993) resulted in a barrage of challenges to the admissibility of fingerprint evidence in federal courts (and in state courts using comparable criteria). The FBI's Onin website (Legal Challenges to Fingerprints, 2005) lists over 40 Daubert and state court rulings in criminal cases since the first one (US v. Byron Mitchell, 2000). The basis of these challenges by the defense has been that the primary method used to compare fingerprints—the Analysis-Comparison-Evaluation-Verification (ACE-V) method--lacks the scientific documentation of its validity required under Daubert. Every court has ruled to admit fingerprint evidence, based on fingerprint's 100 year history of accepted practice, and on the assertions of fingerprint examiners that there is adequate scientific support of the method.

A number of legal scholars have recently published analyses of these Daubert rulings. They focus primarily on what they consider to be misinterpretations of the legal meanings of the Daubert criteria, made both by the government and by the Daubert courts themselves. In the present article we focus explicitly on the appropriate scientific basis to demonstrate validity.

We write this article as research scientists, and describe the available evidence, the kinds of evidence needed, and experiments that would provide that evidence. We conclude that the ACE-V method has not been tested for validity, and until the necessary work is performed to quantify the method and insure that examiners are using the method correctly and consistently, the method cannot be validated.

VALIDITY AND RELIABILITY

Daubert courts often refer to the reliability of scientifically based evidence; we discuss evidence of the validity of the ACE-V method. The distinction is important because a reliable method can consistently produce the same wrong answer, whereas a valid method consistently produces the true answer.

A method is valid when its application produces conclusions in agreement with ground truth. In the case of fingerprint comparisons, ground truth is either certain knowledge that the latent and suspect's fingerprints were made by the same person, or certain knowledge that they were made by two different people. The amount of validity of a method is usually expressed as the converse of an error rate: the percent of time that the conclusion, based upon rigorous application of the method, agrees with ground truth. If scientific evidence shows that the application of the ACE-V method to produce fingerprint individuation evidence results in conclusions that agree with ground truth with a high probability, the method has met the good science requirements of the Daubert decision by the US Supreme Court.

Evidence can also be assessed for its reliability (Cole, 2006). A reliable method is one that produces, for the same comparison, the same result every time it is used, both by many examiners comparing the same set of latent to suspect prints, and by the same examiners (unknowingly) repeating a comparison they had made previously. The amount of reliability is usually expressed as a correlation: the amount of agreement between repeated uses of the method under comparable conditions. A high correlation of reliability indicates that experts using the method reach the same conclusion nearly every time. Daubert should not be directly concerned with reliability. A highly reliable method (producing the same result every time) may still be wrong, that is, be invalid and disagree with ground truth. The flat earth belief was held reliably by nearly every observer for centuries and they were all wrong. However, a method cannot be valid if it is unreliable. If the method produces varying results each time it is used, some of those results are incorrect, and hence the method is invalid. Therefore, the scientific focus to meet the Daubert good science requirements should be on validity—the agreement of the method's conclusions with ground truth.

The Steps of Our Analysis

We divide our analysis of evidence for the validity of the ACE-V method into six parts.

First, we describe the four stages of the ACE-V method. Second, we turn to the kinds of evidence needed to demonstrate the validity of the ACE-V method and describe a prototypic experiment that would provide such evidence, similar to experiments assessing the validity of any method. As part of that description, we show that a search of the research literature fails to uncover any instance of such an experiment applied to ACE-V, and we show that the experimental design requires a set of prerequisites which have yet to be met.

Third, we describe and respond to the arguments presented by the government that ACE-V has already been tested and that it exhibits a zero error rate. Fourth, we analyze the standards that underlie each of the four conclusions examiners draw, based on application of the ACE-V method. We describe the kind of evidence needed to demonstrate the validity of each standard, and for each, describe an experiment that could provide such evidence. From the published literature, we document evidence that application of the standards as presently practiced produces highly variable and therefore invalid results.

Fifth, we discuss the scientific implications of the courts' reliance upon fingerprint examiners rather than research scientists for evaluation of the validity of the method. Sixth, we conclude that there is no scientific evidence for the validity of the ACE-V method, and, that until the prerequisites for specifying the method and its application are met, the ACE-V cannot be tested for its validity.

THE ANALYSIS-COMPARISON-EVALUATION-VERIFICATION METHOD

Fingerprint examiners argue that there are unique and permanent combinations of features on the skin of fingers (and palms and feet, though we confine ourselves in this article to fingers). Further, they argue that images of these patterns (called fingerprints) can be used to individuate people by the proper application of a comparison method. When a perpetrator of a crime touches a surface with a finger and leaves an image of the unique pattern from that finger (called a latent fingerprint), that latent fingerprint image can be found, lifted, and compared to the images of the fingers of a suspect (called exemplar fingerprints, and usually recorded by a trained technician on a ten-print card). Following a method of fingerprint comparison such as the ACE-V, and depending on the examiner's training and experience in the comparison method, the examiner can offer an opinion about ground truth: whether the crime scene latent fingerprint was made by the suspect or by someone else.

The FBI claims in Daubert hearings (Meagher, 1999) that all examiners now use the ACE-V method to make these conclusions, and that there are no other methods in use by fingerprint examiners today. Ashbaugh (2005b) makes a similar claim. Although alternative methods are mentioned in textbooks

(Olsen & Lee, 2001), practicing examiners uniformly refer to the fingerprint comparison method they employ as the ACE-V. Therefore, we restrict our discussion to evidence for the validity of the ACE-V method.

A Description of ACE-V

Because the ACE-V method may not be familiar to some readers outside the fingerprint profession, we provide a brief overview. However, neither the International Association for Identification (IAI) as the professional organization of fingerprint examiners, the FBI, nor any other professional fingerprint organization has provided an official description of the ACE-V method, so our description is based on the most detailed of the published sources.

Huber (1959, 1972) first described the structure of this method, which he applied to every forensic identification discipline, but without suggesting a name. The classic FBI Science of Fingerprints (1958, 1988) contains only a few pages on comparison procedures, but neither refers to that method as ACE-V nor distinguishes among its different steps. Ashbaugh (1999) provides much more detail and examples, in what has become the most influential textbook available. Champod, et al. (2004) offers an even more precise description, spelled out in the form of steps in a flow chart. P. Wertheim (2002) also has a relatively detailed description, which he has used as a model in his training courses. Beeton (2001) gives a shorter version, contrasting some of the differences between Ashbaugh (1999) and P. Wertheim (2002); and Triplett and Cooney (2006) also comment on some of the differences among the accounts.

What follows is a theoretical description, distilled primarily from the authors cited above, and from our own training in IAI-sponsored latent fingerprint courses. We know from examiner testimony offered in Daubert hearings and in trials involving testimony from fingerprint examiners that most practicing fingerprint examiners deviate from this description. (We consider below the implications of the facts that there is no agreed-upon description of the ACE-V method in the fingerprint profession, no professional body has approved any one description as the official ACE-V method, and that individual examiners vary in their practice.)

Analysis stage. The Analysis stage begins when a fingerprint examiner looks at a latent fingerprint and decides whether it contains sufficient quantity and quality of detail so that it exceeds the standard for value. If the quantity and quality of detail does exceed the value standard, then the examiner continues the analysis. If the value decision is negative, the latent fingerprint is not used further. The majority of latent prints found at crime scenes are rejected as of no value (Meagher, 2002).

If the fingerprint examiner continues Analysis of the latent print (he has not yet seen the suspect's exemplar prints), he uses the physical evidence contained in the latent print and that produced by the crime scene investigation to determine which finger made the print, the nature of the surface on which it was deposited, the amount and direction of pressure used in the touch, and the matrix (such as sweat) in which the ridge details of the finger were transferred onto the surface. This analysis is necessary to specify each of the sources of distortion in the latent print that causes the inevitable differences between the latent fingerprint and the patterns of features found on the skin (and on the exemplar image).

The examiner then chooses one feature-rich area of the latent print (preferably near a core or delta). Within this area, he selects the particular features along the various ridge paths in the latent print, in their spatial locations relative to one another, to use to start the comparison between the crime scene latent and the suspect's exemplar prints.

Comparison stage. In the Comparison stage, for the first time, the examiner looks at the suspect's ten exemplar fingerprints. He starts with the most likely finger of the exemplar image, based on what he found during the analysis of the latent print. The examiner goes to the same area of the suspect fingerprint that he had selected in the latent print to determine whether the same patterning of features occurs. If it does not, the examiner concludes that that finger of the suspect cannot be the finger which made the latent print: an exclusion of that finger. He then goes to the next finger of the exemplar image, and repeats this process. If all ten fingers can be excluded as the source of the crime scene latent print, the examiner excludes the suspect as the donor.

If the same pattern of features initially noted in the latent print is found in the corresponding area of one of the suspect's exemplar prints, the examiner goes back to the latent print, selects another area and locates the features there and their relative positions to each other. Then the exemplar is again examined in the new area, to determine whether the corresponding features are also present. This latent-to-exemplar comparison (always in that order) continues until all of the features in the latent print have been compared for agreement of features in corresponding locations in the suspect's exemplar. If substantial agreement is found, the examiner goes to the Evaluation stage.

Throughout Comparison the examiner keeps track of every failure to find a correspondence between the latent and the suspect fingerprint. Any failure in agreement that cannot be accounted for by one of the distortions previously described and labeled in the Analysis stage is a necessary and sufficient condition to exclude the suspect as the perpetrator with the application of the one-unexplained-discrepancy standard. The most common conclusion of the Comparison stage is exclusion (Meagher, 2002).

Evaluation stage. In Evaluation, the examiner applies a sufficiency standard to the amount of corresponding agreement between the latent and the exemplar that dictates his conclusion. If the amount of corresponding agreement exceeds the sufficiency standard, then the examiner concludes that the crime scene latent print can be individuated to the suspect. If the amount of agreement does not exceed the standard, then the conclusion is neither an individuation nor an exclusion—an inconclusive conclusion. Two kinds of sufficiency standards obtain. The first is numeric, in which the amount of agreement is stated as a number, and the threshold for sufficiency is determined by the profession or the crime laboratory. The second is experiential, based on the individual examiner's training and experience.

Verification stage. Verification is employed in larger laboratories for cases in which an examiner has concluded individuation. A second examiner confirms the conclusion of the first. A verification standard describes the rules by which a verifier is selected, informed of the history of the latent print's comparisons, reports his findings, and how conflicting conclusions are resolved.

Scoring the Accuracy of the Four Conclusions

In the overview of ACE-V presented above, the examiner has made four kinds of conclusions: value, exclusion, individuation, or inconclusive. These are described in two reports by the Scientific Working Group on Friction Ridge Analysis, Study and Technology (SWGFAST, 2002a, b). The classification of these four conclusions as either correct or incorrect requires knowledge of the ground truth. The accuracy of the method in reaching true conclusions, or its converse, its error rate, is critical to admissibility under Daubert. We review here the meaning of correct or incorrect for each of the four conclusions. We then present the evidence for ACE-V accuracy with respect to these four conclusions.

Ground truth. Ground truth is certain knowledge that the latent and an exemplar fingerprint came either from the same donor or from two different donors. Ground truth cannot be known in case work, and therefore, research using results from case work cannot be used to establish the validation of the method. In case work, the purpose of investigation, fingerprint comparison, indictment and trial is to find out as much as possible about ground truth. Neither the examiner's opinion nor the jury's verdict is ground truth, though both (or neither) may be consistent with it. The no value and the inconclusive conclusions always miss the correct answer. Whether the prints came from one donor or from two, neither of these conclusions agrees with ground truth. When an examiner reaches either of these conclusions, either an innocent suspect still remains at risk of indictment and conviction, or a guilty perpetrator still remains at large.

The exclusion conclusion is correct when the two prints are from two different donors (and an innocent suspect is released from suspicion), and erroneous when the same donor made both prints (and a guilty perpetrator remains at large). The individuation conclusion is correct when one donor made both fingerprints (and indictment and conviction of the perpetrator is likely), and erroneous when two different donors made the two fingerprints (and an innocent suspect is at risk of indictment and conviction). We now turn to evidence of the validity of the ACE-V method to reach conclusions that agree with ground truth.

Is the ACE-V Method Valid?

Demonstration that an expert used a scientifically validated method (one that has been shown to produce conclusions that agree with ground truth) is intended to assure the court of the accuracy of that method used in the instant case. Similarly, the published error rate information informs the court of the amount of confidence that can be placed in a conclusion based on the method used to reach that conclusion. The validity and the error rate of the method concern the error inherent in the method, as distinct from evidence of any particular individual practitioner's accuracy in applying that method. We return in a later section to arguments that proficiency testing of examiners provides evidence of validity of the method itself. In several Daubert hearings concerning forensic methods other than fingerprints, such as polygraphs (US v. Scheffer, 1999) or ear prints (Washington v. Kunze, 1999), the court has rejected a method specifically on the grounds that the validity of the method has never been demonstrated scientifically.

AN EXPERIMENTAL DESIGN TO TEST THE VALIDITY OF THE ACE-V METHOD

A test of the validity of a methodology (such as the ACE-V) requires an experiment. Because the validity of the method is being tested (the probability that conclusions based on the method agree with ground truth), all other potential sources of error must be controlled or eliminated from that experiment. The subjects are skilled examiners who each compare a number of pairs of prints for which the ground truth is known. For each pair, the examiners apply the ACE-V method, they document the application of the method at each step by completing a response form and report, and they state the conclusion supported by the method. Ideal working conditions are present during the experiment.

The validity of the method (the converse of its error rate) can be computed from the amount of agreement found between the conclusions produced by the method and ground truth, as in Table 1. All deviations from perfect validity in these results (an error rate greater than zero) can be directly attributable to the method, and not to other possible sources of error.

Prerequisite Conditions

With respect to ACE-V, the following four prerequisite conditions must be met in order to perform the validity experiment. Description of the ACE-V method. Definitions of each step in the ACE-V method must be provided.

To do this, the profession has to write and then adopt a manual describing in detail the steps of the ACE-V method. The ACE-V method has yet to be officially described and endorsed as an agreed upon method. The published versions of it differ significantly. Each of the many calls within the profession for the production of a complete, official description of the method implies its absence. For example, the Interpol created a working group in 2000 to define a method for fingerprint identification, and issued a report urging its member countries to adopt a common method (Interpol, 2005). A recent FBI report included a demand for a complete description of the method (Smrz, et al., 2006). At present, the term "ACE-V" refers to some methodological steps which have not been well described, and to differing procedures. Until the method is specified and endorsed, there is no method to test.

Report form for the ACE-V method. Documentation that the examiners are using the ACE-V method in the experiment must be provided. To do this, the profession has to write and then adopt a report form that examiners complete that shows that each step is followed. At present, examiners using the ACE-V method in this country are not required by anyone to record the steps they took to reach their conclusion. Their reports provide only their conclusions, based on the prints examined (see Table 3 for an example of a typical report). Proficiency and certification tests offered by the profession do not require examiners to show their work, so there is no way to determine the extent to which they followed ACE-V correctly and consistently. Nor are examiners required to document their comparison method steps when they testify to an identification in court. In contrast, adequate report forms that show all the steps of an examiner's work are required in other countries, such as Canada (Ashbaugh, 2005a). Unless examiners show their work, their results cannot be used as evidence of the validity of the method employed.

Standardized training. Standardized training programs in the ACE-V method are necessary to ensure that examiners are properly trained, and that the method is uniformly applied. The IAI (2004), TWGFAST (1998), SWGFAST (2006), FBI (Meagher, 2002), and ASCLD (2001) have each provided some guidelines for a training syllabus for the ACE-V method. These recommendations are general, not specific. They are not requirements, and (with the possible exception of the FBI training syllabus, described by Meagher, 2002, but not published), there is no evidence that any of these recommendations are actually written or in practice. The training programs themselves must include assessment procedures, so it can be determined whether individual trainees or working examiners have learned and use the steps of the method correctly. A formalized training program in the official ACE-V method needs to be designed, including specific goals and their assessment. This program must be adopted and required by the profession.

Most examiners receive the majority of their training on-the-job, without either a formal structure of topics covered or formal assessment of success in meeting training goals. Variation in training means that variation in the use of a method must also occur.

Proficiency assessment. The subjects who serve in the validity experiment must have demonstrated high proficiency in the ACE-V method, so that any errors made can be attributed to the method and not to poor performance by the examiners. To do this, subjects must have specific training on the approved manual, and, most importantly, have demonstrated high skill in using the method. However, present assessment of proficiency of examiners in their training and in their case work is incomplete and inadequate (Haber & Haber, 2004) for several reasons.

For example, proficiency tests and procedures have never been assessed for their validity or their reliability (Haber & Haber, 2004). The validity of a proficiency test would be shown by high correlation with other independent measures of skill and ability, such as supervisor ratings, or the quality and quantity of training and experience. The proficiency test manufacturers have never reported any correlations with these independent measures, so nothing is known about the validity of thee tests. Further, no information has ever been reported on the reliability of these tests, the degree to which examiners receive the same score when they take a comparable form of the test again. If not reliable, they cannot be valid. If the present tests were assessed, it is likely that those presently in use, such as the IAI certification test (IAI, 2004), the IAI-ASCLD proficiency test (Koehler, 1999; Cole, 2005), and the FBI (Meagher, 2002) would fail to exhibit acceptable levels of validity or reliability.

In addition, none of the proficiency tests contains fingerprints of known difficulty, because the profession lacks a quantitative measure of print quality (difficulty). One expert observed that the prints used in the FBI proficiency test are so easy they are a joke (Bayles, 2002).

Further, the prints used in proficiency tests do not reflect normal casework. They are predominately or entirely of value, in contrast to case work, in which the majority of latent prints are of no value. These proficiency tests do not include many, if any, exclusions, though, again, the most common outcome in case work is exclusion. When an examiner receives a particular score on such a test, it is impossible to interpret that score other than relative to other examiners who took the same test. The results cannot be generalized to the examiner's performance on the job, or accuracy in court, because the difficulty of the test items is unknown, and the other parameters do not correspond to normal case work.

Consequently, in the absence of a standardized description of ACE-V, a standardized syllabus and training objectives, and measures of performance during each step of a comparison, it is impossible to assess an examiner's level of proficiency (accuracy) in the ACE-V method. If the validity experiment is to address the accuracy of the method, highly skilled examiners must perform the comparisons. At present, there is no objective way to select such a group.

What Is the Evidence for ACE-V Validity?

No such experiment has ever been offered as evidence in any of the Daubert hearings, nor has one ever been published. A recent FBI publication (Budowle, et al., 2006), Haber and Haber (2004), and Cole (2005) reported failure to find a single peer reviewed study that tested the validity of ACE-V. As further evidence of the absence of such an experiment, the National Institute of Justice issued a solicitation for research proposals in 2000 that included testing the validity of the method (NIJ, 2000). That solicitation was never funded, amid controversy (Samuels, 2000; US v. Plaza, 2001). A comparable solicitation was repeated late in 2004 (NIJ, 2004).

A recent experiment by Wertheim, Moenssens, and Langenburg (2006) purported to provide some measure of error rates for examiners using the ACE-V method. The experimenters recorded the accuracy of a large number of latent-to-exemplar comparisons from 108 examiners, collected over several separate week-long training courses. They found that their 92 participants with more than one year of experience made only 81 erroneous individuations out of a total of 6,441: less than a 2% error rate. We provided a detailed critique of this experiment (Haber & Haber, 2006) in which we argued that the experiment cannot be used to assess experimenter error rate (proficiency). This experiment did not use uniformly well-trained examiners, it was carried out in a training and not a case work environment, the difficulty of the latents was adjusted to the participant's skill level by the course instructor to insure success in this part of the course requirement, the conditions did not match normal working conditions, every latent being compared always had a correct match in the packet, no latent print could correctly be judged of no value, extra help in the form of hints was available from the instructor, and participants could determine on which latents they were to be scored, thereby greatly reducing the risk of reaching a wrong conclusion (partially completed packets could be returned without penalty). Finally, no documentation was required as to what steps the participants followed when making comparisons. Therefore, the results cannot be used to estimate examiner error rate when using the ACE-V method.

EVIDENCE OF VALIDITY PRESENTED BY FINGERPRINT EXAMINERS

Examiners in Daubert hearings have asserted that ACE-V has been tested, and that the error rate is very low, or even zero (Meagher, 1999). To support these claims, they offer evidence from history, from adversarial testing, from verification testing, from quality of training and experience, from types of errors that could be made, from publicity, from the scientific method, and from proficiency test results. We summarize and critique there these eight claims that the validity of the ACE-V has been demonstrated.

100 YEARS OF FINGERPRINT HISTORY

Many government witnesses have testified that the validity of the ACE-V method has been tested by its continued acceptance throughout 100 years of history (e, g., Meagher, 2002). However, the hundred years of acceptance provides only "a face" validity which argues that fingerprint individuations must be accurate because people have believed them for a long time. Face validity is not a scientific test of accuracy, only of belief. Further, no one has claimed that the ACE-V method, as distinct from fingerprint evidence, has been accepted for 100 years. As recently as 1988, the FBI's Science of Fingerprints does not refer to the comparison method as ACE-V.

Adversarial Testing

Government witnesses in some Daubert hearings have argued that the results of application of the ACE-V method are tested through the adversarial process during each trial itself (e.g., US v. Havvard, 2001). Since it is claimed in these hearings that no erroneous individuations have ever been uncovered during direct and cross examination, this procedure of testing shows the error rate of the method must be zero. Adversarial testing does not provide a mechanism to assess the error rate of ACE-V. Ground truth is unknown during adversarial proceedings, so the outcome cannot be used to assess validity of the method being used. Further, the vast majority of cases involving fingerprint evidence result in plea bargains, or the fingerprint evidence goes unchallenged and therefore never are subjected to the adversarial process or to second opinions.

Verification Testing

Some government witnesses have argued that verification procedures represent a testing of the method, since a second examiner checks the conclusion of the first (US v. Havvard, 2001), thereby guaranteeing that errors do not occur. Verification testing fails in several ways to provide evidence of validity. In case work verification testing, ground truth is unknown, and agreement between two examiners might mean either that they both were correct in the identification, or that they both made an error either by chance or carelessness, or because some property of the method led both to make the error. Further, most verification testing in crime laboratories is non-blind, which permits contamination and bias to reduce the chances of detecting errors (Haber, 2002; and see our discussion under the Standards Criterion below). Crime laboratories closely guard and do not publish results on the number of verifications they do, the number of those that produced different conclusions, how those differences were resolved, and whether the differences are resolved in ways that reduce errors. The extent to which errors are reduced by current practice is simply unknown. Cole (2005) lists a number of instances of erroneous identifications made in court, nearly all of which had been verified by another examiner, each thereby counting as two erroneous identifications. This list is evidence that errors do occur that get past verifiers.

Skilled Examiners do not Make Errors

The government has claimed that erroneous identifications are only made by poorly trained or inexperienced practitioners. When the method is used by well trained and experienced examiners, no errors are ever made, so that the method itself is error-free (Meagher, 2002; 2003). A number of scientists have noted that if the cause of errors is attributed to practitioners because of their inadequate training and experience, examiner training and experience need to be standardized and tested, and the results of those tests must be known for each examiner before an error occurs. Otherwise, this reasoning is circular (Cole, 2006). The three FBI examiners who concurred in the misidentification of the Madrid bomber (Stacey, 2004) were among the most senior, most experienced, and most trained at the FBI.

Errors involving Value, Exclusion or Inconclusive are not Important

Meagher (2002) has also argued on behalf of the government that the only methodological error rate of relevance to Daubert courts is the percent of erroneous individualizations. Other kinds of errors that can result from the application of ACE-V, including missed identifications based on erroneous exclusions and inconclusive conclusions, are irrelevant and should not be considered in evaluating the validity of ACE-V. If there are no erroneous identifications, then the error rate of the method is zero. This argument fails to consider that erroneous individuations and erroneous exclusions are necessarily highly correlated. If an examiner wants to avoid the possibility of making erroneous individuations, he can simply make fewer identification conclusions. Doing so obviously results in an increase in the rate of perpetrators missed. If the crime laboratory, profession or courts want to minimize erroneous identifications, they can guarantee that outcome by increasing the punishment to examiners for making such an error. From a scientific standpoint, assessment of a method's error rate must reflect the ratio of correct to incorrect conclusions that result from application of that method.

Publicity

As further evidence of a zero error rate, Meagher (2002) reported that in his 35 years working for the FBI, he had never seen or heard of an erroneous identification made by an FBI agent. Since such an error would be widely publicized, a lack of publicity can assure the court that no such errors had ever occurred. When FBI Agent Meagher claimed that no erroneous identifications by FBI agents have occurred because he had never heard of one, he failed to consider that the chances of uncovering an erroneous identification are remote. Most fingerprint identifications are not challenged in court, either because the defendant pled to some other charge, or because the defense did not obtain a second opinion. Further, after conviction, the opportunities for innocent persons to obtain new evidence and have their convictions reviewed and overturned are still extremely rare.

ACE-V is based on the Scientific Method

Several prominent examiners (Ashbaugh, 1999; K. Wertheim, 2003), have argued that that because ACE-V involves hypothesis testing by an examiner, which is a component of the scientific method, that comparability means that the ACE-V method itself is as valid as the scientific method is valid. This argument is flawed. An analogy between the ACE-V method and scientific methods of hypothesis testing does not provide evidence of accuracy. It is an analogy, and nothing more. A method is demonstrably accurate (valid) when its application consistently produces conclusions that agree with ground truth.

Proficiency Test Results

Evidence from practitioner performance testing (proficiency and certification) is claimed by the government (Cole, 2005) to provide estimates of error rates of the ACE-V method. This argument is strange, because (Haber & Haber, 2004) and Cole (2005) have shown that the published proficiency and certification test results show many errors, including erroneous identifications. We would be the first to argue that because the proficiency tests themselves do not reflect casework and have not been validated, they cannot be used as evidence of examiner accuracy—or inaccuracy. More importantly, because none of the published tests in use requires the examiner to document the method he used to reach his conclusions, the results of these tests cannot be used evidence for the validity (accuracy) of ACE-V.

None of the eight arguments offered by the government involves scientific tests and none addresses assessment of the validity of the ACE-V method. Cole (2006) provides a detailed examination of the fallacy of these arguments with respect to scientific testing. He concludes, with us, that none of these arguments bears even indirectly on whether the ACE-V method is a valid procedure on which to base a conclusion.

STANDARDS REQUIRED FOR THE VALIDITY OF A METHOD

An evaluation of a scientific methodology includes evidence that the conclusions reached from the application of the method rest on quantitative standards that themselves have been shown to be valid. Validity of a standard requires that the standard is objective, tested, and has been calibrated against ground truth. We consider the four standards on which ACE-V rests: value, exclusion, individuation and inconclusive (SWGFAST, 2002a, b). Each of these conclusions rests on a separate standard that can be evaluated for its agreement with ground truth. Validation of these standards involves separate experiments, which themselves require prerequisites before they can be run.

Analysis Standard: Value

The first standard the examiner applies concerns whether there is sufficient ridge and feature detail in the latent print to attempt a subsequent comparison: whether the latent print is of value. Value is often expressed in terms of the difficulty of the latent print. If the examiner judges the amount of detail in the latent to exceed this value standard, he continues to analyze the latent fingerprint. If he judges the amount of detail insufficient to meet the value standard, he desists from further work on that latent.

Demonstration of the validity of the value standard requires experimental evidence that highly skilled examiners agree on whether a specific latent print is of value for comparison. If the value standard is to rest on a quantitative basis, as it must for the standard itself to be quantitative, it requires three prerequisites: a quantitative metric of the difficulty of latent prints; evidence that examiners (or computer algorithms) can evaluate the difficulty of individual latents consistently; and a quantitative statement by the profession that latent prints that fail to exceed a specific threshold of difficulty are deemed of no value for comparison. We consider these three prerequisites briefly. A quantitative metric of difficulty requires an experiment. A large number of latent fingerprints are randomly selected from case work. Random selection insures that the proportion of latents of no value in this sample of latents reflects the typical number found in casework. Skilled examiners are then asked to evaluate each latent print in two ways. First, they are to judge the overall difficulty of the latent on a rating scale. Second, they evaluate each print on separate rating scales for a number of specific dimensions expected to predict difficulty.

These include, for example, overall size, level one classification, presence of focal features such as a core or delta, presence of orientation information, contrast, smear, amount of distortion, level two features such as ridges that can be followed or the presence of minutiae on the ridges, and level three features such as pores and ridge edges. Analysis of the rating scale scores across the latent prints would establish the different dimensions of difficulty, so that a total difficulty score could be calculated for each latent, a score that is weighted for the relative importance of each rating scale determined by multivariate analysis of the rating scales. From these data, it is possible to rate objectively the difficulty of any new latent fingerprint on these scales.

The validity of the use of these latent print difficulty scales can then be demonstrated by an experiment. A random sample of examiners can be given a selection of latent prints ranging in rated difficulty (as described above).

If examiners can use these scales accurately, they would agree on the presence of the difficulty variables, and the overall difficulty level of each latent. Once such a metric is established, computer programs could be created so that latent print difficulty could be assessed speedily and uniformly. Yao, et al. (2004) provide an initial attempt to create such a program. No experiment to quantify latent print difficulty has ever been published. No experiment has been performed to determine the consistency with which examiners judge a particular latent print to be of value. However, two kinds of available evidence suggest examiners differ in their value judgments.

First, skilled examiners assert (Ashbaugh, 1999, 2005b; K. Wertheim, 2003) that the value of a latent depends on the training and experience of the examiner, so that one examiner might judge a latent as of no value, whereas another might be willing to proceed to compare it and trust his conclusions. If the standard is training-and-experience dependent, then definitions of difficulty AND definitions of training and experience are each required before an objective value standard can be validated. Second, many of the proficiency tests administered for the IAI and for crime laboratories have required the examiners taking the test to indicate the value of the latent prints. The reported data show that the variation (error rate) on that conclusion is quite high on those tests (Haber & Haber, 2004).

Comparison Standard: One-Unexplained-Discrepancy

During Comparison, the examiner may conclude an exclusion of a suspect's exemplar digit(s) on the basis of the one-unexplained discrepancy standard (Thornton, 1977). This standard is critical to the profession because its absence or violation abrogates the uniqueness principle underlying forensic comparison. Demonstration of the validity of the one discrepancy standard requires experimental evidence that examiners differentiate accurately between two prints that arise from distortions only (a known ground truth of a single donor) and genuine discrepancies (a known ground truth of two donors). No experimental evidence has been published on the one unexplained discrepancy standard.

An experiment to demonstrate the validity of the one discrepancy standard would score examiners on the accuracy with which they apply this standard. The test materials would consist of a set of exemplar-latent pairs, for which all latents are of value, and the ground truth of each pair is known. Starting with the first pair, each examiner would first do a complete analysis of the latent print. The examiner documents a complete record of the analysis (including, for this experiment, special attention to distortions) on a detailed report form. When the examiner completes the analysis by selecting an area of the latent print to use for the beginning of comparison, the same portion of the exemplar print is provided, with the rest of the exemplar print masked off. The examiner's task is to describe the amount of agreement found, note any discrepancies, and for each discrepancy found, account for its cause from the analysis of the distortions in the latent he has already provided. If any of these discrepancies are inexplicable on the basis of the distortions, he should conclude elimination. If every discrepancy can be accounted for by distortions in the latent print, the examiner should put that pair aside, and go onto the next latent print. If examiners are able to use the one-unexplained discrepancy standard, then every exclusion conclusion should be matched with a ground truth of different donors.

However, this experiment cannot be carried out until the following prerequisites are met (Table 4). The profession needs to write and adopt a report form for the Analysis stage which includes a checklist of distortions potentially present in every latent, and a formalized mode of describing the grounds on which each distortion was judged to be explicable or inexplicable during Comparison. Until this report form is available, the validity of the one-unexplained discrepancy standard cannot be demonstrated.

Indirect evidence suggests that examiners differ in their judgments regarding an unexplained discrepancy. Many examiners begin Analysis by looking at the exemplar print and latent print side by side, or, during Comparison, looking back at the latent for features newly discovered in the exemplar. This practice lets the examiner "reshape" the latent. When the analysis of the latent is contaminated by knowledge of the patterning in the exemplar, the one-discrepancy standard is open to errors of bias (Haber & Haber, 2005). Using features from the exemplar to identify ones in the latent creates opportunities for the examiner to overlook inexplicable discrepancies (Ashbaugh, 1999; 2005b).

Evaluation Standard: Numeric Sufficiency

During Evaluation, the examiner may conclude an individuation or inconclusive, on the basis of the number of features in agreement between the two fingerprints. If the amount exceeds the numeric sufficiency standard, the examiner concludes that the two fingerprints have the same donor. If the amount of agreement fails to meet the sufficiency threshold, he concludes that the suspect's finger can neither be included nor excluded as the donor of the latent print.

Demonstration of the validity of the sufficiency standard requires evidence that the conclusions, based on the standard, are consistent with ground truth. Before this experiment can be run, four prerequisites must be met (Table 4). First, the profession must specify the features found in latent fingerprints that are to be counted. Second, the profession needs to specify how the location of each feature in the latent is defined, either in absolute locations, or relative to every other countable feature. Third, the profession needs to specify rules that determine what constitutes agreement of features by kind and location between latent and exemplar, and what constitutes non-agreement. Fourth, evidence is needed to determine whether examiners apply these rules consistently.

Once these prerequisites are fulfilled, then an experiment to demonstrate the validity of the numeric sufficiency standard can be carried out. The purpose of the experiment is to establish a quantitative metric of agreement between latent and exemplar, with a concomitant probability of matching ground truth. Skilled examiners are given a large number of latent prints, all of value but of varying difficulty. Each latent print is to be fully analyzed. Then, each latent print is paired with an exemplar, some of same donor ground truth, some of different donor, and the pair is to be compared and evaluated. The examiner records every corresponding point in agreement by type and location between the two prints.

It is to be expected that as the number of features in agreement increases, the probability of ground truth being a single donor increases. For example, it might be found that when ten points are found in agreement, 80% of the pairs have a ground truth of one donor, and 20% of the pairs with ten points in agreement have two donors. It might be further found that 95% of the pairs with 25 points in agreement have a ground truth of one donor, with only 5% having two donors. The profession could then decide what value to choose for the sufficiency standard, depending on the percentage of erroneous identifications the profession is willing to accept when the examiner concludes identification.

The available evidence suggests that skilled examiners vary in the number of features they perceive in a latent, and in the number of features they count as in agreement in latent and exemplar. Langenburg (2004) reported that even skilled examiners differed as to the number of minutiae they labeled in latent prints. Evett and Williams (1996) reported data showing substantial variation in judgments from highly experienced examiners in the amount of agreement between each of 10 latent-exemplar pairs (all of value). On one pair, judgments ranged from 14 to 51 points of agreement. The Langenburg and Evett and Williams results show the critical importance to the profession of specifying what constitutes a feature, and what constitutes agreement. The current variability among examiners means that a quantitative test of a sufficiency standard based on number of features in agreement will fail to meet reasonable standards of validity. Until 1973, examiners in the United States based their conclusions on a numeric sufficiency standard. In 1973 the IAI ruled that all numeric standards based on a point count of correspondences be abandoned, because there was no scientific evidence to support any particular numeric standard (IAI, 1973; Stacey, 2005; Budowle, et al., 2006). No experiment such as the one we describe has been carried out.

EVALUATION STANDARD: TRAINING AND EXPERIENCE

In 1973, the numeric sufficiency standard was replaced by a subjective, personal "training and experience" sufficiency standard (Ashbaugh, 2005b). The examiner asserts the equivalent of the following under oath: "Based on my training and my experience, I have never seen this amount of agreement between two fingerprints obtained from two different people—therefore I am absolutely confident that this amount of agreement means that the same person made both fingerprints" (wording suggested by Ashbaugh, 2005b). However, when an examiner refers to his own experience as the source of the standard, he forgets that his experience (case work) has no access to ground truth. Hence, there is no way for him to determine whether some amount of agreement is an identification grounded in truth. An experiment to test the validity of a training and experience sufficiency standard has several prerequisites. Training needs to be specified in quantitative units, and experience needs to be specified in quantitative units. Neither of these exists today.

Given that current training is not standardized, and that experience is heterogeneous, and that the profession has not grappled with defining either training or experience quantitatively, neither of those definitions can be specified today. Therefore, the subjective personal sufficiency standard based on training and experience in its present form cannot be validated. Once training and experience are quantitatively defined, an experiment to demonstrate the validity of this sufficiency standard could be performed. Examiners of varying training and experience are asked to compare pairs of latent and exemplar prints of known ground truth, representative of case work in difficulty. The presumed results are that as training and experience increase, so do agreement with ground truth. If sufficient accuracy were found, the profession could then adopt a training and experience sufficiency criterion: a given amount of quantified training and quantified experience assures resulting conclusions with a known probability of being correct.

Because practicing examiners differ greatly in the amount and kinds of training and experiences they have undergone, it is expected that different examiners will reach different conclusions when comparing the same set of prints. There is ample evidence that this is in fact the case, both among highly trained and experienced examiners, and across the entire spectrum of training and experience (Evett & Williams, 1996; the Mitchell FBI survey reported in US v. Mitchell, 2000; and certification examination results, Grieve, 1996). Further, neither the court nor the examiner himself has access to any measure of ground truth on which to estimate the accuracy of the examiner's conclusion. It is simply his opinion, based on his personal experience and training. This subjective and opinion-based standard differs from a numerical standard, in that with the latter, the validity of the standard can be tested. With the former, which is now the standard required by the IAI, the probability of error is untested, and cannot be tested in current practice.

Verification Standard: Blind Replication

The final standard is stipulated by the fingerprint profession itself, and is under the control of the crime laboratory, not the individual examiner. During Verification of an individuation conclusion, a second examiner compares the latent and exemplar prints. The blind component of the standard (Smrz, et al., 2006; Stacey, 2005; SWGFAST, 2006) requires that in the absence of knowledge of any prior conclusion or who made it, the verifying examiner must reach the same conclusion as the first examiner. However, current practice in nearly all crime laboratories in which verification is performed is non-blind.

Validation of the verification standard requires evidence of the accuracy of the replication process: the percent of errors made by the first examiner that are uncovered by the verifier. Research on quality control (Arvizo, 2002, 2003; Boone, et al., 1982) shows that non-blind verification catches relatively few errors, whereas blind verification, and especially double blind verification, catches many more errors. Blind verification is standard procedure in governmental testing laboratories (for example, FDA Guidelines, 2001), and is a requirement for publication in virtually all scientific journals.

Fingerprint examiners have testified for the government that non-blind verification catches most errors (Meagher, 2002). This claim can be tested in an experiment by comparing non-blind with single blind procedures. "Problem" pairs of latent-exemplar prints known to have resulted in mistakes in training, on proficiency tests, or in court are given to a number of examiners who are asked to complete an ACE-V procedure and offer a conclusion. Ground truth for these pairs may not be known, but the selection of pairs is restricted to those for which expert examiners have consistently agreed which of the pairs are identifications and which exclusions. Half of the pairs (of both identifications and exclusions) are distributed among the examiners with the notation that this pair has already been individuated (or excluded) by a skilled examiner and they are to confirm the conclusion (non-blind). The other half of the pairs (of both identifications and exclusions) is distributed among the examiners as if each is a new pair, and each examiner is asked to use ACE-V to reach a conclusion (blind). The results would compare, between the non-blind and blind conditions, the number of times the examiners differed from the experts' classification of the pair.

This experiment has never been reported in this form. However, Dror and his colleagues have reported several experiments to show that skilled examiners can be induced to change their initial conclusion after presentation of biasing information. Dror, Charlton and Peron (2006) selected five senior examiners (who had previously volunteered to be tested from time to time for research purposes) and asked them to do a complete ACE-V on a single latent-exemplar pair. The pair for each examiner was selected from that examiner's file (unbeknownst to the examiner), in which the examiner had made an identification conclusion at least five years earlier. The pair in each case was re-presented to the examiners as an example of the erroneous identification of Brandon Mayfield made by the FBI (Stacey, 2004). While the Mayfield case had just broken and the five examiners were familiar with the facts, none of them had yet seen the Mayfield latent or exemplar print and none remembered the pair they had personally individuated years earlier. One of the 5 examiners changed his previous identification conclusion to inconclusive, and 3 changed their previous identification to exclusion. Only one verified his previous identification. We equate the biasing information presented in this experiment with non-blind replications: the examiner has information about the expected outcome of his work.

In a follow up experiment, Dror and Charlton (2006) tested six examiners on four pairs they had previously concluded identification and four pairs they had previously excluded. When the 48 pairs were now presented a second time (again, unbeknownst to the examiners), half were accompanied by biasing information (for the earlier exclusions, the examiners were now told the suspect confessed; for the earlier identifications, they were told that the suspect had been in jail at the time of the crime). For the 24 pairs on which bias was introduced, four instances of identification were changed to exclusion. No exclusions were changed to identification. For the 24 pairs on which no bias was introduced, one examiner changed an identification to an exclusion conclusion, and one changed an exclusion to an identification. This finding shows variability in accuracy even in the absence of bias. Dror's work shows that non-blind replication led to change consistent with the introduced bias. Even more troubling in the context of the validity of ACE-V, Dror's data show that highly skilled examiners reach different conclusions with themselves over time, even in blind replications of their own work. See also Dror, Peron, Hind & Charlton, 2005) for another example, and Dror & Rosenthal (2006) for further statistical analyses of the experiments discussed here.

EVIDENCE PRESENTED BY THE GOVERNMENT ON THE VALIDITY OF THE FOUR STANDARDS

The Daubert ruling does not refer explicitly to validation of the standards inherent in a method. We, as research scientists, have described the kinds of evidence needed to demonstrate the validity of the four conclusions a fingerprint examiner may draw from application of the ACE-V method. Fingerprint examiners, testifying on behalf of the government in Daubert hearings, have not addressed them, other than negatively to note that the numeric sufficiency standard was abandoned in 1973 due to lack of validation.

WHICH COMMUNITY OF EXPERTS: EXAMINERS OR RESEARCHERS?

The 40+ courts in which the validity of fingerprint evidence was challenged have unanimously ruled that this evidence was admissible. This decision reflects the testimony of fingerprint examiners, who affirm that the ACE-V method is valid, and that fingerprint evidence has been accepted for one hundred years. One of the criteria used under Daubert concerns evidence that the members of relevant scientific communities accept the method as valid (Saks, 2003; Faigman, et al., 1997, 2002). Fingerprint examiners certainly comprise a relevant community. However, with rare exceptions, they are untrained in experimental science and lack the qualifications to evaluate the validity of a method.

When other scientifically based methods have undergone the scrutiny of a Daubert hearing, most, if not all of the testimony has rested on evidence presented by scientists trained to evaluate the research evidence of the validity (or non-validity) of the method in question. These scientists all belong to a community of researchers who have the training and skills to carry out and publish the empirical studies that demonstrate the method's validity. However, in the Daubert hearings on the acceptance of the validity of the ACE-V method, fingerprint examiners argued that the fingerprint profession is qualified to speak to the validity of the method they use, and that only they are qualified to provide the scientific evidence of the validity of the method used to make the forensic comparisons (See Cole, 2005, for a review of these claims). The fingerprint community is the scientific group intended under Daubert (Meagher, 2002). In some Daubert hearings, the dissenting testimony by scientists has been ruled inadmissible (US vs. Havvard, 2001).

The uniform rulings to accept fingerprint examiners as the relevant scientific community is surprising, given the 1993 Daubert Court ruling. Furthermore, in accepting practitioners as the relevant community, these courts have failed to recognize a dichotomy in the backgrounds and training of the experts providing the testimony in the hearings. For example, no research scientist familiar with comparative procedures underlying ACE-V has ever testified in a Daubert hearing or submitted an Amicus brief on the government side. No research scientist who has testified in a Daubert hearing on the validity of the ACE-V method has accepted that the validity of the ACE-V method has been demonstrated.

As a different example, legal scholars familiar with the scientific evidence on the validity of ACE-V have never testified or submitted an amicus brief in a Daubert hearing on the government side. When legal scholars have testified in Daubert hearings regarding fingerprint evidence, they have been almost unanimous in attesting to the absence of scientific validity evidence of the ACE-V method. A substantial legal scholarship published literature attacks the fingerprint profession for its failure to test the validity of the ACE-V method (Saks, 2003; Faigman, et al, 1997; 2002). As a final example of this dichotomy, in a recent Amicus Brief submitted to the Massachusetts Supreme Judicial Council (Massachusetts v. Patterson, 2005), five Ph.D. scientists and 12 legal scholars signed the brief opposing the admission of fingerprint evidence based on problems with its validity. In the same case, the government briefs were contributed only by fingerprint examiners or prosecutors, and did not include a single contribution from a research scientist or a legal scholar.

Fingerprint evidence has a 100 year history of court acceptance, but the ACE-V has not been systematically tested for validity. Belief in its accuracy by the professionals who practice the method does not constitute evidence of acceptance by a relevant scientific community.

10

Recognition of Fingerprint

INTRODUCTION

WHAT IS A FINGERPRINT?

A fingerprint is the feature pattern of one finger. It is believed with strong evidences that each fingerprint is unique. Each person has his own fingerprints with the permanent uniqueness. So fingerprints have being used for identification and forensic investigation for a long time.

Fig. A fingerprint image acquired by an Optical Sensor

A fingerprint is composed of many ridges and furrows. These ridges and furrows present good similarities in each small local window, like parallelism and average width. However, shown by intensive research on fingerprint recognition, fingerprints are not distinguished by their ridges and furrows, but by Minutia, which are some abnormal points on the ridges. Among the variety of minutia types reported in literatures, two are mostly significant and in heavy usage: one is called termination, which is the immediate ending of a ridge; the other is called bifurcation, which is the point on the ridge from which two branches derive.

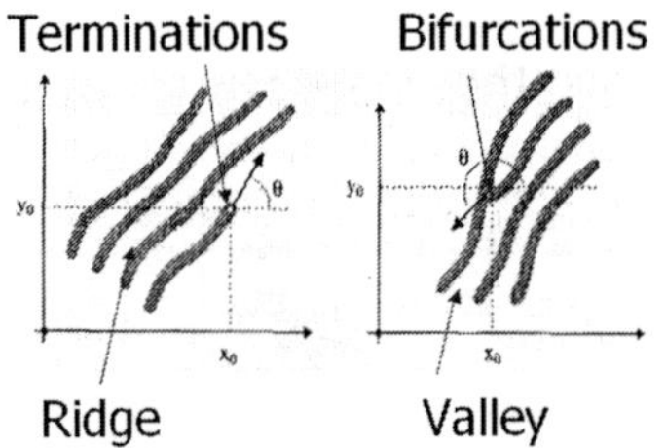

Fig. Minutia. (Valley is also referred as Furrow, Termination is also called Ending, and Bifurcation is also called Branch)

WHAT IS FINGERPRINT RECOGNITION?

The fingerprint recognition problem can be grouped into two sub-domains: one is fingerprint verification and the other is fingerprint identification. In addition, different from the manual approach for fingerprint recognition by experts, the fingerprint recognition here is referred as AFRS (Automatic Fingerprint Recognition System), which is program-based.

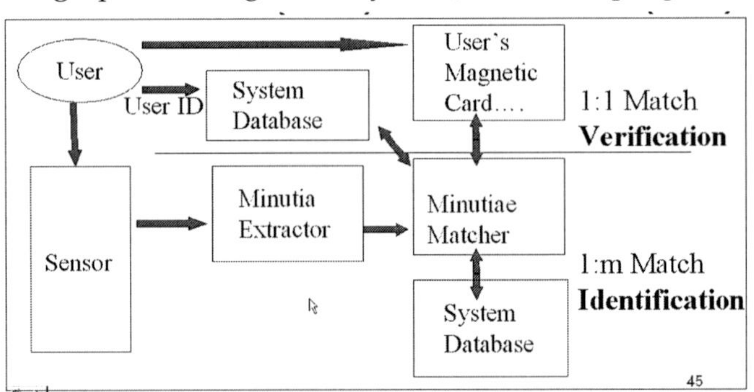

Fig. Verification vs. Identification

Fingerprint verification is to verify the authenticity of one person by his fingerprint. The user provides his fingerprint together with his identity information like his ID number. The fingerprint verification system retrieves the fingerprint template according to the ID number and matches the template with the real-time acquired fingerprint from the user. Usually it is the underlying design principle of AFAS (Automatic Fingerprint Authentication System).

Fingerprint identification is to specify one person's identity by his fingerprint(s). Without knowledge of the person's identity, the fingerprint identification system tries to match his fingerprint(s) with those in the whole fingerprint database. It is especially useful for criminal investigation cases. And it is the design principle of AFIS (Automatic Fingerprint Identification System). However, all fingerprint recognition problems, either verification or identification, are ultimately based on a well-defined representation of a fingerprint. As long as the representation of fingerprints remains the uniqueness and keeps simple, the fingerprint matching, either for the 1-to-1 verification case or 1-to-m identification case, is straightforward and easy.

TWO APPROACHES FOR FINGERPRINT RECOGNITION

Two representation forms for fingerprints separate the two approaches for fingerprint recognition. The first approach, which is minutia-based, represents the fingerprint by its local features, like terminations and bifurcations. This approach has been intensively studied, also is the backbone of the current available fingerprint recognition products. I also concentrate on this approach in my honors project. The second approach, which uses image-based methods, tries to do matching based on the global features of a whole fingerprint image. It is an advanced and newly emerging method for fingerprint recognition. And it is useful to solve some intractable problems of the first approach. But my project does not aim at this method, so further study in this direction is not expanded in my thesis.

HISTORY OF FINGERPRINT RECOGNITION

Fingerprint imaging technology has been in existence for centuries. The use of fingerprints as a unique human identifier dates back to second century B.C. China, where the identity of the sender of an important document could be verified by his fingerprint impression in the wax seal .

The first modern use of fingerprints occurred in 1856 when Sir William Herschel, the chief magistrate of the Hooghly district in Jungipoor, India, had a local businessman, Rajyadhar Konai, impress his handprint on the back of a contract. Later, the right index and middle fingers were printed next to the signature on all contracts made with the locals. The purpose was to frighten the signer of repudiating the contract because the locals believed that personal contact with the document made it more binding. As his fingerprint collection grew Sir Herschel began to realize that fingerprints could prove or disprove identity .

The 19th century introduced systematic approaches to matching fingerprints to certain individuals. One systematic approach, the Henry Classification System, based on patterns such as loops and whorls, is still used today to organize fingerprint card files . In the late 1960's NEC worked with the FBI and the Home Office in London, which had been working on a system for New Scotland Yard from the late 1960's, to eventually develop a minutia-based fingerprint identification system. It was initially installed in Tokyo in 1981 and in San Francisco in 1983.

The first country to adopt a national computerized form of fingerprint imaging was Australia in 1986, which implemented fingerprint imaging technology into its law enforcement system . In 1996 after nearly a year of study, the National Institute of Standards and Technology has been convinced that minutia is an acceptable way to store fingerprint biometric data on smart cards. With the NIST acceptance of minutia it became inevitable this would set an industry standard.

MINUTIA-BASED ALGORITHM

Minutia-based algorithms extract information such as ridge ending, bifurcation, and short ridge from a fingerprint image.

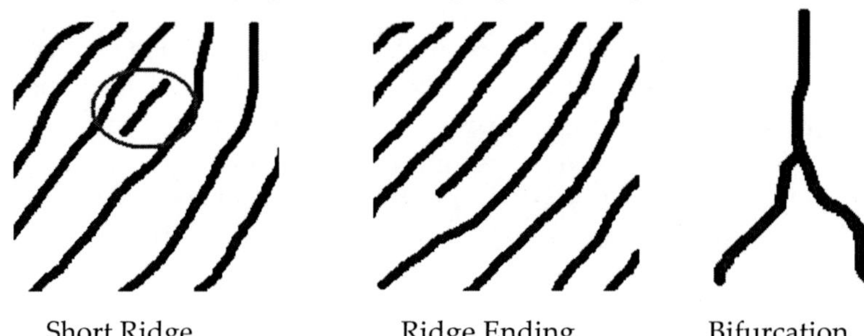

Short Ridge Ridge Ending Bifurcation

These features are then stored as mathematical templates. The identification or verification process compares the template of the live image with a database of enrolled templates (identification), or with a single enrolled template (authentication).

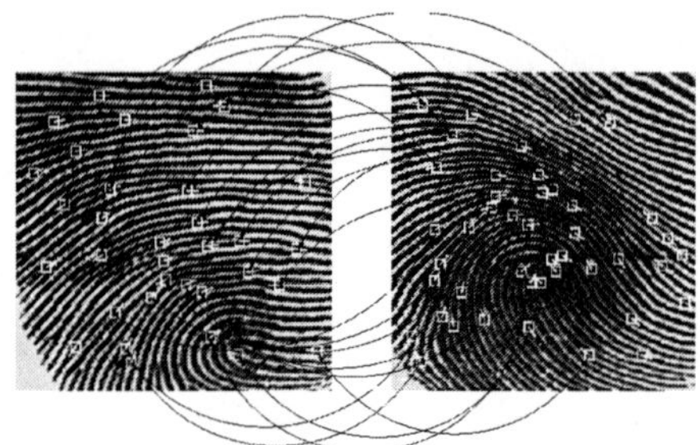

(Image source National Institute of Standards and Technology)

People with no or few minutia points (special skin conditions) cannot enroll or use the system effectively. This is exemplified by the fingerprint immigration programs where finger moistening peripherals are standard. Moreover, a low number of minutia points can be a limiting factor for security of the algorithm. This can lead to false minutia points (areas of obfuscation that appear due to low-quality enrollment, imaging, or fingerprint ridge detail). In an application environment, enrollment without assistance may take several attempts due to poor position or lack of pressure. While not quantified, user frustration will certainly have a negative impact on technology acceptance.

Moreover, the widely used minutiae-based representation does not utilize a significant component of the rich discriminatory information available in the fingerprints.

Spectrum Analysis

Utilizing research from Nagoya Institute of Technology Graduate School in Japan, DDS has developed an algorithm based on Spectrum Analysis. This technique captures cross sections of a sliced fingerprint pattern and converts them to waves. Spectrum analysis uses the spectral series of the waves as feature information, finding the maximum correlations in the wave and verifies the identity of the fingerprint.

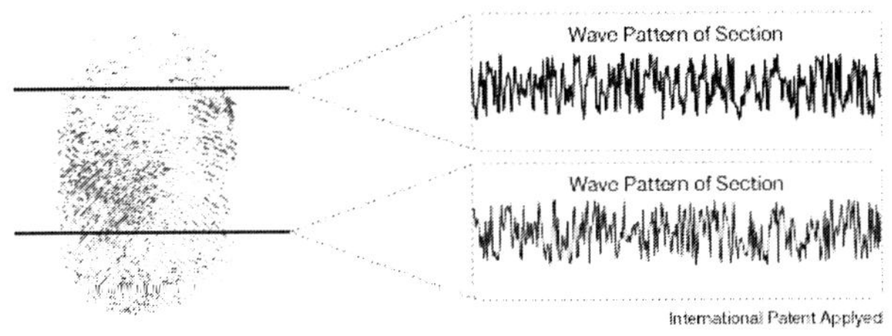

This algorithm of spectrum analysis works extremely well because this algorithm extracts characteristics from the concavo-convex information of a fingerprint without being influenced by the position of the characteristic points used in the conventional Minutia and Pattern-matching method. In the course of verification under the spectrum analysis algorithm, it is not necessary to store the fingerprint image itself in the system which eliminates the possibility of exposure or leakage of fingerprint images. In principle, it is impossible to regenerate original fingerprint image from the extracted characteristics of images. This addresses issues raised by the IEEE on fingerprint reconstruction of minutiae based systems.

The algorithm performs extremely well in controlled environments where positioning of the finger in enrollment and verification are similar. However, with disparate fingerprint positioning for enrollment and verifications results can be less than desired. This requirement limits the application developer to more controlled ergonomic environments and may reduce some commercial viability.

The Hybrid technology

To accentuate the strengths of both the Spectrum Analysis and Minutiae-based algorithm and limited the inherent weaknesses, the company has combined both algorithms and created a Hybrid algorithm. The combined technology provides for rapid and accurate enrollment and verification in difficult environments. This Hybrid algorithm extends beyond combining scores from both techniques to form a single decision; instead the technology utilizes a proprietary technique of "Shading". Shading analyzes the scores of both results and places and associates an importance value with each. Using an algorithm based upon a database of past results, a final score is calculated by using the individual results as a function of importance.

With Shading, if one score is high and one is low, more decision weighting is placed upon the higher score. If both scores are low then information from both are weighed more equally and the results are combined together for a final decision. This ability allows for both algorithms to have low scores and still be accurate.

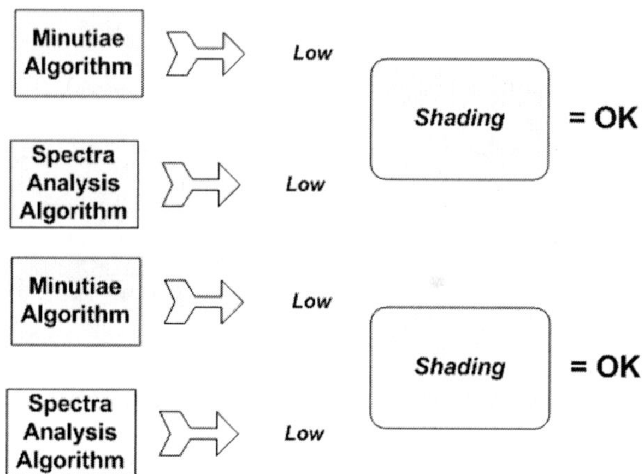

The technology addresses both the strengths of both as well as the challenges. Spectrum analysis can work well with poor image quality and difficult to read fingerprints and minutiae-based algorithms can work very

Hybrid Enrollment Examples

Conclusion

By combining the Spectrum Analysis algorithm and Minutiae based algorithms together, the Hybrid technology brings better performance by utilizing the benefits of both algorithms. Moreover, by applying a shading algorithm the technology enhances difficult to read fingerprints caused by poor print quality and poor application ergonomics.

With a more flexible enrollment and matching capability, the system works well with difficult fingerprints, user ergonomics, and partial captures. With this increased robustness, application developers are less restricted with end user applications. Thus is produced a lower cost of support and lower costs of sales as well as less resistance to technology.

SYSTEM LEVEL DESIGN

A fingerprint recognition system constitutes of fingerprint acquiring device, minutia extractor and minutia matcher.

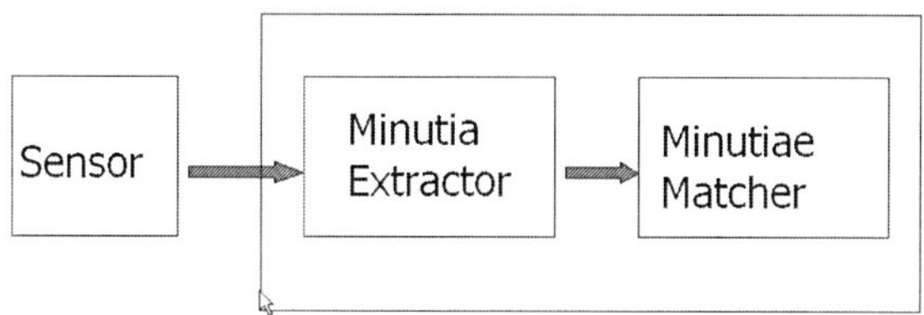

Fig. Simplified Fingerprint Recognition System

For fingerprint acquisition, optical or semi conduct sensors are widely used. They have high efficiency and acceptable accuracy except for some cases that the user's finger is too dirty or dry. However, the testing database for my project is from the available fingerprints provided by FVC2002 (Fingerprint Verification Competition 2002). So no acquisition stage is implemented. The minutia extractor and minutia matcher modules are explained in detail in the next part for algorithm design and other subsequent sections.

ALGORITHM LEVEL DESIGN

To implement a minutia extractor, a three-stage approach is widely used by researchers. They are preprocessing, minutia extraction and postprocessing stage.

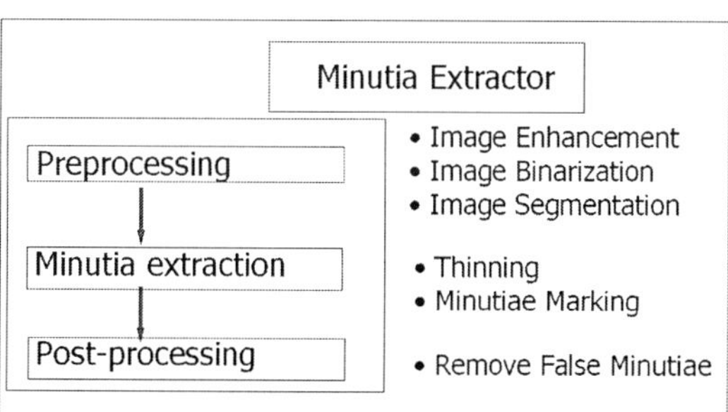

Fig. Minutia Extractor

For the fingerprint image preprocessing stage, I use Histogram Equalization and Fourier Transform to do image enhancement . And then the fingerprint image is binarized using the locally adaptive threshold method . The image segmentation task is fulfilled by a three-step approach: block direction estimation, segmentation by direction intensity and Region of Interest extraction by Morphological operations. Most methods used in the preprocessing stage are developed by other researchers but they form a brand new combination in my project through trial and error. Also the morphological operations for extraction ROI are introduced to fingerprint image segmentation by myself.

For minutia extraction stage, three thinning algorithms are tested and the Morphological thinning operation is finally bid out with high efficiency and pretty good thinning quality. The minutia marking is a simple task as most literatures reported but one special case is found during my implementation and an additional check mechanism is enforced to avoid such kind of oversight. For the postprocessing stage, a more rigorous algorithm is developed to remove false minutia based. Also a novel representation for bifurcations is proposed to unify terminations and bifurcations.

Minutia Matcher

- Ridge correlation to specify reference minutia pair
- Align two fingerprint images
- Minutiae Match

Fig. Minutia Matcher

The minutia matcher chooses any two minutia as a reference minutia pair and then match their associated ridges first. If the ridges match well , two fingerprint images are aligned and matching is conducted for all remaining minutia.

FINGERPRINT IMAGE PREPROCESSING

FINGERPRINT IMAGE ENHANCEMENT

Fingerprint Image enhancement is to make the image clearer for easy further operations. Since the fingerprint images acquired from sensors or other medias are not assured with perfect quality, those enhancement methods, for increasing the contrast between ridges and furrows and for connecting the false broken points of ridges due to insufficient amount of ink, are very useful for keep a higher accuracy to fingerprint recognition. Two Methods are adopted in my fingerprint recognition system: the first one is Histogram Equalization; the next one is Fourier Transform.

Histogram Equalization:

Histogram equalization is to expand the pixel value distribution of an image so as to increase the perceptional information. The original histogram of a fingerprint image has the bimodal type, the histogram after the histogram equalization occupies all the range from 0 to 255 and the visualization effect is enhanced.

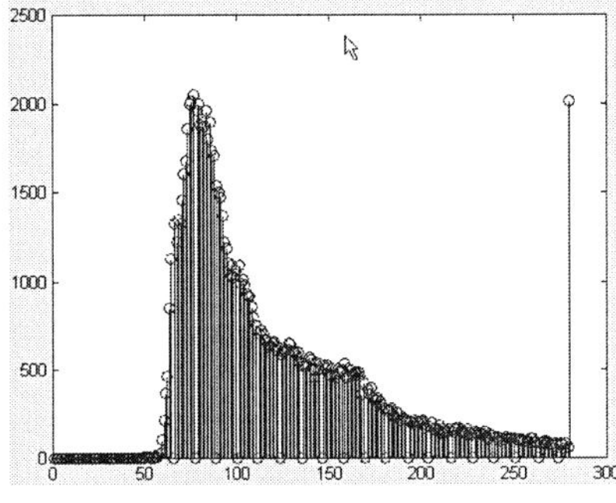

The Original histogram of a fingerprint image

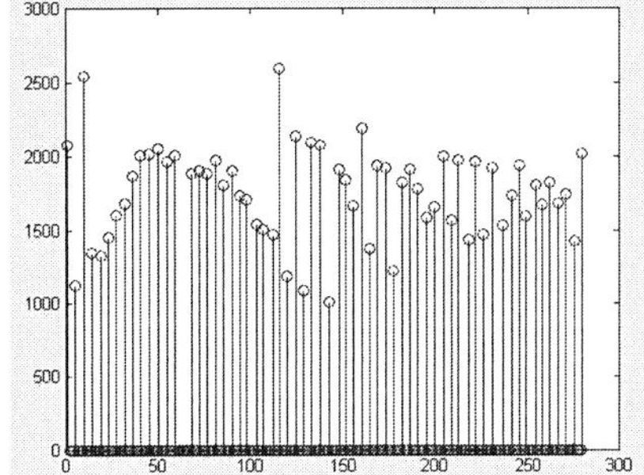

Histogram after the Histogram Equalization

The right side of the following figure is the output after the histogram equalization.

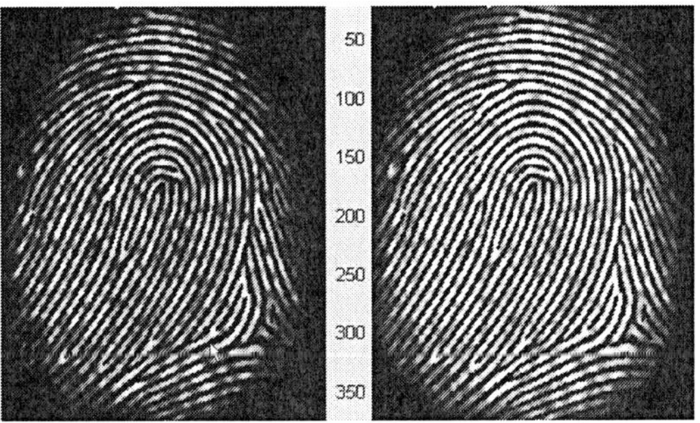

Fig. Histogram Enhancement.

Original Image (Left). Enhanced image (Right)

Fingerprint Enhancement by Fourier Transform

We divide the image into small processing blocks (32 by 32 pixels) and perform the Fourier transform according to:

$$F(u,v) = \sum_{x=0}^{M-1}\sum_{y=0}^{N-1} f(x,y) \times \exp\left\{-j2\pi \times \left(\frac{ux}{M} + \frac{vy}{N}\right)\right\}$$

for u = 0, 1, 2, ..., 31 and v = 0, 1, 2, ..., 31.

In order to enhance a specific block by its dominant frequencies, we multiply the FFT of the block by its magnitude a set of times. Where the magnitude of the original FFT = abs(F(u,v)) = |F(u,v)|.

Get the enhanced block according to

$$g(x,y) = F^{-1}\left\{F(u,v) \times |F(u,v)|^{K}\right\},$$

where F-1(F(u,v)) is done by:

$$f(x,y)=\frac{1}{MN}\sum_{x=0}^{M-1}\sum_{y=0}^{N-1}F(u,v)\times\exp\left\{j2\pi\times\left(\frac{ux}{M}+\frac{vy}{N}\right)\right\}$$

for x = 0, 1, 2, ..., 31 and y = 0, 1, 2, ..., 31.

The k in formula is an experimentally determined constant, which we choose k=0.45 to calculate. While having a higher "k" improves the appearance of the ridges, filling up small holes in ridges, having too high a "k" can result in false joining of ridges. Thus a termination might become a bifurcation.

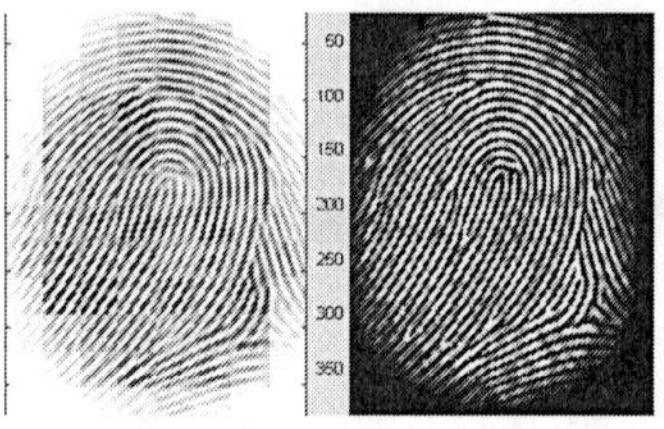

Fig. Fingerprint enhancement by FFT

Enhanced image (left), Original image (right): The enhanced image after FFT has the improvements to connect some falsely broken points on ridges and to remove some spurious connections between ridges. The shown image at the left side of figure above is also processed with histogram equalization after the FFT transform. The side effect of each block is obvious but it has no harm to the further operations because I find the image after consecutive binarization operation is pretty good as long as the side effect is not too severe.

FINGERPRINT IMAGE BINARIZATION

Fingerprint Image Binarization is to transform the 8-bit Gray fingerprint image to a 1-bit image with 0-value for ridges and 1-value for furrows. After the operation, ridges in the fingerprint are highlighted with black color while furrows are white.

A locally adaptive binarization method is performed to binarize the fingerprint image. Such a named method comes from the mechanism of transforming a pixel value to 1 if the value is larger than the mean intensity value of the current block (16x16) to which the pixel belongs.

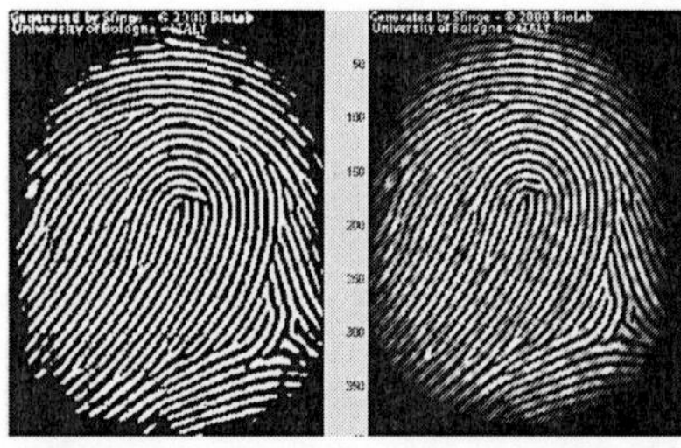

Fig. The Fingerprint image after adaptive binarization

Binarized image(left), Enhanced gray image(right)

FINGERPRINT IMAGE SEGMENTATION

In general, only a Region of Interest (ROI) is useful to be recognized for each fingerprint image. The image area without effective ridges and furrows is first discarded since it only holds background information. Then the bound of the remaining effective area is sketched out since the minutia in the bound region are confusing with those spurious minutia that are generated when the ridges are out of the sensor. To extract the ROI, a two-step method is used. The first step is block direction estimation and direction variety check , while the second is intrigued from some Morphological methods.

Block direction estimation

Estimate the block direction for each block of the fingerprint image with WxW in size(W is 16 pixels by default). The algorithm is:

I. Calculate the gradient values along x-direction (gx) and y-direction (gy) for each pixel of the block. Two Sobel filters are used to fulfill the task.

II. For each block, use Following formula to get the Least Square approximation of the block direction.

$tg2\beta = 2\ \Sigma\Sigma\ (gx*gy)/\Sigma\Sigma\ (gx^2-gy^2)$ for all the pixels in each block.

The formula is easy to understand by regarding gradient values along x-direction and y-direction as cosine value and sine value. So the tangent value of the block direction is estimated nearly the same as the way illustrated by the following formula.

$$tg2\theta = 2\sin\theta\ \cos\theta/(\cos 2\theta\ -\sin 2\theta)$$

After finished with the estimation of each block direction, those blocks without significant information on ridges and furrows are discarded based on the following formulas:

$$E = \{2\Sigma\Sigma\ (gx*gy) + \Sigma\Sigma\ (gx^2-gy^2)\}/\ W*W*\Sigma\Sigma\ (gx^2 + gy^2)$$

For each block, if its certainty level E is below a threshold, then the block is regarded as a background block. The direction map is shown in the following diagram. We assume there is only one fingerprint in each image.

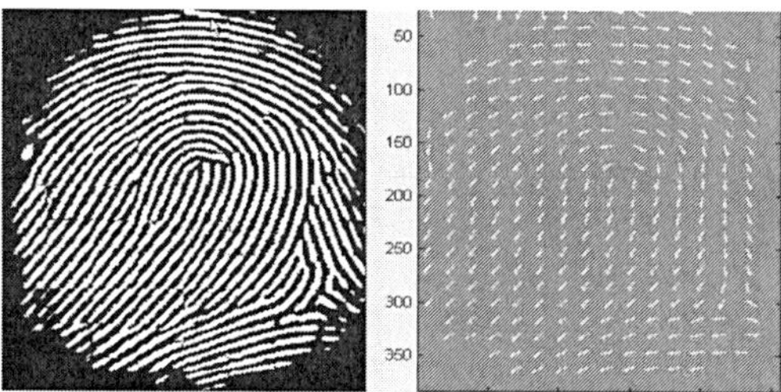

Fig. Direction map.

Binarized fingerprint (left), Direction map (right)

ROI extraction by Morphological operations

Two Morphological operations called 'OPEN' and 'CLOSE' are adopted. The 'OPEN' operation can expand images and remove peaks introduced by background noise. The 'CLOSE' operation can shrink images and eliminate small cavities.

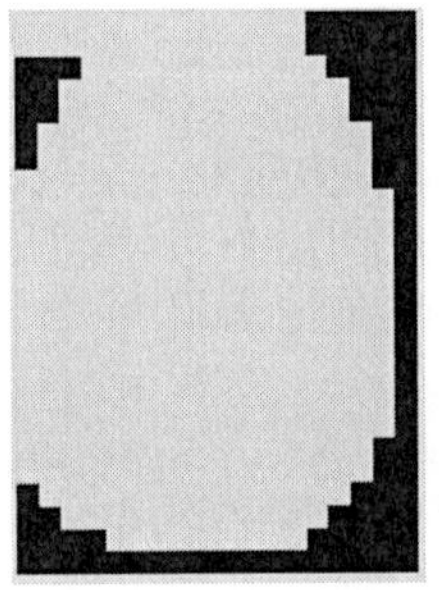

Fig. Original Image Area

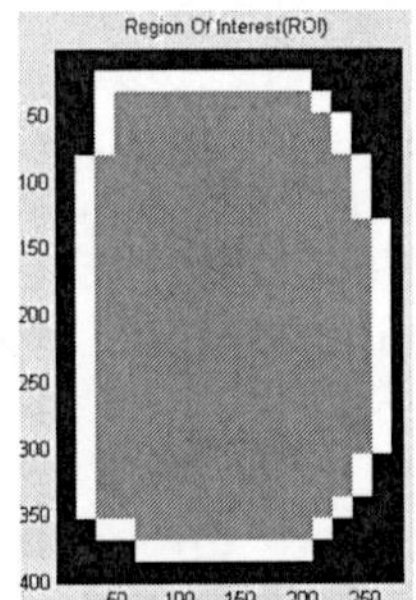

Fig. After CLOSE operation

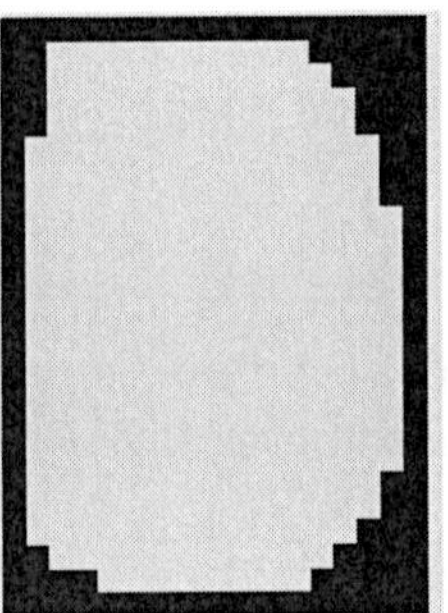

Fig. After OPEN operation

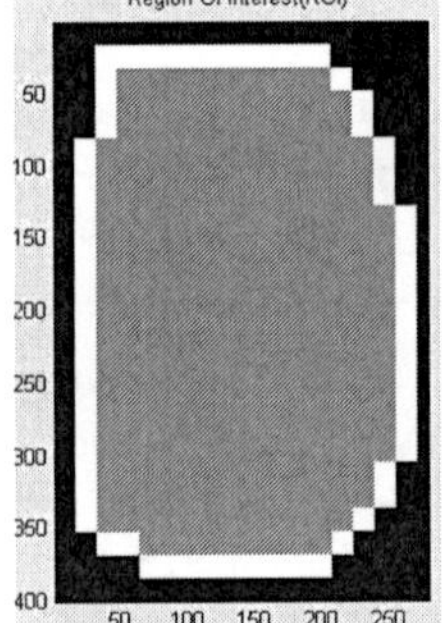

Fig. ROI + Bound

The interest fingerprint image area and its bound. The bound is the subtraction of the closed area from the opened area. Then the algorithm throws away those leftmost, rightmost, uppermost and bottommost blocks out of the bound so as to get the tightly bounded region just containing the bound and inner area.

FINGERPRINT RIDGE THINNING

Ridge Thinning is to eliminate the redundant pixels of ridges till the ridges are just one pixel wide. uses an iterative, parallel thinning algorithm. In each scan of the full fingerprint image, the algorithm marks down redundant pixels in each small image window (3x3). And finally removes all those marked pixels after several scans. In my testing, such an iterative, parallel thinning algorithm has bad efficiency although it can get an ideal thinned ridge map after enough scans. uses a one-in-all method to extract thinned ridges from gray-level fingerprint images directly. Their method traces along the ridges having maximum gray intensity value. However, binarization is implicitly enforced since only pixels with maximum gray intensity value are remained. Also in my testing, the advancement of each trace step still has large computation complexity although it does not require the movement of pixel by pixel as in other thinning algorithms. Thus the third method is bid out which uses the built-in Morphological thinning function in MATLAB.

The thinned ridge map is then filtered by other three Morphological operations toremove some H breaks, isolated points and spikes.

MINUTIA MARKING

After the fingerprint ridge thinning, marking minutia points is relatively easy. But it is still not a trivial task as most literatures declared because at least one special case evokes my caution during the minutia marking stage.

In general, for each 3x3 window, if the central pixel is 1 and has exactly 3 one-value neighbors, then the central pixel is a ridge branch. If the central pixel is 1 and has only 1 one-value neighbor, then the central pixel is a ridge ending.

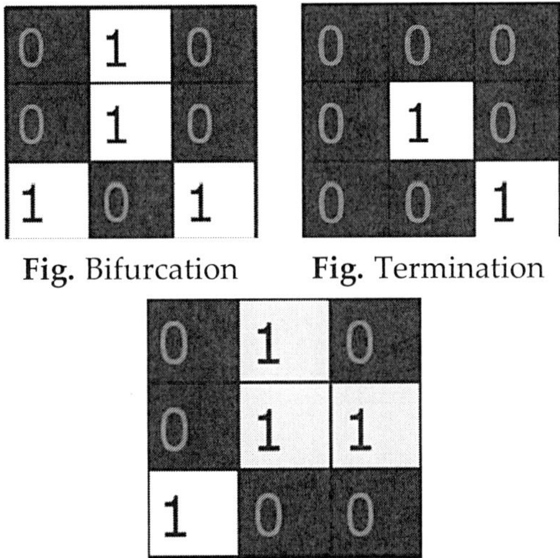

Fig. Bifurcation **Fig.** Termination

Fig. Triple counting branch

A special case that a genuine branch is triple counted. Suppose both the uppermost pixel with value 1 and the rightmost pixel with value 1 have another neighbor outside the 3x3 window, so the two pixels will be marked as branches too. But actually only one branch is located in the small region. So a check routine requiring that none of the neighbors of a branch are branches is added.

Also the average inter-ridge width D is estimated at this stage. The average inter-ridge width refers to the average distance between two neighboring ridges. The way to approximate the D value is simple. Scan a row of the thinned ridge image and sum up all pixels in the row whose value is one. Then divide the row length with the above summation to get an inter-ridge width. For more accuracy, such kind of row scan is performed upon several other rows and column scans are also conducted, finally all the inter-ridge widths are averaged to get the D.

Together with the minutia marking, all thinned ridges in the fingerprint image are labeled with a unique ID for further operation. The labeling operation is realized by using the Morphological operation: BWLABEL.

MINUTIA POSTPROCESSING

FALSE MINUTIA REMOVAL

The preprocessing stage does not totally heal the fingerprint image. For example, false ridge breaks due to insufficient amount of ink and ridge cross-connections due to over inking are not totally eliminated. Actually all the earlier stages themselves occasionally introduce some artifacts which later lead to spurious minutia. These false minutia will significantly affect the accuracy of matching if they are simply regarded as genuine minutia. So some mechanisms of removing false minutia are essential to keep the fingerprint verification system effective.

Seven types of false minutia are specified in following diagrams:

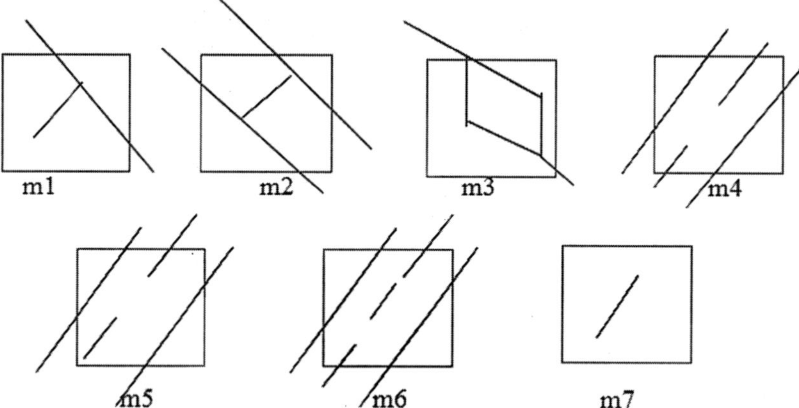

Fig. False Minutia Structures. m1 is a spike piercing into a valley. In the m2 case a spike falsely connects two ridges. m3 has two near bifurcations located in the same ridge. The two ridge broken points in the m4 case have nearly the same orientation and a short distance. m5 is alike the m4 case with the exception that one part of the broken ridge is so short that another termination is generated. m6 extends the m4 case but with the extra property that a third ridge is found in the middle of the two parts of the broken ridge. m7 has only one short ridge found in the threshold window.

My procedures in removing false minutia are:

1. If the distance between one bifurcation and one termination is less than D and the two minutia are in the same ridge(m1 case) . Remove both of them. Where D is the average inter-ridge width representing the average distance between two parallel neighboring ridges.
2. If the distance between two bifurcations is less than D and they are in the same ridge, remove the two bifurcations. (m2, m3 cases).
3. If two terminations are within a distance D and their directions are coincident with a small angle variation. And they suffice the condition that no any other termination is located between the two terminations. Then the two terminations are regarded as false minutia derived from a broken ridge and are removed. (case m4,m5, m6).
4. If two terminations are located in a short ridge with length less than D, remove the two terminations (m7).

My proposed procedures in removing false minutia have two advantages. One is that the ridge ID is used to distinguish minutia and the seven types of false minutia are strictly defined comparing with those loosely defined by other methods. The second advantage is that the order of removal procedures is well considered to reduce the computation complexity. It surpasses the way adopted by that does not utilize the relations among the false minutia types. For example, the procedure3 solves the m4, m5 and m6 cases in a single check routine. And after procedure 3, the number of false minutia satisfying the m7 case is significantly reduced.

UNIFY TERMINATIONS AND BIFURCATIONS

Since various data acquisition conditions such as impression pressure can easily change one type of minutia into the other, most researchers adopt the unification representation for both termination and bifurcation. So each minutia is completely characterized by the following parameters at last: 1) x-coordinate, 2) y-coordinate, and 3) orientation.

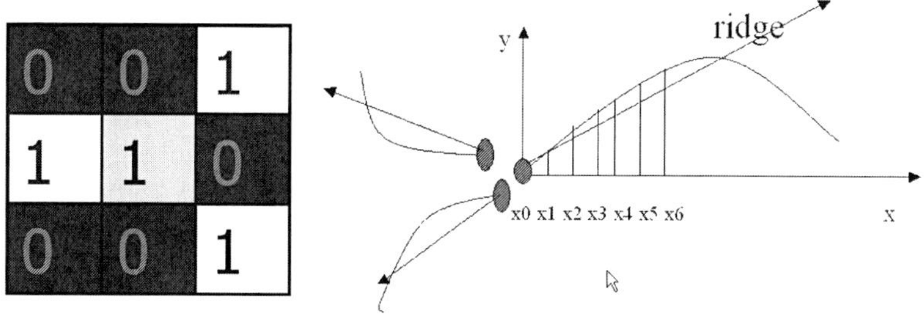

Fig. A bifurcation to three terminations

- Three neighbors become terminations (Left)
- Each termination has their own orientation (Right)

The orientation calculation for a bifurcation needs to be specially considered. All three ridges deriving from the bifurcation point have their own direction, represents the bifurcation orientation using a technique. simply chooses the minimum angle among the three anticlockwise orientations starting from the x-axis. Both methods cast the other two directions away, so some information loses. Here I propose a novel representation to break a bifurcation into three terminations. The three new terminations are the three neighbor pixels of the bifurcation and each of the three ridges connected to the bifurcation before is now associated with a termination respectively.

And the orientation of each termination (tx,ty) is estimated by following method :

Track a ridge segment whose starting point is the termination and length is D. Sum up all x-coordinates of points in the ridge segment. Divide above summation with D to get sx. Then get sy using the same way. Get the direction from: atan((sy-ty)/(sx-tx))

MINUTIA MATCH

Given two set of minutia of two fingerprint images, the minutia match algorithm determines whether the two minutia sets are from the same finger or not.

An alignment-based match algorithm partially derived from the is used in my project. It includes two consecutive stages: one is alignment stage and the second is match stage.

1. Alignment stage. Given two fingerprint images to be matched, choose any one minutia from each image, calculate the similarity of the two ridges associated with the two referenced minutia points. If the similarity is larger than a threshold, transform each set of minutia to a new coordination system whose origin is at the referenced point and whose x-axis is coincident with the direction of the referenced point.
2. Match stage: After we get two set of transformed minutia points, we use the elastic match algorithm to count the matched minutia pairs by assuming two minutia having nearly the same position and direction are identical.

ALIGNMENT STAGE

1. The ridge associated with each minutia is represented as a series of x-coordinates (x_1, $x_2 \ldots x_n$) of the points on the ridge. A point is sampled per ridge length L starting from the minutia point, where the L is the average inter-ridge length. And n is set to 10 unless the total ridge length is less than 10*L.

   So the similarity of correlating the two ridges is derived from:

   $$S = \Sigma^m_{i=0} x_i X_i / [\Sigma^m_{i=0} x_i^2 X_i^2]^\wedge 0.5,$$

   where (xi~xn) and (X_i~X_N) are the set of minutia for each fingerprint image respectively. And m is minimal one of the n and N value. If the similarity score is larger than 0.8, then go to step 2, otherwise continue to match the next pair of ridges.

2. For each fingerprint, translate and rotate all other minutia with respect to the reference minutia according to the following formula:

$$\begin{pmatrix} xi\_new \\ yi\_new \\ \theta i\_new \end{pmatrix} = TM * \begin{bmatrix} (xi - x) \\ (yi - y) \\ (\theta i - \theta) \end{bmatrix},$$

where (x,y, θ) is the parameters of the reference minutia, and TM is

$$TM = \begin{pmatrix} \cos\theta & -\sin\theta & 0 \\ \sin\theta & \cos\theta & 0 \\ 0 & 0 & 1 \end{pmatrix}$$

The following diagram illustrate the effect of translation and rotation:

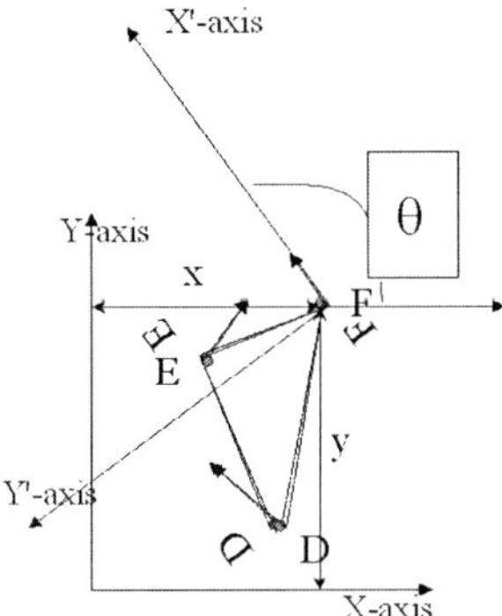

Fig. The new coordinate system is originated at minutia F and the new x-axis is coincident with the direction of minutia F. No scaling effect is taken into account by assuming two fingerprints from the same finger have nearly the same size.

My method to align two fingerprints is almost the same with the one used by but is different at step 2. Lin's method uses the rotation angle calculated from all the sparsely sampled ridge points. My method use the rotation angle calculated earlier by densely tracing a short ridge start from the minutia with length D. Since I have already got the minutia direction at the minutia extraction stage, obviously my method reduces the redundant calculation but still holds the accuracy.

Also Lin's way to do transformation is to directly align one fingerprint image to another according to the discrepancy of the reference minutia pair. But it still requires a transform to the polar coordinate system for each image at the next minutia match stage. My approach is to transform each according to its own reference minutia and then do match in a unified x-y coordinate. Therefore, less computation workload is achieved through my method.

MATCH STAGE

The matching algorithm for the aligned minutia patterns needs to be elastic since the strict match requiring that all parameters (x, y, ?) are the same for two identical minutia is impossible due to the slight deformations and inexact quantizations of minutia. My approach to elastically match minutia is achieved by placing a bounding box around each template minutia. If the minutia to be matched is within the rectangle box and the direction discrepancy between them is very small, then the two minutia are regarded as a matched minutia pair. Each minutia in the template image either has no matched minutia or has only one corresponding minutia.

The final match ratio for two fingerprints is the number of total matched pair over the number of minutia of the template fingerprint. The score is 100*ratio and ranges from 0 to 100. If the score is larger than a pre-specified threshold, the two fingerprints are from the same finger. However, the elastic match algorithm has large computation complexity and is vulnerable to spurious minutia.

EXPERIMENTATION RESULTS

EVALUATION INDEXES FOR FINGERPRINT RECOGNITION

Two indexes are well accepted to determine the performance of a fingerprint recognition system: one is FRR (false rejection rate) and the other is FAR (false acceptance rate). For an image database, each sample is matched against the remaining samples of the same finger to compute the False Rejection Rate. If the matching g against h is performed, the symmetric one (i.e., h against g) is not executed to avoid correlation. All the scores for such matches are composed into a series of Correct Score. Also the first sample of each finger in the database is matched against the first sample of the remaining fingers to compute the False Acceptance Rate. If the matching g against h is performed, the symmetric one (i.e., h against g) is not executed to avoid correlation. All the scores from such matches are composed into a series of Incorrect Score.

EXPERIMENTATION RESULTS

A fingerprint database from the FVC2000 (Fingerprint Verification Competition 2000) is used to test the experiment performance. My program tests all the images without any fine-tuning for the database. The experiments show my program can differentiate imposturous minutia pairs from genuine minutia pairs in a certain confidence level. Furthermore, good experiment designs can surely improve the accuracy. Further studies on good designs of training and testing are expected to improve the result.

Here is the diagram for Correct Score and Incorrect Score distribution:

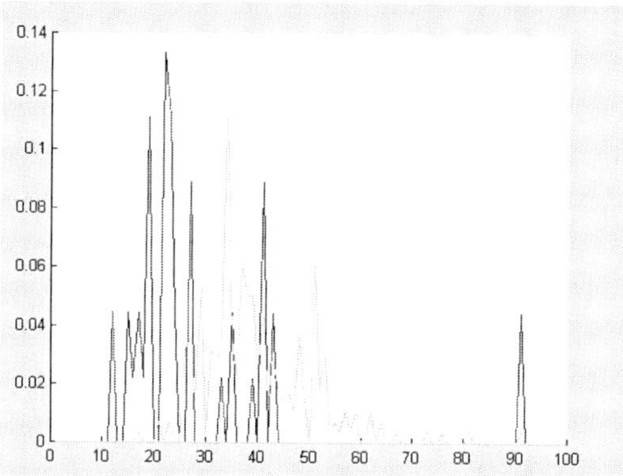

Fig. Distribution of Correct Scores and Incorrect Scores

- Red line: Incorrect Score
- Green line: Correct Scores

It can be seen from the above figure that there exist two partially overlapped distributions. The Red curve whose peaks are mainly located at the left part means the average incorrect match score is 25. The green curve whose peaks are mainly located on the right side of red curve means the average correct match score is 35. This indicates the algorithm is capable of differentiate fingerprints at a good correct rate by setting an appropriate threshold value.

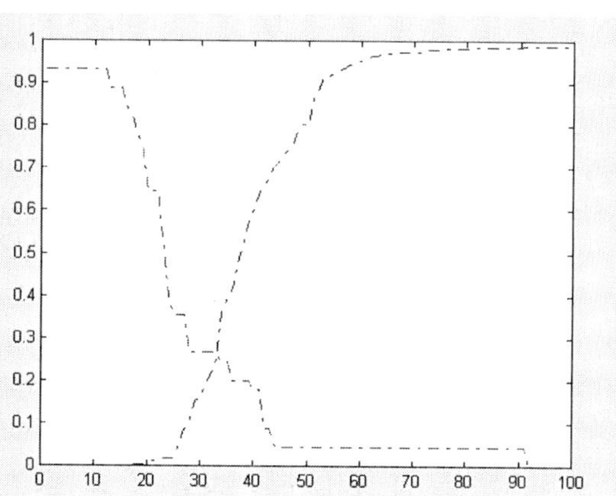

Fig. FAR and FRR curve

- Blue dot line: FRR curve
- Red dot line: FAR curve

The above diagram shows the FRR and FAR curves. At the equal error rate 25%, the separating score 33 will falsely reject 25% genuine minutia pairs and falsely accept 25% imposturous minutia pairs and has 75% verification rate. The high incorrect acceptance and false rejection are due to some fingerprint images with bad quality and the vulnerable minutia match algorithm.

MULTIMODAL BIOMETRICS SYSTEM FOR EFFICIENT HUMAN RECOGNITION

"Biometrics" means "life measurement", but the term is usually associated with the use of unique physiological characteristics to identify an individual. One of the applications which most people associate with biometrics is security. However, biometrics identification has eventually a much broader relevance as computer interface becomes more natural. It is an automated method of recognizing a person based on a physiological or behavioral characteristic. Among the features measured are; face fingerprints, hand geometry, handwriting, iris, retinal, vein, voice etc. Biometric technologies are becoming the foundation of an extensive array of highly secure identification and personal verification solutions . As the level of security breaches and transaction fraud increases, the need for highly secure identification and personal verification technologies is becoming apparent. In recent years, biometrics authentication has seen considerable improvements in reliability and accuracy, with some of the traits offering good performance. However, even the best biometric traits till date are facing numerous problems; some of them are inherent to the technology itself. In particular, biometric authentication systems generally suffer from enrollment problems due to non-universal biometric traits, susceptibility to biometric spoofing or insufficient accuracy caused by noisy data acquisition in certain environments.

One way to overcome these problems is the use of multi-biometrics. Driven by lower hardware costs, a multi biometric system uses multiple sensors for data acquisition. This allows capturing multiple samples of a single biometric trait (called multi-sample biometrics) and/or samples of multiple biometric traits (called multi source or multimodal biometrics). This approach also enables a user who does not possess a particular biometric identifier to still enroll and authenticate using other traits, thus eliminating the enrollment problems and making it universal. A unimodal biometric system consists of three major modules: sensor module, feature extraction module and matching module. The performance of a biometric system is largely affected by the reliability of the sensor used and the degrees of freedom offered by the features extracted from the sensed signal. Further, if the biometric trait being sensed or measured is noisy (a fingerprint with a scar or a voice altered by a cold, for example), the resultant matching score computed by the matching module may not be reliable. This problem can be solved by installing multiple sensors that capture different biometric traits. Such systems, known as multimodal biometric systems , are expected to be more reliable due to the presence of multiple pieces of evidence. These systems are also able to meet the stringent performance requirements imposed by various applications. However, multimodal systems address the problem of non-universality: it is possible for a subset of users who do not possess a particular biometric. For example, the feature extraction module of a fingerprint authentication system may be unable to extract features from fingerprints associated with specific individuals, due to the poor quality of the ridges. In such instances, it is useful to acquire multiple biometric traits for verifying the identity. Multimodal systems also provide anti-spoofing measures by making it difficult for an intruder to spoof multiple biometric traits simultaneously. By asking the user to present a random subset of biometric traits, the system ensures that a live user is indeed present at the point of acquisition. However, an integration scheme is required to fuse the information presented by the individual modalities.

MULTIMODAL BIOMETRICS SYSTEM

Multimodal biometric systems are those that utilize more than one physiological or behavioral characteristic for enrollment, verification, or identification. In applications such as border entry/exit, access control, civil identification, and network security, multi-modal biometric systems are looked to as a means of reducing false non-match and false match rates, providing a secondary means of enrollment, verification, and identification if sufficient data cannot be acquired from a given biometric sample, and combating attempts to fool biometric systems through fraudulent data sources such as fake fingers.

Ross and Jain (2003) have presented an overview of Multimodal Biometrics and have proposed various levels of fusion, various possible scenarios, the different modes of operation, integration strategies and design issues. A multimodal system can operate in one of three different modes: serial mode, parallel mode, or hierarchical mode. In the serial mode of operation, the output of one modality is typically used to narrow down the number of possible identities before the next modality is used. Therefore, multiple sources of information (e.g., multiple traits) do not have to be acquired simultaneously. Further, a decision could be made before acquiring all the traits. This can reduce the overall recognition time. In the parallel mode of operation, the information from multiple modalities is used simultaneously in order to perform recognition. The levels fusion proposed for multimodal systems are broadly categorized into three system architectures which are according to the strategies used for information fusion as shown in Figure 1:

- Fusion at the Feature Extraction Level
- Fusion at the Matching Score Level
- Fusion at the Decision Level

In Fusion at the Feature Extraction Level, information extracted from the different sensors is encoded into a joint feature vector, which is then compared to an enrollment template (which itself is a joint feature vector stored in a database) and assigned a matching score as in a single biometric system.

In Fusion at the Matching Score Level, feature vectors are created independently for each sensor and are then compared to the enrollment templates which are stored separately for each biometric trait. Based on the proximity of feature vector and template, each subsystem computes its own matching score. These individual scores are finally combined into a total score which is passed to the decision module.

In Fusion at the Decision Level, a separate authentication decision is made for each biometric trait. These decisions are then combined into a final vote. This architecture is rather loosely coupled system architecture, with each subsystem performing like a single biometric system. A substantial amount of work has been carried out on the combination of multiple classifiers. Most of such work focuses on fusing 'weak' classifiers for the purpose of increasing the overall performance (Tolba & Rezq, 2000) . A hybrid fingerprint matcher which fuses minutiae and reference point location classifiers has been proposed by Ross, Jain & Riesman (2003). It has been reported that the performance of the hybrid matcher is better than individual classifiers.

Apart from fusion of multi classifiers, much work has also been done to combine traits/different modalities at various levels. Yunhong, Tan & Jain (2003) proposed the fusion of iris and face modalities and reported that besides improving verification performance, the fusion of these two has several other advantages. Dass, Nandakumar & Jain (2005) have proposed an approach to score level fusion in multimodal biometrics systems . Experimental results have been presented on face, fingerprint and hand geometry using product rule and coupla method. It is found that both fusion rules show better performance than individual recognizers. Common theoretical framework for combining classifiers using sum rule, median rule, max and min rule are analyzed by Kittler et al. (1998) under the most restrictive assumptions and have observed that sum rule outperforms other classifiers combination schemes.

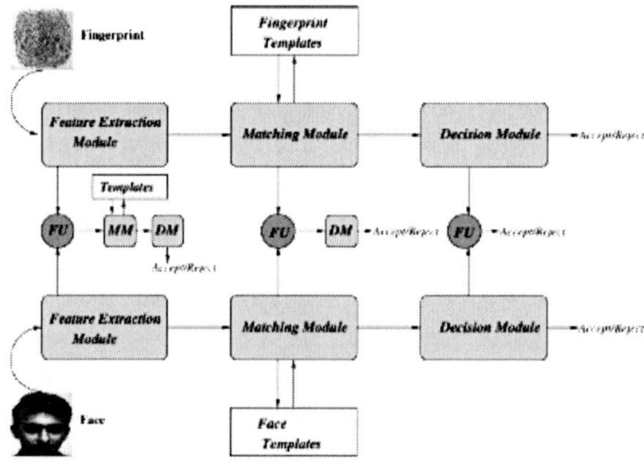

Fig. Multimodal System using three levels of Fusion (taken from Ross & Jain, 2003)

Guiyu Feng et. al. (2004) presents a novel fusion strategy for personal identification using face and palmprint biometrics . The work considers the feature level fusion scheme. The purpose of the proposed paper is to investigate whether the integration of face and palmprint biometrics can achieve higher performance that may not be possible using a single biometric indicator alone. Both Principal Component Analysis (PCA) and Independent Component Analysis (ICA) are considered in this feature vector fusion context. It is found that the performance improved significantly.

MULTIMODAL BIOMETRICS SYSTEM AT IITK

The multimodal biometric system at IITK is developed using four traits i.e., face, fingerprint, iris and signature (as shown in Figure 2). In Face Recognition, the input face image is recognized using Elastic Bunch Graph matching algorithm. In Fingerprint Verification, the input image is enhanced to bring out obscure information based on Gabor filtering and matching is done by combination of Reference Point and Minutiae matching algorithms. In Iris Recognition, the input image is localized by finding the pupillary and outer iris boundary and is matched using combination of Haar Wavelet and Circular Mellin operator. In Signature Verification, feature vector consists of Global and Local features of signature image and is matched using Euclidean Distance. The modules based on the individual traits returns an integer value after matching the database and query feature vectors. First of all the fusion is done at classifier level i.e., for face, fingerprint and iris, multiple classifiers are combined at matching score level followed by fusion at multiple modalities level. The final score is generated by using sum of score technique at matching score level which is passed to the decision module. The brief description of various recognition algorithms are presented below:

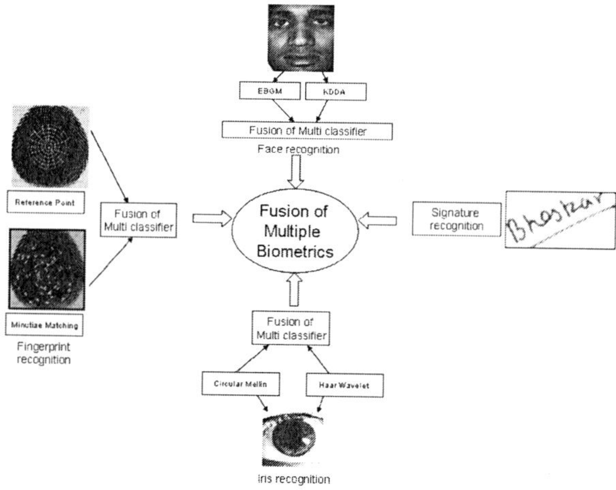

Fig. Multimodal Biometric System at IITK

Face Recognition

Face Recognition is a noninvasive process where a portion of the subject's face is photographed and the resulting image is reduced to a digital code. Facial recognition records the spatial geometry of distinguishing features of the face . The recognition algorithm takes facial image, measures the unique characteristics and computes the template corresponding to each face. Using templates, the algorithm then compares that image with another image and produces a score that measures how similar the images are to each other.

FEATURE EXTRACTION USING EBGM AND KDDA

Elastic Bunch Graph Matching (EBGM)

Face recognition using elastic bunch graph matching is based on recognizing novel faces by estimating a set of novel features using a data structure called a bunch graph. Similarly for each query image, the landmarks are estimated and located using bunch graph. Then the features are extracted by convolution with the number of instances of Gabor filters followed by the creation of face graph. The matching score (MSEBGM) is calculated on the basis of similarity between face graphs of database and query image. The diagrammatic representation of EBGM algorithm is shown in Figure 3.

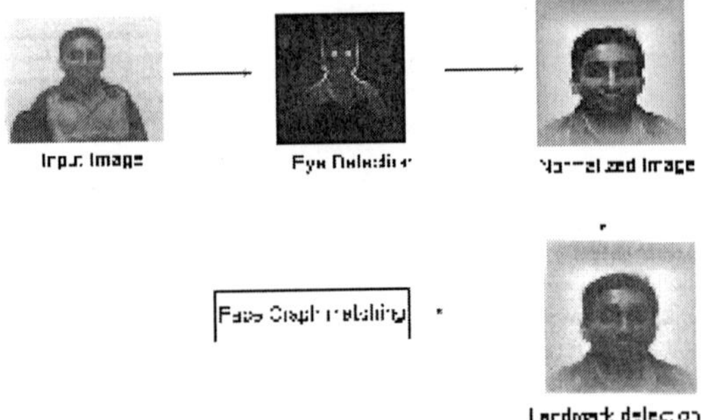

Fig. Steps involved in face recognition

Kernel Direct Discriminant Analysis (KDDA)

Face recognition using KDDA is based on computation of feature space F (from training set) and projection of input pattern into the feature space to calculate significant discriminant features. For each of the m features in the database and n features in the query image, reference features are chosen depending on the distance and rotation between the positions of features in the feature space. The matching score for each transformation of database and query feature vectors are calculated with respect to reference feature chosen using bounding box technique. MSKDDA is defined by the maximum of all matching scores divided by the maximum number of features (among the query and the database).

COMBINATION OF EBGM AND KDDA

The matching scores from the above two classifiers are converted from distance to similarity score and are combined at matching score level using sum of score technique which significantly increases the accuracy of the face recognition system.

Fingerprint Recognition

The fingerprint recognition system has been developed by the fusion of Reference Point and Minutiae Matching Techniques . The key steps involved are fingerprint enhancement, feature extraction using Reference point Algorithm and Minutiae Matching approach and computation of matching score. The goal of fingerprint enhancement is to increase the clarity of ridge structure so that minutiae and the reference points can be easily and correctly extracted.

Feature Extraction using Reference point and Minutiae matching approach

Reference Point Algorithm gracefully handles local noise in a poor quality fingerprint. The detection should necessarily consider a large neighborhood in the fingerprint image. For an accurate localization of the reference point, the input image is segmented to remove any kind of noise present in the image. Further Sobel Operator is applied to obtain gradient of segmented image. The Orientation Field is estimated along with the Y component. A specific pattern in which the value of Y-Component is maximum is Reference point (the point of maximum curvature).

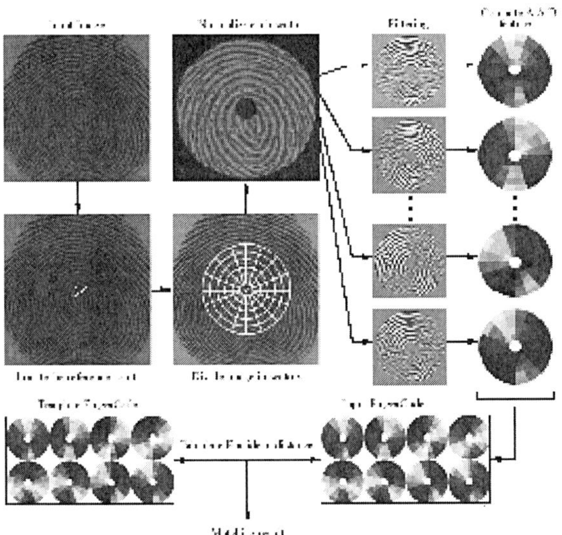

Fig. Diagrammatic representation of Reference point Location algorithm

The finger code is generated by drawing concentric circles of fixed radius centered at reference point (as shown in Figure 4). The image is segmented into 5 tracks and 16 sectors from the detected reference point. The size of the feature vector is 512 values. The distance (DRef) for the database and query feature vectors is calculated using Euclidean distance method.

Minutiae Matching

The input fingerprint image is enhanced using Gabor Filters The enhanced image is further binarized and thinned using a morphological operation that successively erodes away the foreground pixels until they are one pixel wide. The thinned image is used to detect minutiae points by locating ridge ending and bifurcations using Crossing Number (CN) method. The matching score MSMIN between the database and query image is computed using Elastic matching approach . Figure 5 shows various steps involved in minutiae extraction.

COMBINATION OF REFERENCE POINT AND MINUTIAE MATCHING ALGORITHM

The matching scores from the above two classifiers are converted from distance to similarity score and are combined at matching score level using sum of score technique which significantly increases the accuracy of the fingerprint system.

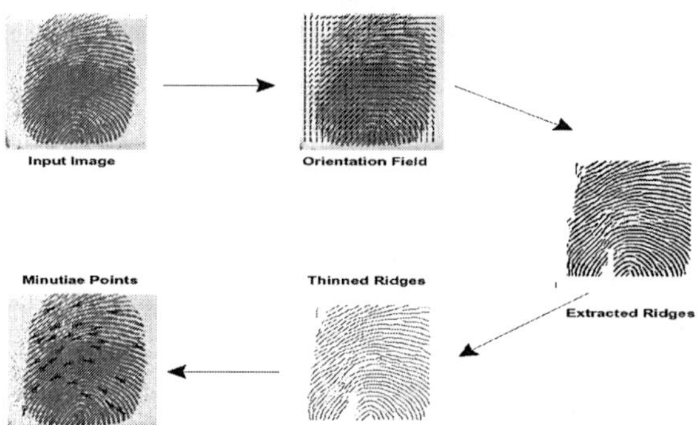

Fig. Steps involved in minutiae extraction

Iris Recognition

The iris image acquired from a 3CCD camera is localized by finding the center of pupil from the spectrum image. The radius of the pupil is the distance between the pupil center and nearest non-zero pixel. The outer iris boundary is detected by drawing concentric circles of different radii from the pupil center and the intensities lying over the perimeter of the circle are summed up. Among the candidate iris circles, the circle having a maximum change in intensity with respect to the previous drawn circle is the outer iris boundary (shown in Figure 6). The annular region lying between pupil and iris boundary is transformed to polar co-ordinates to take into consideration the possibility of pupil dilation and appearing of different size in different images. From the normalized strip the eyelids are detected and removed.

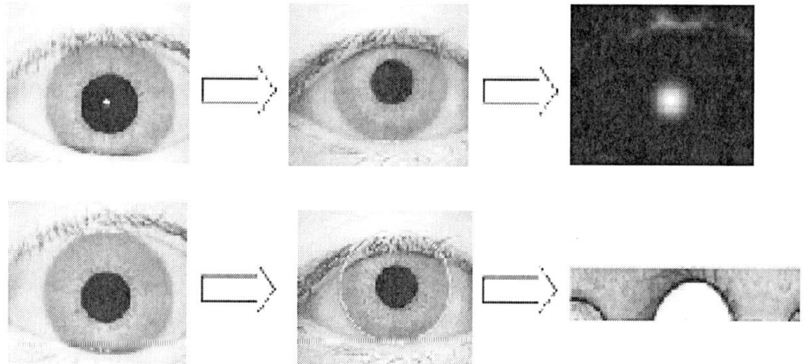

Fig. Steps involved in iris preprocessing and normalization

FEATURE EXTRACTION USING HAAR WAVELET AND CIRCULAR MELLIN OPERATOR

Haar Wavelet

Haar wavelet is widely used in texture recognition algorithms . The input signal S (polarized iris image) is decomposed into approximation, vertical, horizontal and diagonal coefficients using the wavelet transformation and coefficients for the fourth and fifth levels are chosen to reduce space complexity and discard the redundant information. The iris code is generated by assigning one to the positive coefficient values and zero to negative values.

Circular Mellin operators

These "Circular Mellin" operators are invariant to both scale and orientation of the target and represent the spectral decomposition of the image scene in the polar-log coordinate system. Features in iris images are extracted based on the phase of convolution of polarized iris image with mellin operators. The iris code is one for positive phase values and zero for negative phase values.

The iris codes generated using Haar Wavelet and Circular Mellin operators are matched using Hamming Distance approach.

COMBINATION OF HAAR WAVELET AND CIRCULAR MELLIN OPERATOR

The individual matching scores generated by above mentioned classifiers are converted from distance to similarity score and are fused at matching score level for better performance of iris recognition.

Signature Verification

In biometrics terminology, the signature is a behavioral characteristic of a person and can be used to identify/verify a person's identity. The signature recognition algorithm consists of three major modules i.e., preprocessing and noise removal, feature extraction and computation of Euclidean distance. Offline signature acquisition is carried out statically, unlike online signature acquisition, by capturing the signature image using a high resolution scanner. A scanned signature image may require morphological operations (shown in Figure 7) like normalization, noise removal by eliminating extra dots from the image, conversion to grayscale, thinning and extraction of high pressure region.

FEATURE EXTRACTION USING GLOBAL AND LOCAL FEATURES

The features of the signature images can be classified into two categories - global and local . Global features include the global characteristics of an image. Ismail and Gad have described global features as characteristics which identify or describe the signature as a whole. Examples include: width/height (or length), baseline, area of black pixels etc. They are less responsive to small distortions and hence are less sensitive to noise as well, compared to local features which are confined to a limited portion of the signature. In contrast to global features, they are susceptible to small distortions like dirt but are not influenced by other regions of the signature. Hence, though extraction of local features requires a huge number of computations, they are much more precise. However, the grid size has to be chosen very carefully. It can neither be too gross nor be too detailed. Examples include local gradients, pixel distribution in local segments etc. Many of the global features such as global baseline, center of gravity, and distribution of black pixels have their local counterparts as well.

The difference/distance (DSign) between the two feature sets are computed using weighted Euclidean distance measure.

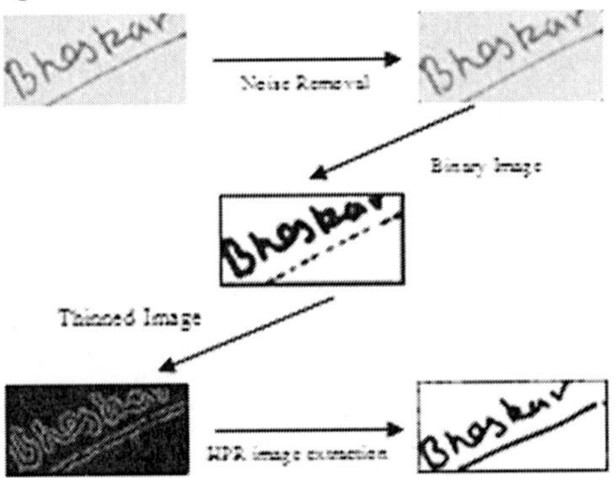

Fig. Preprocessing and noise removal

Fusion

The different biometrics systems can be integrated at multi-classifier and multi-modality level to improve the performance of the verification system. However, it can be thought as a conventional fusion problem i.e. can be thought to combine evidence provided by different biometrics to improve the overall decision accuracy.

The multimodal biometric system at IITK is developed at multi-classifier and multi-modalities level. At multi- classifier level, multiple algorithms are developed and combined for traits like face, fingerprint and iris. The following steps are performed for fusion at classifier level:

- S1: Given a query image as input, features are extracted by the individual recognizers and then an individual comparison algorithm for each recognizer compares the set of features and calculates the matching scores or distances corresponding to each recognizer for various traits.
- S2: The scores/distances obtained in S1 are normalized to a common range between 0 to 1.
- S3: These scores are then converted from distance to similarity score by subtraction from 1 if it is a dissimilarity score. For example the dissimilarity scores, in case of fingerprint recognition using reference point algorithm (DRef), iris recognition using Haar Wavelet (DHaar) and Circular Mellin operator (DMellin) are converted to similarity scores (MSRef, MSHaar, MSMellin)
- S4: The matching scores are further rescaled so that threshold value becomes same for each recognizer.
- S5: Then the combined matching score is calculated by fusion of the matching scores of multiple classifiers using sum rule technique.

$$MS = \frac{\alpha \times MS + \beta \times MS}{2}, \quad MS = \frac{\alpha \times MS + \beta \times MS}{2}$$

$$MS = \frac{\alpha \times MS + \beta \times MS}{2}$$

where α and β are the weights assigned to individual classifiers. Currently equal weightage is given to each classifiers and the value of α and β is one.

The multimodal biometric system at IITK is developed by integrating four traits i.e., face, fingerprint, iris and signature at matching score level. Based on the proximity of feature vector and template, each subsystem computes its own matching score. These individual scores are finally combined into a total score, which is passed to the decision module. The same steps for fusion at classifiers level are followed for multiple modalities level i.e., matching scores are computed for each trait (face, fingerprint, iris and signature) followed by normalization to the common scale and distance to similarity score conversion for all the four traits. The matching scores are further rescaled so that the threshold value becomes common for all the subsystems. Finally, the sum of score technique is applied for combining the matching scores of four traits i.e., face, fingerprint, iris and signature. Thus the final score MSFinal is given by,

$$MS_{Final} = \frac{1}{4}(a \times MS_{Face} + b \times MS_{Finger} + c \times MS_{Iris} + d \times MS_{Sign})$$

where MSFace = matching score of face, MSFinger = matching score of fingerprint, MSIris = matching score of iris, and MSSign = matching score of signature and a, b, c and d are the weights assigned to the various traits. Currently, equal weightage is assigned to each trait so the value of a, b, c and d is one. The final matching score (MSFinal) is compared against a certain threshold value to recognize the person as genuine or an imposter.

11

Biometric Fingerprinting and Retinal Scanning

BIOMETRICS

Biometrics (or biometric authentication) refers to the identification of humans by their characteristics or traits. Biometrics is used in computer science as a form of identification and access control. It is also used to identify individuals in groups that are under surveillance. Biometric identifiers are the distinctive, measurable characteristics used to label and describe individuals. Biometric identifiers are often categorized as physiological versus behavioral characteristics. Physiological characteristics are related to the shape of the body. Examples include, but are not limited to fingerprint, face recognition, DNA, Palm print, hand geometry, iris recognition, retina and odour/scent. Behavioral characteristics are related to the pattern of behavior of a person, including but not limited to: typing rhythm, gait, and voice. Some researchers have coined the termbehaviometrics to describe the latter class of biometrics. More traditional means of access control include token-based identification systems, such as a driver's license or passport, and knowledge-based identification systems, such as a password or personal identification number. Since biometric identifiers are unique to individuals, they are more reliable in verifying identity than token and knowledge-based methods; however, the collection of biometric identifiers raises privacy concerns about the ultimate use of this information.

BIOMETRIC FUNCTIONALITY

Many different aspects of human physiology, chemistry or behavior can be used for biometric authentication. The selection of a particular biometric for use in a specific application involves a weighting of several factors. Jain *et al.* (1999) identified seven such factors to be used when assessing the suitability of any trait for use in biometric authentication. Universality means that every person using a system should possess the trait. Uniqueness means the trait should be sufficiently different for individuals in the relevant population such that they can be distinguished from one another. Permanence relates to the manner in which a trait varies over time. More specifically, a trait with 'good' permanence will be reasonably invariant over time with respect to the specific matching algorithm. Measurability (collectability) relates to the ease of acquisition or measurement of the trait. In addition, acquired data should be in a form that permits subsequent processing and extraction of the relevant feature sets. Performancerelates to the accuracy, speed, and robustness of technology used (see performance section for more details). Acceptability relates to how well individuals in the relevant population accept the technology such that they are willing to have their biometric trait captured and assessed. Circumvention relates to the ease with which a trait might be imitated using an artifact or substitute.

No single biometric will meet all the requirements of every possible application. The basic block diagram of a biome in the following two modes. In verification mode the system performs a one-to-one comparison of a captured biometric with a specific template stored in a biometric database in order to verify the individual is the person they claim to be. Three steps involved in person verification. In the first step, reference models for all the users are generated and stored in the model database. In the second step, some samples are matched with reference models to generate the genuine and impostor scores and calculate the threshold. Third step is the testing step. This process may use a smart card, username or ID number (e.g. PIN) to indicate which template should be used for comparison. 'Positive recognition' is a common use of verification mode, "where the aim is to prevent multiple people from using same identity".

In Identification mode the system performs a one-to-many comparison against a biometric database in attempt to establish the identity of an unknown individual. The system will succeed in identifying the individual if the comparison of the biometric sample to a template in the database falls within a previously set threshold. Identification mode can be used either for 'positive recognition' (so that the user does not have to provide any information about the template to be used) or for 'negative recognition' of the person "where the system establishes whether the person is who she (implicitly or explicitly) denies to be". The latter function can only be achieved through biometrics since other methods of personal recognition such as passwords, PINs or keys are ineffective.

The first time an individual uses a biometric system is called *enrollment*. During the enrollment, biometric information from an individual is captured and stored. In subsequent uses, biometric information is detected and compared with the information stored at the time of enrollment. Note that it is crucial that storage and retrieval of such systems themselves be secure if the biometric system is to be robust. The first block (sensor) is the interface between the real world and the system; it has to acquire all the necessary data.

Most of the times it is an image acquisition system, but it can change according to the characteristics desired. The second block performs all the necessary pre-processing: it has to remove artifacts from the sensor, to enhance the input (e.g. removing background noise), to use some kind of normalization, etc. In the third block necessary features are extracted. This step is an important step as the correct features need to be extracted in the optimal way. A vector of numbers or an image with particular properties is used to create a *template*. A template is a synthesis of the relevant characteristics extracted from the source. Elements of the biometric measurement that are not used in the comparison algorithm are discarded in the template to reduce the filesize and to protect the identity of the enrollee.

If enrollment is being performed, the template is simply stored somewhere (on a card or within a database or both). If a matching phase is being performed, the obtained template is passed to a matcher that compares it with other existing templates, estimating the distance between them using any algorithm (e.g. Hamming distance). The matching program will analyze the template with the input. This will then be output for any specified use or purpose (e.g. entrance in a restricted area). Selection of biometrics in any practical application depending upon the characteristic measurements and user requirements. We should consider Performance, Acceptability, Circumvention, Robustness, Population coverage, Size, Identity theft deterrence in selecting a particular biometric. Selection of biometric based on user requirement considers Sensor availability, Device availability, Computational time and reliability, Cost, Sensor area and power consumption

MULTIMODAL BIOMETRIC SYSTEM

Multimodal biometric systems use multiple sensors or biometrics to overcome the limitations of unimodal biometric systems. For instance iris recognition systems can be compromised by aging irides and finger scanning systems by worn-out or cut fingerprints. While unimodal biometric systems are limited by the integrity of their identifier, it is unlikely that several unimodal systems will suffer from identical limitations. Multimodal biometric systems can obtain sets of information from the same marker (i.e., multiple images of an iris, or scans of the same finger) or information from different biometrics (requiring fingerprint scans and, using voice recognition, a spoken pass-code).Multimodal biometric systems can integrate these unimodal systems sequentially, simultaneously, a combination thereof, or in series, which refer to sequential, parallel, hierarchical and serial integration modes, respectively. The interested reader is pointed to Choubisa for detailed tradeoffs of response time, accuracy, and costs between integration modes.

Broadly, the information fusion is divided into three parts, pre-mapping fusion, midst-mapping fusion, and post-mapping fusion/late fusion.In pre-mapping fusion information can be combined at sensor level or feature level.

Sensor-level fusion can be mainly organized in three classes: (1) single sensor-multiple instances, (2) intra-class multiple sensors, and (3) inter-class multiple sensors. Feature-level fusion can be mainly organized in two categories: (1) intra-class and (2) inter-class. Intra-class is again classified into four subcategories: (a) Same sensor-same features, (b) Same sensor-different features, (c) Different sensors-same features, and (d) Different sensors-different features.

PERFORMANCE

The following are used as performance metrics for biometric systems:

- False acceptance rate or false match rate (FAR or FMR): the probability that the system incorrectly matches the input pattern to a non-matching template in the database. It measures the percent of invalid inputs which are incorrectly accepted. In case of similarity scale, if the person is imposter in real, but the matching score is higher than the threshold, then he is treated as genuine that increase the FAR and hence performance also depends upon the selection of threshold value.
- False rejection rate or false non-match rate (FRR or FNMR): the probability that the system fails to detect a match between the input pattern and a matching template in the database. It measures the percent of valid inputs which are incorrectly rejected.
- Receiver operating characteristic or relative operating characteristic (ROC): The ROC plot is a visual characterization of the trade-off between the FAR and the FRR. In general, the matching algorithm performs a decision based on a threshold which determines how close to a template the input needs to be for it to be considered a match. If the threshold is reduced, there will be fewer false non-matches but more false accepts. Correspondingly, a higher threshold will reduce the FAR but increase the FRR. A common variation is the *Detection error trade-off (DET),* which is obtained using normal deviate scales on both axes. This more linear graph illuminates the differences for higher performances (rarer errors).
- Equal error rate or crossover error rate (EER or CER): the rate at which both accept and reject errors are equal. The value of the EER can be easily obtained from the ROC curve. The EER is a quick way to compare the accuracy of devices with different ROC curves. In general, the device with the lowest EER is most accurate.
- Failure to enroll rate (FTE or FER): the rate at which attempts to create a template from an input is unsuccessful. This is most commonly caused by low quality inputs.
- Failure to capture rate (FTC): Within automatic systems, the probability that the system fails to detect a biometric input when presented correctly.

- Template capacity: the maximum number of sets of data which can be stored in the system.

HISTORY OF BIOMETRICS

The earliest cataloging of fingerprints dates back to 1891 when Juan Vucetich started a collection of fingerprints of criminals in Argentina. The History of Fingerprints.

ADAPTIVE BIOMETRIC SYSTEMS

Adaptive biometric Systems aim to auto-update the templates or model to the intra-class variation of the operational data. The two-fold advantages of these systems are solving the problem of limited training data and tracking the temporal variations of the input data through adaptation. Recently, adaptive biometrics have received a significant attention from the research community. This research direction is expected to gain momentum because of their key promulgated advantages. First, with an adaptive biometric system, one no longer needs to collect a large number of biometric samples during the enrollment process. Second, it is no longer necessary to re-enroll or retrain the system from scratch in order to cope with the changing environment. This convenience can significantly reduce the cost of maintaining a biometric system. Despite these advantages, there are several open issues involved with these systems. For mis-classification error (false acceptance) by the biometric system, cause adaptation using impostor sample. However, continuous research efforts are directed to resolve the open issues associated to the field of adaptive biometrics. More information about adaptive biometric systems can be found in the critical review by Rattani et al.

CURRENT, EMERGING AND FUTURE APPLICATIONS OF BIOMETRICS

India's Universal ID Program

India's Universal ID (UID) program seeks to provide a unique identity to all 1.2 billion residents. Although it is in early stages, the UID program is already the largest biometric identification program in the world with more than 200 million people enrolled as of January 2012. For more information, see Countries applying biometrics.

RECENT ADVANCES IN EMERGING BIOMETRICS

In recent times, biometrics based on brain (electroencephalogram) and heart (electrocardiogram) signals have emerged. The research group at University of Wolverhampton led by Ramaswamy Palaniappan has shown that people have certain distinct brain and heart patterns that are specific for each individual. The advantage of such 'futuristic' technology is that it is more fraud resistant compared to conventional biometrics like fingerprints.

However, such technology is generally more cumbersome and still has issues such as lower accuracy and poor reproducibility over time.

Proposal calls for biometric authentication to access certain public networks

John Michael (Mike) McConnell, a former vice admiral in the United States Navy, a former Director of US National Intelligence, and Senior Vice President of Booz Allen Hamilton promoted the development of a future capability to require biometric authentication to access certain public networks in his Keynote Speech at the 2009 Biometric Consortium Conference.

A basic premise in the above proposal is that the person that has uniquely authenticated themselves using biometrics with the computer is in fact also the agent performing potentially malicious actions from that computer. However, if control of the computer has been subverted, for example in which the computer is part of a botnet controlled by a hacker, then knowledge of the identity of the user at the terminal does not materially improve network security or aid law enforcement activities.

Recently, another approach to biometric security was developed, this method scans the entire body of prospects to guarantee a better identification of this prospect. This method is not globally accepted because it is very complex and prospects are concerned about their privacy. Very few technologists apply it globally.

ISSUES AND CONCERNS

Privacy and discrimination

It is possible that data obtained during biometric enrollment may be used in ways for which the enrolled individual has not consented. For example, biometric security that utilizes an employee's DNA profile could also be used to screen for various genetic diseases or other 'undesirable' traits.

There are three categories of privacy concerns:

1. Unintended functional scope: The authentication goes further than authentication, such as finding a tumor.
2. Unintended application scope: The authentication process correctly identifies the subject when the subject did not wish to be identified.
3. Covert identification: The subject is identified without seeking identification or authentication, i.e. a subject's face is identified in a crowd.

Danger to owners of secured items

When thieves cannot get access to secure properties, there is a chance that the thieves will stalk and assault the property owner to gain access. If the item is secured with a biometric device, the damage to the owner could be irreversible, and potentially cost more than the secured property. For example, in 2005, Malaysian car thieves cut off the finger of a Mercedes-Benz S-Class owner when attempting to steal the car.

Cancelable biometrics

One advantage of passwords over biometrics is that they can be re-issued. If a token or a password is lost or stolen, it can be cancelled and replaced by a newer version. This is not naturally available in biometrics. If someone's face is compromised from a database, they cannot cancel or reissue it. Cancelable biometrics is a way in which to incorporate protection and the replacement features into biometrics. It was first proposed by Ratha et al.

Several methods for generating new exclusive biometrics have been proposed. The first fingerprint based cancelable biometric system was designed and developed by Tulyakov et al. Essentially, cancelable biometrics perform a distortion of the biometric image or features before matching. The variability in the distortion parameters provides the cancelable nature of the scheme. Some of the proposed techniques operate using their own recognition engines, such as Teoh et al. and Savvides et al.,whereas other methods, such as Dabbah et al., take the advantage of the advancement of the well-established biometric research for their recognition front-end to conduct recognition. Although this increases the restrictions on the protection system, it makes the cancellable templates more accessible for available biometric technologies

Soft biometrics

Soft biometrics traits are physical, behavioural or adhered human characteristics, which have been derived from the way human beings normally distinguish their peers (e.g. height, gender, hair color). Those attributes have a low discriminating power, thus not capable of identification performance, additionally they are fully available to everyone which makes them privacy-safe.

International sharing of biometric data

Many countries, including the United States, are planning to share biometric data with other nations.

In testimony before the US House Appropriations Committee, Subcommittee on Homeland Security on "biometric identification" in 2009, Kathleen Kraninger and Robert A Mocny commented on international cooperation and collaboration with respect to biometric data, as follows:

> "To ensure we can shut down terrorist networks before they ever get to the United States, we must also take the lead in driving international biometric standards. By developing compatible systems, we will be able to securely share terrorist information internationally to bolster our defenses. Just as we are improving the way we collaborate within the U.S. Government to identify and weed out terrorists and other dangerous people, we have the same

> obligation to work with our partners abroad to prevent terrorists from making any move undetected. Biometrics provide a new way to bring terrorists' true identities to light, stripping them of their greatest advantage—remaining unknown."

According to an article written in 2009 by S. Magnuson in the National Defense Magazine entitled "Defense Department Under Pressure to Share Biometric Data" the United States has bi-lateral agreements with other nations aimed at sharing biometric data. To quote that article:

> "Miller [a consultant to the Office of Homeland Defense and America's security affairs] said the United States has bi-lateral agreements to share biometric data with about 25 countries. Every time a foreign leader has visited Washington during the last few years, the State Department has made sure they sign such an agreement."

Governments are unlikely to disclose full capabilities of biometric deployments

Certain members of the civilian community are worried about how biometric data is used but full disclosure may not be forthcoming. In particular, the Unclassified Report of the Defense Science Board Task Force on Defense Biometrics states that it is wise to protect, and sometimes even to disguise, the true and total extent of national capabilities in areas related directly to the conduct of security-related activities. This also potentially applies to Biometrics. It goes on to say that this is a classic feature of intelligence and military operations. In short, the goal is to preserve the security of 'sources and methods'.

BIOLOGICAL CHARACTERISTICS OF BIOMETRICS

Biometrics refers to the use of physiological or biological characteristics to measure the identity of an individual. These features are unique to each individual and remain unaltered during a person's lifetime. These features make biometrics a promising solution to the society. The access to the secured area can be made by the use of ID numbers or password which amounts to knowledge based security. But such information can easily be accessed by intruders and they can breach the doors of security. The problem arises in case of monetary transactions and highly restricted to information zone. Thus to overcome the above mentioned issue biometric traits are used. A biometric system is essentially a pattern recognition system which makes a personal identification by determining the authenticity of a specific physiological or behavioral characteristic possessed by the user. Biometric technologies are thus defined as the automated methods of identifying or authenticating the identity of a living person based on a physiological or behavioral characteristic.

BIOMETRIC TECHNOLOGIES

Biometric technologies are being adopted across North America by law enforcement officials and surveillance advocates as identification technologies *par excellence*. Utopian descriptions laud their potential range of applications, from the assertion that they will eliminate racial profiling by replacing the discriminatory gaze of the state official with the neutral gaze of the biometric camera to the suggestion that they will help to eradicate crime by stabilizing the otherwise mercurial identities of criminalized individuals and allowing their precise individual identification. Although developed for law enforcement use, in the early 1990s state governments campaigned for the addition of biometric identifiers to existing welfare systems, resulting in the widespread biometric fingerprinting of welfare recipients. By 2000, biometric identifiers were being used in welfare programs in more than twelve states.

The biometric identification of welfare recipients includes both biometric fingerprinting and retinal scanning. Fingerprinting has historically and culturally been a marker of interaction with and surveillance by the criminal justice system. Biometric fingerprinting and other biometric forms of identity marking join a long history of surveillance over recipients of state welfare. These "new" forms of identity capture claim an allegedly greater level of technological sophistication.

BIOMETRIC TECHNOLOGIES: TECHNOLOGIZED VISION ENVISIONING THE BODY

Biometrics is the science of using biological information for the purposes of identification. Here we focus primarily on two related biometric technologies: finger imaging and retinal scanning, the two most commonly used methods in identifying welfare recipients. Finger imaging takes a picture of the fingerprint. Retinal scanning takes a picture of the blood vessels at the back of the eye. Widescale deployment of retinal scanning is hindered by its invasive nature. Unlike iris scanning, which simply takes a picture of your iris, retinal scanning uses infrared light to penetrate the eye and to produce an image of the vessels at the back. This quality has made many reluctant to use it. Thus, retinal scanning has been most successful when used coercively, as in the mandatory scanning of welfare recipients.

RETINAL SCANNING AND FINGER IMAGING

Both retinal scanning and finger imaging are converted into digital information that may be encoded onto a passport or a smartcard for purposes of identification or stored in an information database. The distinction between finger imaging and retinal scanning is fundamentally artificial. In a neo-liberal market of increasing consolidation biometric identification companies are merging, resulting in multi-modal biometric solutions that enable scans that simultaneously check fingerprints and retinal patterns.

RETINAL SCAN

A retinal scan, commonly confused with the more appropriately named "iris scanner", is a biometric technique that uses the unique patterns on a person's retina to identify them. It is not to be confused with another ocular-based technology, iris recognition. The biometric use of this scan is used to examine the pattern of blood vessels at the back of the eye. The human retina is a thin tissue composed of neural cells that is located in the posterior portion of the eye. Because of the complex structure of the capillaries that supply the retina with blood, each person's retina is unique. The network of blood vesselsin the retina is not entirely genetically determined and thus even identical twins do not share a similar pattern. Although retinal patterns may be altered in cases of diabetes, glaucoma or retinal degenerative disorders, the retina typically remains unchanged from birth until death. Due to its unique and unchanging nature, the retina appears to be the most precise and reliable biometric, aside from DNA. Advocates of retinal scanning have concluded that it is so accurate that its error rate is estimated to be only one in a million. A biometric identifier known as a retinal scan is used to map the unique patterns of a person's retina. The blood vessels within the retina absorb light more readily than the surrounding tissue and are easily identified with appropriate lighting. A retinal scan is performed by casting an unperceived beam of low-energy infrared light into a person's eye as they look through the scanner's eyepiece. This beam of light traces a standardized path on the retina. Because retinal blood vessels are more absorbent of this light than the rest of the eye, the amount of reflection varies during the scan. The pattern of variations is converted to computer code and stored in a database.

History

The idea for retinal identification was first conceived by Dr. Carleton Simon and Dr. Isadore Goldstein and was published in the New York State Journal of Medicine in 1935.The idea was a little before its time, but once technology caught up, the concept for a retinal scanning device emerged in 1975. In 1976, Robert "Buzz" Hill formed a corporation named EyeDentify, Inc., and made a full-time effort to research and develop such a device. In 1978, specific means for a retinal scanner was patented, followed by a commercial model in 1981.

In popular culture

The relative obscurity and "high tech" nature of retinal scans means that they are a frequent device in fiction to suggest that an area has been particularly strongly secured against intrusion. Some notable examples include:

In the movie *Star Trek II: The Wrath of Khan* (1982), Admiral Kirk gains access to top secret computer files by use of a retinal scan. In the movie *Batman*

(1966), Batman describes to Robin how the tiny vessels in the retina are unique to the individual and utilizing the portable retina scan device in the Batmobile they could confirm the identity of the Penguin. Characters in the films *GoldenEye* (1995), *Mission: Impossible* (1996), *Barb Wire* (1996), *Entrapment* (1999), *Minority Report* (2002) and *Paycheck* (2003) utilize or try to deceive retinal scanners. In the *Splinter Cell* series, retinal scanners are used to identify agents within Third Echelon and guards within military/business complexes.

Uses[edit]

Retinal scanners are typically used for authentication and identification purposes. Retinal scanning has been utilized by several government agencies including the FBI, CIA, and NASA. However, in recent years, retinal scanning has become more commercially popular. Retinal scanning has been used in prisons, for ATM identity verification and the prevention of welfare fraud.Retinal scanning also has medical application. Communicable illnesses such as AIDS, syphilis, malaria, chicken pox and Lyme disease as well as hereditary diseases like leukemia, lymphoma, and sickle cell anemia impact the eyes. Pregnancy also affects the eyes. Likewise, indications of chronic health conditions such as congestive heart failure, atherosclerosis, and cholesterol issues first appear in the eyes.

Pros and cons

Advantages

- Low occurrence of false positives
- Extremely low (almost 0%) false negative rates
- Highly reliable because no two people have the same retinal pattern
- Speedy results: Identity of the subject is verified very quickly

Disadvantages

- Measurement accuracy can be affected by a disease such as cataracts
- Measurement accuracy can also be affected by severe astigmatism
- Scanning procedure is perceived by some as invasive
- Not very user friendly
- Subject being scanned must be close to the camera optics
- High equipment cost

BIOMETRICS: RETINAL SCANNING

Retinal scans use infrared light to survey the unique pattern of blood vessels of the retina, which is the nerve tissue that lines the back of the eye. The first company to exploit the idea, developed in the 1950s, was Eydentify, founded by Robert Hill in 1976. Hill was an electrical engineer who happened on the idea of using retinal scans as a form of identification when he was helping his father, an ophthalmologist, detect eye disease through photographs.

There is little chance that retinal patterns can be replicated or forged. Retinal scans are therefore considered to be among the least violable of biometric security measures. (Fingerprints, by comparison, are relatively easy to forge.) Nevertheless, retinal scanners are not at present good candidates for widespread use. First they are expensive. Second, in order to work, users must permit light beams to be shone directly into their eyes for 10 to 15 seconds. The sensation is unpleasant and intrusive enough to make widespread acceptance among the general public unlikely. Additionally, diseases such as cataracts can cause the retina to change over time.

EXPLAINER: RETINAL SCAN TECHNOLOGY

Retinal scans map the unique patterns of a person's retina. The blood vessels within the retina absorb light more readily than the surrounding tissue and are easily identified with appropriate lighting. A retinal scan is performed by casting an unperceived beam of low-energy infrared light into a person's eye as they look through the scanner's eyepiece. This beam of light traces a standardized path on the retina. Once the scanner device captures a retinal image, specialized software compiles the unique features of the network of retinal blood vessels into a template. Retinal scan algorithms require a high-quality image and will not let a user enroll or verify until the system is able to capture an image of sufficient quality. The retina template generated is typically one of the smallest of any biometric technology.

Retinal scan is a highly dependable technology because it is highly accurate and difficult to spoof, in terms of identification. The technology, however, has notable disadvantages including difficult image acquisition and limited user applications. Often enrollment in a retinal scan biometric system is lengthy due to requirement of multiple image capture, which can cause user discomfort. However, once user is acclimated to the process, an enrolled person can be identified with a retinal scan process in seconds. Retinal scan technology has robust matching capabilities and is typically configured to do one-to-many identification against a database of users. However, because quality image acquisition is so difficult, many attempts are often required to get to the point where a match can take place.

While the algorithms themselves are robust, it can be a difficult process to provide sufficient data for matching to take place. In many cases, a user may be falsely rejected because of an inability to provide adequate data to generate a match template. Because retinal blood vessels are more absorbent of log-energy infrared light than the rest of the eye, the amount of reflection varies during the scan. The pattern of variations is converted to computer code and stored in a database. Retinal scans should therefore not be confused with another ocular-based technology, iris recognition, which is described as the process of recognizing a person by analyzing the random pattern of the iris. The retina's intricate network of blood vessels is a physiological characteristic that remains stable throughout the life of a person. As with fingerprints and iris patterns, genetic factors do not determine the exact pattern of blood vessels in the retina. This allows retinal scan technology to differentiate between identical twins and provide robust identification. The retina contains at least as much individual data as a fingerprint, but, unlike a fingerprint, is an internal organ and is less susceptible to either intentional or unintentional modification. Certain eye-related medical conditions and diseases, such as cataracts and glaucoma, can render a person unable to use retina-scan technology, as the blood vessels can be obscured.

Retinal scan devices are mainly used for physical access applications and are usually used in environments requiring exceptionally high degrees of security and accountability such as high-level government, military, and corrections applications. Retinal scanning has been utilized by several U.S. government agencies including the Federal Bureau of Investigation (FBI), theCentral Intelligence Agency (CIA), and NASA. Retinal scanning is also used for medical diagnostic applications. Examining the eyes using retinal scanning can aid in diagnosing chronic health conditions such as congestive heart failure and atherosclerosis. Diseases such as AIDS, syphilis, malaria, chicken pox and Lyme disease, as well as hereditary diseases, such as leukemia, lymphoma, and sickle cell anemia, also impact the eyes and can be detected using retinal scan technology.

INTRODUCING BIOMETRICS TO WELFARE ENTITLEMENT

Interest in, and availability of, biometric imaging systems for welfare recipients emerged at a time when welfare programs were particularly visible. Biometrics companies were interested in expanding markets for their newly developed products, allowing "yesterday's technological exotica" to be translated into today's "everyday tool". Biometric companies' need for expanding markets drove the implementation of the technologies. In one of the earliest demonstrations of the application of biometrics to welfare, Unisys Corporation developed biometric fingerprint imaging for San Diego county at no cost. In return, San Diego granted Unisys the right to use the project as a testing opportunity to develop similar systems for other state and local governments.. The system Unisys designed was one of the first "fingerprint imaging system[s]... designed to detect welfare fraud". When companies such as Digital Biometrics Inc. and Identix Inc. became involved in the biometric identification of welfare recipients they reported "their first profitable quarters ever".

In the U.S. the inclusion of biometric identification technologies in welfare programs was part of a campaign of sweeping reforms, begun one year before Bill Clinton pledged to end welfare "as we know it" during his election campaign of 1992. As has been documented by John Gilliom (2001) and Jullily Kohler-Hausmann (2007), the unpopularity of state-subsidized welfare in the

United States dates to the 1970s. Welfare is a gendered program and - with respect to aid for families with children - is in fact the state's sole (and limited) way of recognizing that mothering and homemaking have "social and economic value outside of the patriarchal family". Connections continue to be drawn between racialization, gendered identity, (un)willingness to work and state assistance. The backlash resulting in the reification of these connections accompanied the expansion of aid to racialized women – which occurred only as a result of the institutionalization of civil rights laws that prohibited "discrimination in the administration of welfare benefits". These changes allowed women, who previously had been prevented by racist and sexist policies from benefiting from state assistance, to receive aid for the first time. Prior policies had often prevented women of colour from receiving aid: welfare was frequently "withheld from black women in the South when field laborers were needed".

Contempt for welfare recipients stems from racist, sexist and classist stereotypes of the undeserving poor. Martin Gilens writes that the news media continually distorts welfare, depicting "overly racialized images of poverty" and associates these images with the suggestion that the poor are unwilling to work. Moreover, he argues that "Americans who think most welfare recipients (or poor people) are Black express more negative views about people on welfare and are more likely to blame poverty on a lack of effort rather than on circumstances beyond the control of the poor". The exaggerated stereotype of the African American welfare queen exploited by Ronald Reagan in 1976, and utilized by conservative politicians from Clarence Thomas to George W. Bush, epitomizes the sexist, racist and classist nexus that has been essential to the expansion of the criminalization of poverty to the criminalization of welfare; it is these oppressive categorizations that have been used to justify the rollbacks to federal and state assistance.

Significant cuts to welfare began in the early 1970s under the Quality Control movement, which provided states with incentives for "accuracy" and imposed penalties for errors in welfare administration. Thus, the 1973 Quality Control regulations dramatically intensified the surveillance of welfare recipients. This trend continued with President Reagan's cuts to welfare in the 1980s. To achieve his goal of welfare reform, Reagan granted states a freer hand in their ability to tailor aid programs, beginning the elimination of federal responsibility for state wellbeing. This strategy was furthered by President Clinton through the *Personal Responsibility and Work Opportunity Reconciliation Act (PRWORA)* of 1996, signed by most states in 1997. *PRWORA* gave states a free hand in the administration of welfare programs. Major reforms to welfare included work requirements for aid recipients, time limits to welfare, mandatory job training and the use of social programs such as health care and child care as 'carrots' to encourage welfare recipients to work. The Clinton-era reforms allowed states the flexibility to extend biometric identification programs to welfare. Parallel cuts to welfare were made in Canada during the 1980s. A conservative majority significantly reduced the welfare state through cuts and changes to regulation, including a dramatic reduction in federal responsibility for welfare. In 1990, the federal government imposed limits on its funding to welfare programs in Canada's three richest provinces, containing over half the nation's poor. Like the US under Reagan, the Canadian government summarily asked the provinces to pick up the remainder of the welfare tab without federal help. Furthermore, as in the US, welfare was increasingly tied to work-incentive programs, despite substantial evidence that these programs cost more than they saved. Federal cuts to welfare, coupled with the requirement that provinces pick up the tab, paved the way for steep welfare cuts by provincial governments. In Ontario, a conservative majority headed by Premier Mike Harris made workfare and learnfare mandatory for many recipients under the *Ontario Works Act* of 1997 and resulted in large numbers being struck from the rolls. Sean Hier documents how the devastating potential of the cuts was actualized when "college student Kimberly Rogers died eight months' pregnant while under 'house arrest' in her apartment". She is thought to have succumbed to the extreme temperatures in her apartment as a result of a late summer heat wave. Significantly, the *Ontario Works Act* also called for the use of biometric fingerprinting technologies to avoid fraud, but, although millions were spent developing the technologies, privacy legislation in Canada prevented their implementation.

In the United States, two of the primary programs targeted for the inclusion of biometric technologies were *Home Relief* (for single adults with no dependents) and *Aid to Families with Dependent Children (AFDC)*, now *Temporary Aid to Needy Families (TANF)*. In keeping with the attachment to quick-fix solutions, the use of biometric technologies was described as a straightforward answer to the problem of welfare, capable of ensuring biometric accountability and eliminating fraud. The purported "scientific simplicity" of biometrics made them extremely popular as a solution to the complex and highly contested nature and administration of welfare. Biometrics are a trigger point for those interested in the single-issue policies that have come to dominate the so-called "compassionate conservatism" of the New Right: i.e. single-issue policies aimed at gaining "popular support for economic policies favorable to the economic elite", and that fundamentally undermine the policies and spirit of the welfare state.

The unequal distribution of wealth and power between cities and suburbs, cities are particularly vulnerable to cuts to the New Deal. The first location to be targeted for a biometrics program aimed at "getting tough on welfare fraud" was Los Angeles, followed shortly by New York. The able-bodied poor without children are traditionally among the first welfare clients to be subject to cutbacks. LA was no exception; it was here that the first recipients on *Home*

Relief were finger-imaged. Biometric fingerprinting was quickly expanded to other counties in California. In 1994, general assistance welfare clients in San Francisco, Alameda and Contra Costa counties in California were biometrically fingerprinted. The majority of those scanned were single young men. Men (particularly able-bodied men) are culturally constructed as independent and self-sustaining individuals, and they remain suspect as recipients of "caretaking" from the state. It is thus not surprising that their benefits were among the first to be policed by biometrics. In 1994, LA's biometric identification program was expanded to families receiving *AFDC*, making it the first place in the country to biometrically fingerprint families with children. This pattern of *Home Relief* recipients being the first to be fingerprinted, followed by the expansion of fingerprinting to families with children was repeated in New York in 1995.

There is a long history of sorting the poor into those deserving and those undeserving of relief. Although the basis for classification changes, the imperative to discover who is worthy of aid persists. Michael Katz (1986) describes a study done in the early 19th century in Massachusetts that strove to classify the "impotent" from the "able" bodied poor. These attempts at classification have endured as those seeking aid from the state are divided by age, gender, race, mental ability, (dis)ability and parental status in order to evaluate the legitimacy of their claim. Biometric technologies are the latest technology of power utilized in this quest, as they automate this process of social sorting. Yet, like all of their predecessors, biometrics classification systems are prone to error. For example, when George W. Bush was Governor of Texas, he called for the expansion of biometric identification to families with children. Ironically, elderly, ill Americans became the prime targets of the new biometric tests as Bush selected Medicaid as the target program requiring biometric identification. Once biometric technologies for welfare clients were established in one county, they usually spread to other counties in the same state, as in California. In Wisconsin in 1996, largely as a result of Republican governor and notorious welfare-reformer Tommy Thompson's efforts , biometric scanning of welfare clients was expanded to include retinal scanning.

Biometrically identifying welfare recipients fits what Lawrence Mead has termed the "new paternalism": a "supervisory approach to poverty" that advocates intense scrutiny of those receiving aid such as *General Relief, Home Relief* or *AFDC/TANF*. Paternalists emphasize that "some intrusion" into the lives of welfare clients is both necessary and acceptable ; thus, welfare reform under new paternalism has led to an intensification of the process to "peer into the lives of those on assistance, looking for the smallest of reasons to deny them benefits". This approach to poverty facilitates the introduction of biometrics to welfare programs and ignores suggestions that there are privacy concerns that need to be addressed. After all, state politicians conclude, biometric technologies only target the guilty: "'I'm not after the people who deserve welfare, God bless them,' said Joseph Rizzo, a Republican county legislator from Islip Terrace and a sponsor of the [biometrics] bill. 'I'm after the people who are ripping off the system'". The rhetoric of "nothing to fear, nothing to hide" is an oft-repeated refrain of the promoters of biometrics technologies that obfuscates the way that they are reliably used against those living at the margins of the state. In a letter to the editor published in *The Seattle Times*, one writer expressed irritation at the idea that welfare recipients might have the right to privacy: "[to] the majority of us who have to pay the tab for welfare programs, requiring the recipients of our generosity to be fingerprinted as proof of eligibility doesn't seem unreasonable in the least. If nothing else, it may be the first time some of them ever got their hands dirty putting a dollar in their pocket" (1996).

Paternalistic welfare reforms make the state into the surrogate (and suspicious) father of those receiving aid. New biometric technologies make this paternalism scientific. A particularly dramatic example of making "new paternalism" scientific is the newfound emphasis on discovering actual paternity, demonstrating the confluence of biological paternity testing and biometric identification. In one program, a "hospital-based paternity ID program" was implemented in order to "ensure future child support". Critical-race feminist and queer theorist Anna Marie Smith has dubbed this newfound emphasis on discovering the paternity of the children born to welfare mothers part of a system of neo-eugenics – in which women receiving welfare are subject to extreme forms of sexual regulation. Smith includes among neo-eugenic practices invasive procedures that compel welfare mothers to disclose their sexual histories, and open their homes and their DNA to the government, as well as policies that force women fleeing from violent biological fathers to place themselves at risk (2007). In Canada, pre-1987 'spouse in the house' rules that deprived women of their welfare checks if they were cohabiting with a partner qualify as an extreme form of sexual regulation.

Making welfare scientific has been a continual goal of reform. Efforts to eliminate caseworker discretion and shift "the administration of welfare toward a more bureaucratic model that emphasize[s] adherence to rules and procedures" are ongoing. Mary Jo Bane and David Ellwood cite a "Department of Public Welfare director of labor relations as saying 'We've been trying to get the people who think like social workers out and the people who think like bank tellers in'". Scientific jargon is regularly imported into the aid administration process. The last ten years of reform have produced a discourse of scientific rhetoric that emphasizes "compliance" to new welfare standards. Allowing your biometric information to be captured by the state becomes an essential component of compliance – a necessary designation to continue to receive benefits. Biometrics help to make the business of getting benefits appear scientific by producing an image purporting to be "the only true form of identification.... And you carry it with you wherever you go". Biometrics not only facilitate state surveillance of welfare clients and help to determine whether compliance has been achieved, but claim to replace the discretionary eye of the caseworker with the neutral eye of the scanner.

Some professionals have been uncomfortable with the inclusion of imaging technologies, since they transfer the site of expertise into the virtual hands of a machine. In the case of welfare, however, administration officials have welcomed the transfer of authority. Most welfare professionals do not want the responsibility of striking clients off the rolls – authority that, as the case of Kimberly Rogers reminds us, may be equated with the power of life and death. Others appreciate the change because it allows them to justify their decisions with scientific rhetoric. Faced with the impossible goal of reducing their caseloads, one can understand how some welfare workers would be grateful to defer this decision to a machine. Biometric technologies have a double advantage: not only do they fit the new emphasis on making welfare scientific, they claim to convert a subjective individual choice into one made by an "objective" machine. State officials argue that the addition of biometric technologies to welfare helps prevent soft-hearted caseworkers from providing benefits beyond the terms of the welfare system. At the same time, these additions justify a hard line approach to denying benefits with the rhetoric of the scientific gaze. Thus, the 'scientific neutrality' of biometric technologies is essential to their popularity as a tool of welfare reform. In many cases biometrics have become the "central feature of.... [the] campaign to reduce welfare spending".

Although reformers claim that making welfare more scientific has made the system more objective and therefore more just, careful study of the impact of applying a quasi-scientific model to welfare reveals a mass of contradictions. Biometric technologies were introduced to welfare in the guise of "pilot programs": preliminary programs used to "test" a particular reform. The scientific theory behind this process was that the success of the pilot projects would then be reviewed – as in a lab setup in which a hypothesis is formulated, tested, then revised based on the success of the experiment. However, the impact of pilot programs on welfare reform has been considerable. Rogers-Dillon documents that these programs have served to "restructure social policy outside of the legislative process", rather than to scientifically test welfare reforms. Additionally, pilot programs made a wider range of ideas politically viable , including fingerprinting every aid recipient, an idea that previously had been untenable. The "pilot" nature of the projects was used to make the case that these systems were only being "tested" rather than implemented, and thus the pilot programs succeeded where other attempts at instituting "reforms" had failed.

Rogers-Dillon additionally demonstrated that although the scientific discourse of experimentation was invoked, those actually responsible for implementing welfare reform "did not have a scientist's view of experimentation". Biometric programs demonstrate the failure of the so-called 'scientific' process. Pilot projects involving biometrics were both expensive and politically popular, and failure became an unacceptable test outcome; the projects simply had to work. The "review" stage necessary to scientific experimentation was completely undermined by political officials responsible for the implementation of biometric technologies. An audit done in California made clear that the biometric Statewide Fingerprinting Imaging System used to test welfare clients for fraud was implemented without knowing how "much fraud actually existed in those aid programs, making it impossible to know whether the $31 million fingerprint program was necessary". The report concluded that the additional $11.4 million a year to operate the program was far too high, given that most of the fraud detected "turned out to be errors made by county staff, and the level of detected duplicate-aid has been small". Yet, although this review was essential to the scientific process used to justify the implementation of a biometrics program, then-governor Gray Davis ultimately rejected the report produced by the review process. The Governor's office asserted that it was in "disagreement with many of the report's fundamental findings... which we find either inaccurate or unsupportable", though it cited no evidence to support this claim. Thus, while science was invoked to justify the implementation of the program, it was disdained during the review process.

ASPECTS OF BIOMETRIC IDENTIFICATION

A biometric device or biometric identification device is one that captures a physiological 'image' and uses that image to permit or deny access. The access being controlled could be to a computer account or to a room. The goal of these devices is to provide a more accurate and secure method for physical or logical access. Everyone has a story or has heard one of someone forgetting his or her password and not being able to 'log in' to an account. Almost everyone also has a story about losing a set of keys. Biometric devices are meant to be an improvement over keys or passwords, since these latter devices can be lost or forgotten. In theory, your biometric cannot be lost or stolen because it is specific to you. Note that this is a different usage of the word biometrics. Biometrics, as the statistics community commonly uses it, is a statistical or mathematical analysis of biological phenomena, e.g. the journal *Biometrics*.

There are a wide variety of biometrics that are being used. A biometric is the physiological image that is used to determine identity. A partial list (in no particular order) would include: fingerprints, hand geometry, finger geometry, hand vein patterns, ear geometry, face recognition, voice recognition, retinal scans, iris patterns, handwriting, keystroke dynamics and walking gait.

The goal of a biometric device is to accurately determine whether or not you are who you say you are. There are several factors that go into a 'good' biometric device. Jain et al.(1998) suggest that a biometric should possess the following characteristics: universality, uniqueness, permanence, collectability, performance, acceptability, and circumvention. Universality means that as many people as possible should have the biometric in question. Not every person has a right index finger, so that a biometric device based solely on this will not be universal. Next, uniqueness implies that each person should have a different version of the biometric. Fingerprints are generally thought to be unique. Permanence is the condition that the biometric should not change over time. A biometric device based on facial recognition is not ideal in this sense because people change their hair, they grow beards and they get wrinkles. The ease with which a biometric can be captured is it's collectability. It might be possible to create a biometric device based upon your EEG, but it would be difficult to capture that information quickly and easily. On the other hand, a fingerprint or an iris is fairly exposed and, therefore, easily collectible. Performance measures how easy a particular biometric is to use and implement. Acceptability is the degree to which the there is public acceptance of the biometric for identification purposes. Fingerprints are a prime example of a biometric with high acceptability, since they have been used for centuries as a method of identification. Finally, circumvention is the amount of work need to fool the system. Signatures are notoriously easy to reproduce, whereas creating a copy of a fingerprint is far more difficult.

The basic biometric system contains five subsystems: data collection, transmission, signal processing, a decision-making algorithm and a database (Wayman, 1998). The data collection mechanism is a sensor of some kind. For facial recognition, the data collection mechanism is a camera. For keystroke dynamics, the data collection mechanism is a keyboard. The information from the data collection mechanism is then transmitted to the signal-processing unit. As part of the transmission, techniques such a signal compression may be implemented on the presented biometric. The signal-processing unit then extracts the relevant details from the transmitted image and compares that image to one or more stored images, or templates, of the biometric from the database. The decision-making phase then decides whether or not the presented biometric is 'close enough' to the stored template to be considered a match. Several aspects of this process, particularly the decision-making step, have a statistical flavor.

One of the most important aspects of any biometric device is its matching performance. The matching performance is usually measured in terms of false accept and false reject rates. (Within the biometrics industry, these are sometimes referred to as the false match and false non-match rates, respectively, (U.K. Biometrics Working Group, 2000)). We will refer to users that are enrolled in the database as genuine users and we will refer to users who are not enrolled in the database as imposters. Thus, the matching performance describes how well the system allows access to genuine users and denies access to imposters.

When an individual presents their biometric, the 'image' is processed and matched against one or more stored templates from the database. The number of comparisons depends upon the mode that the device uses. There are two basic modes of operation. The first is verification or one-to-one mode. In this mode, some identifier such as a name or an ID number is given to the system and it verifies that your biometric matches the biometrics stored under your name. The second mode of operation is identification or one-to-many mode. Under this scenario, the biometric system compares the presented biometric to the entire database looking for a match. Though these systems have very different methodologies, their performance is measured in the same way.

In either of these modes, the result of the decision-making algorithm is a match score, T. A comparison is then made between the match score, T, and a threshold, t. If $T>t$, then system decides that a match has been made and permits access. This is called an accept. If T t, then the system decides that a match has not been made and denies access. This is generally referred to as a reject. The statistical aspects of biometric device performance focus on the rate at which errors are made in this process. These errors are akin to the Type I and Type II errors that are encountered in hypothesis testing. A Type I error is a false reject and a Type II error is a false accept.

BIOMETRICS: THE SCIENCE OF MEASURING PHYSICAL-BIOLOGICAL TRAITS

BIOMETRICS NOW APPLIES ALMOST EXCLUSIVELY TO THE MEASUREMENT OF BIOLOGICAL TRAITS FOR SECURITY PURPOSES

The term biometrics is now widely known as "the science of measuring physical characteristics to verify a person's identity; this includes voice recognition, iris and face scans, and fingerprint recognition." This definition represents a recently created application used in the industrial-tech world.

Since biometrics is a system for measuring unique biological traits for the purpose of identification; it includes utilization of time clocks, the "easy way" to track and to report employees authentication to increase security, and the enhancemeant of access with the convenience of hand readers or finger prints; so, there is no further need for ID badges or time cards and the biometric system also eliminates the "buddy punching" of time cards or employees clocking each other in. When some recognition systems verify the identities of individuals by the size and shape of the hand, they do so without the fingerprints or palm prints being utilized. Fingerprint recognition has emerged as one of the most popular and convenient biometric technologies because it is more accurate than voice recognition and cheaper than iris scanning, supporters say. Now, more and more everyday gadgets are coming equipped with fingerprint scanners, including some cell phones. A few airports and government agencies, such as the FBI, have dabbled with biometrics to identify employees. In recent years, a wave of new users, from schools to banks, have adopted the technology. The goal: tighten security, reduce security costs, and meet stricter laws imposed after the 2001 terrorist attacks.

Biometrics Plays an Important Role in Physical Access Control

Biometrics identifies a person via a unique human characteristic: the size and shape of a hand, a fingerprint, one's face or several aspects of the eye. If the goal of an access control system is to control where people, not credentials, can and cannot go, then only a biometric device truly provides this capability to the end user.

As a result, biometrics is used on the front doors of thousands of businesses around the world, at the doors to the tarmacs of major airports, and at the entrances of other facilities where the combination of security and convenience are desired. More than 900 biometric hand readers control client and employee access to special areas of Italian banks and more than 100 units perform similar functions in Russia. In the united Kingdom, Her Majesty's prisons rely on biometrics for prisoner and visitor tracking. Universities use hand readers for the on-campus meal program and to safeguard access to dormitories and to protect their computer centers. Hospitals utilize the biometric devices for access control and payroll accuracy.

Since 1991, biometric systems have produced millions of verifications at San Francisco International airport (SFO), with more than 50,000 produced on high volume days. Hand readers span the entire airport, securing more than 180 doors and verifying the identity of more than 18,000 employees. The use of biometrics at San Francisco is airport-wide and fully integrated into the primary access control system.

The Benefits of Biometrics in Access Control

The goal of any access control system is to let authorized people, not just their credentials, into specific places. Only with the use of a biometric device can this goal be achieved. A card-based access system will control the access of authorized pieces of plastic, but not who is in possession of the card. Systems using PINs require an individual only know a specific number to gain entry; but who actually entered the code cannot be determined. On the contrary, biometric devices verify who people are by what they are, whether by hand, eye, fingerprint, or voice recognition. Biometric reductions in errors have lowered the capital costs of ID cards in recent years and the true benefit of eliminating them is realized through reduced administrative efforts. For example, a lost card must be replaced and reissued by someone. Just as there is a price associated with the time spent to complete this seemingly simple task, when added together, the overall administration of a card system is costly.

Contrary to using badges, sign-ins or other ways of tracking employees, a biometric time clock assures that no employee can punch in for another, eliminating time fraud and reducing payroll costs. Because every person's biometric characteristic (hand, fingerprint, eyes, face, etc.) is unique, a biometric time clock provides a quick, accurate, and reliable way to record in- and out-punches for each employee. That's why so many companies now employ biometrics.

Potential Problems with Biometric Systems

One of the most crucial factors in the success of a biometric system is user acceptance of the device. It must cause no discomfort or concern for the user. If people are afraid to use the device, they probably will not use it properly, which may result in users not being granted access. The biometric device must work correctly. When it functions properly, it does two things: it keeps unauthorized people out and lets authorized people in. No device is perfect. In the biometric world, the probability of letting the wrong person in and right person out, is characterized by the "false accept" and the "false reject" error rates. This contrast and the frustration of dealing with a high number of false rejects will have authorized users and management alike looking for a way to replace the biometric system with something else if these factors are not considered up front.

"Smart" Passports for U.S. Citizens

The U.S. governments planned to issue "smart" passports, featuring embedded microchips that store a compressed image of the owner's face, to U.S. citizens in October, 2004. Designed to prevent tampering, the digital passports were set to include cryptographically signed digital images to guarantee their authenticity. Although civil liberties groups have expressed concerns about the government using the new passports as a monitoring tool, Frank Moss, deputy assistant secretary for Passport Services at the U.S. Department of State, maintained that information will only be forwarded to centralized databases if there is a query over the authenticity of a passport. What is more, Moss says the passports would only include basic passport information.

Some technical experts have also warned that smart passports do not guarantee safety; adding that the new passports would only help to identify known suspects or people who have forged passports. Richard Clayton, a hardware security expert at Cambridge University in the United Kingdom, adds that everyone involved in the September 11 terrorist attacks had a photo ID. Meanwhile, the European Union planned to spend 140 million euros to develop an interoperable biometric system, which would enable passports to carry fingerprints and iris scan biometric data. Such biometric information would be much easier to cross-reference than photographs of an individual with different hair styles and facial hair.

TESTING BIOMETRIC TECHNOLOGIES

In other cases, pilot programs testing biometric technologies were not only institutionalized after the test period (even when audits questioned their effectiveness), but were expanded even when clearly illegal. Suffolk County, New York, voted to extend a pilot program requiring fingerprinting for welfare recipients even though it had not been ratified by the State Legislature. "If they don't think it's legal, let them take us to court" blustered Joseph Rizzo, the Republican county legislator responsible for proposing the program. This institutionalization occurred despite the fact that the program was questioned by Republican Governor Cuomo, originally responsible for enactment of the biometrics program,

Notwithstanding the questionable "scientific" process to which welfare recipients are subjected, science continues to be used as a primary method of justification for these programs. Scientific rhetoric continues to be admired by the American public and positivist approaches to social problems remain popular. Sadly, attempting to quantify welfare reforms reveals a common error: basing studies in "social science or humanities on an ideal version of those in the natural sciences". Scientific methods were imported into welfare reform, transformed, misused, and misapplied. The reforms were then justified and made permanent, backed by claims that they were supported by "good science". Scientific methods only gilded the bars of the iron cage of bureaucratic patriarchy responsible for administering welfare benefits.

Thus, those receiving aid become implicated in a powerful regime of "technobiopower", in which "informatics, biologics, and economics" intersect in order to police the most vulnerable citizen-subjects. Biometrics becomes the system by which chip and gene can be joined, as the borders between the natural and the artificial are imploded and then manipulated by consolidated governmental powers driven by the hope of profit and electoral gain. This regime of technobiopower, despite protests to the contrary, makes more than a gesture toward criminalization.

INTENSIFICATION IN THE CRIMINALIZATION OF WELFARE RECIPIENTS AS A RESULT OF BIOMETRIC REFORM

Those receiving welfare have long borne the stigma customarily associated with unlawful acts. Investigators "routinely order [welfare] applicants to empty their pockets, then flip through their wallets and personal possessions, demanding to know the identity of every name they come across". In 1995, Governor Weld of Massachusetts claimed a link between "welfare, fatherlessness and crime," arguing that there are "a lot of kids who come out of fatherless families who seem to have ice water in their veins and no milk of human kindness". The Clinton administration's reforms brought new procedures that explicitly criminalized welfare. In the state of Florida, panels were instituted to review the cases of welfare clients who had been found to be "noncompliant". Although the review boards were established as "independent, community-based panel[s] to review cases," and could have been run quite informally, in practice, they were set up to reproduce a judicial hearing.. Rogers-Dillon found that those aid recipients who most represented 'deviant femininity' were most heavily policed. Women were more likely to be found "guilty" by these panels if they failed to meet normative standards of femininity – this raises questions as to how queer women might have been policed by pseudo-scientific reforms to welfare.

Finger imaging was invented in a climate of expanding *technologies of criminalization*. Refined over a period of "20 years for its obvious first customer, law enforcement agencies" , biometric-fingerprinting development was driven by the FBI, and was identical to the finger-imaging technology used to verify the identity of prison inmates. Despite these clear links to criminalization, biometrics companies and government officials continue to try to shake this association. One way is to distinguish biometric finger "imaging" from manual fingerprinting: "That's manual fingerprinting with ink. This is a very hi-tech system that is a clean process and a dignified process". Another is to claim that the process is entirely different because welfare bureaus only take two prints while prisons take ten. Yet the relationship between the finger imaging of welfare recipients and the criminalizing of welfare is unmistakable: "Opponents argue that finger imaging equates welfare recipients with criminals and may intimidate legitimate welfare recipients from applying for benefits they deserve" ; "'There's an assumption of guilt that goes with fingerprinting,' said Democratic Assemblyman Herman D. Farrell Jr.. 'Why do we choose this class of people to fingerprint?'". The connection between the two is only deepened as "function creep" - the process by which a tool designed for one purpose then applied to a new (usually larger) set of problems - took a frightening turn. Despite earlier claims that the "fingerprints will not be provided to any law enforcement agency" , the fingerprints of welfare clients are being made available to other state agencies.

In Massachusetts, Governor Weld proposed sharing prints taken for welfare authentification with the judicial system: "Weld's plan would also go further than current law by allowing law-enforcement officials investigating crimes to subpoena welfare fingerprint records". This type of proposal is reality in New York, where "state law also allows social service officials to pass on to law enforcement officials cases of fraud revealed through the finger-imaging program, which they had not been allowed to do under the earlier law initiating the program for *Home Relief* recipients". Thus, in the context of receiving aid from the state (a process that had already been criminalized), biometrics became a powerful *technology of criminalization*: engraving on the body a number that easily could be tracked by the eye of the state. In those states where this information can be shared, biometrics additionally transformed welfare fraud into a crime that is easy to prosecute; law enforcement agencies now have the necessary identifying information regarding the accused. Offices that biometrically fingerprint welfare recipients represent the first successful attempt of the US state to take the fingerprints of citizens and residents before they have committed crimes – making them the US's very own pre-crime unit.

FAILURES OF BIOMETRICS PROGRAMS

The expansion of biometrics programs into other areas of civic participation has continued unabated, despite the numerous failures induced by this system of 'bio-benefits'. Although the difficulties ensuing from the application of biometrics programs to welfare are many, we focus here on three particularly problematic outcomes resulting from the marriage of technology and state assistance.

Economic Issues

A cursory cost-benefit analysis reveals that biometric-technology programs have failed in their objective of saving the US state money. As with the claim of scientific method to justify the addition of biometrics to welfare, a quasi-economics has been used to defend the expenditure of millions of dollars on biometrics programs. In California, cost savings in LA were extrapolated to the rest of California. A later audit revealed both the initial savings figures and the projections to be flawed economics: "Auditors say the state erred in assuming that conditions in Los Angeles would hold true elsewhere" (Fingerprint Failure Fix or End Flawed Welfare ID program 2003). Yet the California model frequently is used to rationalize the expansion of biometrics programs in other states. In two counties of New York, $500,000 was saved from the "4.3 percent of *Home Relief* [single, childless] who chose not to reapply for welfare after being informed they would have to be fingerprinted. When that figure was extrapolated across the entire state, the study projected an annual savings of $46.2 million in welfare benefits" ; again without any evidence and in the absence of traditional economic models that emphasize the quality of the initial sample. New York City spent $40-50 million to implement increased biometric enforcement to save a projected $250 million. This target most certainly could not be met, given the unexpectedly low incidence of fraud in New York State at less than 3%. Even Governor Cuomo of New York doubted the results of a study commissioned by his own department regarding possible welfare savings for the rest of New York based on savings in one county.

In other states, the economic rationalization supposedly driving implementation of expensive biometrics systems barely was justified. In California, Governor Wilson asserted that the best economic outcome that could hoped for was that biometric technologies would save enough in fraud detection to pay for their implementation ; "I'm absolutely confident we're going to save enough to pay for the system" said Michael Genest of the Department of Social Service. If real savings were not anticipated, we must ask why biometrics were introduced as welfare reform in the first place. Huge spending on "fraud-proof" biometrics additions to welfare usually was justified without reference to exactly how much fraud would be halted. As noted, in California, a biometrics program costing $31 million dollars to implement and $11.4 million dollars per year to maintain was approved without information regarding the extent of welfare fraud in California.

Nor can the vast expenditures on biometrics programs be justified given the finding that most welfare recipients are telling the truth. Fewer than 0.3% of welfare clients in New York state have been found to be committing welfare fraud. In California most cases of fraud were found to result from administrative error. This suggests that biometrics programs and their costs were wholly unnecessary.

Part of the reason that these expenditures on "fraud-busting" biometrics programs were useless is because fraud is not well-defined. Most welfare fraud occurs when recipients work for pay 'under the table' in addition to collecting their benefits. This type of fraud is not discoverable using biometric fingerprinting systems, designed to detect somebody signing up for welfare benefits more than once. Thus, the projected figure of fraud at 3%, used as a multiplier to calculate savings, is highly inaccurate. In fact, many states found that 3% was too high an estimate for other types of fraud. Moreover, considerable welfare fraud is committed not by clients, but by service providers. The *New York Times* asserted that "most welfare fraud is not done by welfare recipients, but by providers, including doctors and landlords. As a result of the high incidence of provider fraud, lawmakers in Connecticut debated an "amendment that would have required doctors and others who supply goods and services to welfare recipients to be fingerprinted". Not surprisingly, this proposal was rejected – revealing that more powerful citizens are able to reject a criminal classification, while those living at the margins are not.

"Fraud" is not the only term lacking a clear conceptualization within the broader discourse of welfare restructuring. The "success" of a biometrics program is similarly difficult. As with the term "fraud", the definitional nature of "success" has implications for the administration and justification of biometric-identification programs. Success of biometric-identification programs is described in only one way: caseload decline. The accomplishments of biometrics programs have been defined by their ability to 'get people off the rolls'. Commonly, those dropped from the rolls are ousted not because they have found work and upward mobility, but because the overseers of the welfare system have not marked (i.e. fingerprinted) them. This occurs for one of three reasons: a welfare client has not understood the new process to be followed in order to continue to receive benefits. Yet this range of reasons is described by the all-inclusive term 'refusal' (to be fingerprinted). Thus, Los Angeles is considered a biometrics success because 3324 people were removed from the welfare rolls. Importantly, only 314 applicants were dropped for fraud while 3010 were dropped for 'refusing' to be fingerprinted. Here, refusal is clearly confounded, wrongly, with fraud. The use of biometric programs in welfare services is widely identified as a disincentive to fraud: "officials say that the program's greatest value is as a deterrent".

Biometrics and the Construction of Disability

Biometrics programs to test welfare clients for fraud create a discourse that works to manipulate the category of disability. Here, we understand disability as a construct that may be manipulated by hegemonic systems of representation. One becomes a disabled person if one is afraid of being fingerprinted. The impact of biometric-fingerprinting on those with mental health issues is highlighted by advocates, who assert that plans "to use new fingerprint technology to reduce welfare fraud may be unnecessarily frightening for people with psychiatric illnesses who depend on welfare". They further argue that "many of these people are already paranoid... To ask these people to surrender their fingerprints to the welfare bureaucracy could put them over the edge. Many would simply refuse and drop off the system altogether". Other community workers note that psychiatric patients "have enough trouble just getting out of bed in the morning"; to require them to follow the complicated steps required by biometric fingerprinting is ridiculous. Nor can fears about the ramifications of being finger-imaged be labeled wholly paranoid, given the ways in which this highly sensitive biometric information is shared between welfare and law enforcement agencies.

No provisions are made for those welfare clients who are afraid to be fingerprinted. Suggestions of working hard to "identify" people with disabilities in an attempt to "steer the mentally ill into SSI disability grants" seem halfhearted and unlikely given the caseloads of most welfare workers. Nor is this disproportionate impact of welfare reform on disabled persons exceptional. Targeting people with disabilities using biometrics programs repeats their targeting by other welfare reforms. In her study of a "non-compliant" community of welfare clients in Florida, Rogers-Dillon found that "noncompliant often means non-functioning." She noted that a "sizeable portion of the 'noncompliant' population may have had serious functional difficulties". Rogers-Dillon found that noncompliance was explicitly connected to disability by the aid recipients themselves. A disproportionate percentage of the noncompliant population "was, or considered itself to be, disabled".

Biometrics and Immigration

Immigrants also are targeted by biometric additions to welfare programs. Officials assume that anyone who fails to re-enroll on welfare following the implementation of a biometrics program has previously committed or intends to commit welfare fraud: a "1994 study by the Cuomo administration found that fingerprinting had saved Rockland and Onondaga nearly $500,000 by trimming 4.3 percent of *Home Relief* recipients from welfare rolls. That 4.3 percent consisted of people who chose not to reapply for welfare after being informed that they would be fingerprinted, and the study assumed that "most of them had hoped to cheat the system or were already cheating it". This means that no records were kept of those who refused re-enrollment because they "were afraid their fingerprints would be used for some other reason, such as to challenge their immigration status". Forms in English also made the process difficult for immigrants who were not fluent. For refugees used to harassment or intimidation by government officials in their countries of origin, biometric fingerprinting can serve as a frightening reminder of old perils. To label these fears wholly paranoid is problematic in a post-9/11 climate of increased harassment and intimidation of US residents, particularly racialized immigrants.

This is illustrated in the following case pertaining to welfare reform's impact on the application process for benefits: "The strict rules can be hard on people who do not speak English, who are homeless or who do not have traditional living arrangements. Yakov Gavritoc, a 61-year-old Russian immigrant who speaks no English and has a heart condition, was living with his son in an apartment on 63rd Street in Bensonhurst, Brooklyn, when he applied for Home Relief benefits in April, according to his son, Gavril Gavrilov. When investigators came to the apartment to check his address, the older man did not understand the forms they slipped under the door, and he did not call the number listed. A few weeks later the son said, Mr. Gavrilov's application was rejected for improper residence. 'When we lived in Michigan last year, they gave him welfare," the son added. "He does not understand why they will not do that here'". The same hazards for people hoping to claim benefits are introduced by biometric reforms to welfare – which are dependent upon reading dense documentation about the new procedure and following a complicated process to go and get fingerprinted.

If biometric technologies are not saving the state the vast amounts of money that politicians promised (the primary justification for their implementation), and if they make the lives of the most vulnerable people within the welfare system increasingly perilous, why do they continue to be such a popular approach to welfare reform?

Index

M

P

R

S

T

U

V

W